Parametric Landscape Architecture Planning and Design

Yangyang Yuan

Parametric Landscape Architecture Planning and Design

Springer

Yangyang Yuan
Department of Landscape Architecture
Southeast University
Nanjing, Jiangsu, China

ISBN 978-981-95-8010-1 ISBN 978-981-95-8011-8 (eBook)
https://doi.org/10.1007/978-981-95-8011-8

Jointly published with Southeast University Press
The print edition is not for sale in China (Mainland). Customers from China (Mainland) please order the print book from: Southeast University Press.

Translation from the Chinese Simplified language edition: "《参数化风景园林规划设计》" by Yangyang Yuan, © Southeast University Press 2023. Published by Southeast University Press. All Rights Reserved.

This Springer imprint is published by the registered company Springer Nature Singapore Pte Ltd.
The registered company address is: 152 Beach Road, #21-01/04 Gateway East, Singapore 189721, Singapore

Foreword

With the advent of the digital era, both the form and connotation of design have undergone significant transformations, accompanied by continuous innovations in design theories and methods. As a tool and means, digital technology has played a crucial role in advancing the discipline of landscape architecture, gradually influencing the thinking and decision-making processes of landscape professionals. Traditional landscape planning and design methods have mainly been experience-based, relying heavily on designers' intuition, insight, and accumulated knowledge. These approaches often lack scientific rigor and teachability, resulting in outcomes that tend to be contingent and uncertain. As an ancient yet young discipline, landscape architecture can no longer rely solely on a priori methods to address the complex realities of contemporary society.

From the history of scientific development, we observe that by the late 19th century, human inquiry gradually shifted from the idealism of the Enlightenment era toward positivism, aided by advances in mathematics and logic, thus embarking on a path of quantification and empirical validation. The development of digital technology has steered landscape research from qualitative to quantitative approaches, providing robust support for the scientific advancement of the field. Originally belonging to mathematics, statistics, and computer science, the concept of "parametrization" has recently emerged as a key innovation in planning and design research through its integration with design. Parametric landscape planning and design focuses on regulation and optimization within a systemic framework, transcending mere parametric form generation. Instead, it utilizes parametric design mechanisms to achieve an integrated study of landscape planning and design—from environmental composition and construction processes to presentation forms—realizing overall optimization of the landscape system under human intervention through multi-level and multi-objective coupling.

This book primarily presents the author's doctoral research results, completed under the guidance of Professor Cheng Yuning. Centered on "Parametric Landscape Planning and Design," it analyzes the parametric design mechanism from a methodological perspective, explores the landscape planning and design process guided by coupling principles, constructs a parametric landscape planning and design system, and establishes parametric evaluation and planning design models for landscape architecture. The book integrates theoretical and practical outcomes of applying parametric methods comprehensively in landscape planning and design, demonstrating the practical value of parametric technologies in actual design work. Leveraging digital technology, the author strives to demystify the "grey box" of design, clearly delineate the landscape planning and design process, and ultimately build theories and methods for landscape planning and design in the digital age, thereby promoting the scientific development of landscape architecture.

Nanjing, China Yangyang Yuan

Preface to the Series

Landscape architecture is a vital component of disciplines related to the human living environment, embodying both the attributes of humanistic art and scientific technology. It is grounded in scientific cognition and expressed through artistic form, representing not only the crystallization of human wisdom but also an ideal paradigm for reshaping and mediating the relationship between humans and nature. The scientific nature of landscape architecture imparts it with characteristics of quantification and logic, making digital landscapes an inherent extension of the discipline. Simultaneously, its artistic dimension is inseparably linked with "digital" elements—multi-dimensional representation, virtual reality, digital twins and biomimicry, and digital construction techniques—all of which have greatly enriched the expressive power of landscape art.

At the core of modern landscape design lies the effective resolution of ecological and morphological issues. In the digital era, scientific analysis and evaluation enable precise understanding of ecological environments. Algorithms can even be used to simulate and predict landscape dynamics—what is referred to as "digital simulation." Digital landscape technologies significantly expand designers' capabilities in perceiving and shaping form, leading to a morphological diversity unprecedented in history. The relationship between landscape form and ecology is inherently "synchronic"; digital landscape methods allow for the coordinated optimization of ecology and morphology within the same spatiotemporal dimension. In essence, these methods enable the integrated and synergistic development of ecological systems and spatial forms in the human living environment.

Traditional disciplines of planning and design often rely on field surveys, questionnaires, and other manual methods to collect data, which is then analyzed and synthesized to form a secondary cognition of the objective world, serving as the basis for planning and design decisions. However, throughout this process, due to repeated translation of information across different stages of cognition and the variability of individual experience, issues such as information loss, randomness, and misjudgment are inevitable—ultimately compromising the quality of planning and design outcomes. In contrast, the digital landscape approach utilizes a comprehensive data stream that is not only efficient but also highly accurate. Environmental and site data acquisition, evaluation, and analysis technologies—represented by RS, GIS, and EI/RSEI—serve to support and guide the logic of planning and design generation. This greatly enhances design efficiency, minimizes omissions and errors, and reduces subjective biases, thereby improving overall precision. As a result, digital landscape approaches have already initiated a paradigm shift in both planning and design methods as well as implementation and control strategies in landscape architecture.

The fundamental goal of planning and designing human living environments—including landscape environments—is to create sustainable and livable spaces. But how can the quality of a design scheme be assessed? How can the effectiveness of planning and design be evaluated? It is clear that purely experience-based judgments are no longer sufficient. By employing digital landscape methods and technologies, the performance of the landscape environment can be precisely measured across its entire life cycle. This enables the entire planning and design process to be quantitatively described and dynamically adjusted, thereby significantly enhancing human capabilities in understanding, planning, and designing livable landscape environments.

This book series is primarily authored by my research team, drawing upon our respective doctoral dissertations. Adopting an open framework, each volume focuses on specific fields or topics, all revolving around the central theme of landscape planning and design. This has led to a series of studies covering parametric design, landscape color, plant landscapes, landscape hydrology, sponge cities, human-centered landscape environments, spatial configuration, and heritage parks. These works collectively showcase our achievements in applying digital landscape theories, methods, and technologies in planning and design, and reflect both the academic advancement and practical relevance of our research. Over the past two decades, digital landscape research has gone through two major development phases, and is now entering the 3.0 era, characterized by the emergence of digital twin technologies. Under the national strategy of "Digital China," the ongoing innovation in digital landscape theory, methods, and technologies is poised to lead the landscape architecture profession toward new horizons and comprehensively enhance the quality of the human living environment.

Nanjing, China Yangyang Yuan

Contents

Chapter 1
Parametric Design in Landscape Architecture

Since the mid-eighteenth century, with the Industrial Revolution and particularly in the last century, urbanization has accelerated globally, and urban issues have become increasingly prominent. With the development of theories such as systems theory, cybernetics, and interdisciplinary research, scholars from various countries have expanded the scope of urban studies. The Greek architect C. A. Doxiadis introduced the concept of "Ekistics: The Science of Human Settlements," proposing the study of human habitats as a unified whole. In recent years, as China's urbanization has rapidly advanced, and in response to China's national context, Professor Wu Liangyong officially established the concept of "The Sciences of Human Settlements" in 1993. In his book Introduction to the Sciences of Human Settlements, he states: "The Sciences of Human Settlements focuses on the study of places where human settlements occur and represents an interdisciplinary field that integrates various academic disciplines. Architecture, landscape, and urban planning form the 'core disciplines' within the broader system of human settlements science."[1]As a key component of the science of systems and complexity, Landscape Architecture upholds these characteristics and is closely related to other relevant disciplines.

Chinese landscape architecture has a history of several thousand years, and modern landscape architecture has developed for over 60 years in China. Rooted in the aesthetic appreciation of natural landscapes, it integrates humanistic spirit as a source of creative inspiration. In his book *Yuan Ye* (The Craft of Garden Making), Ji Cheng states, "The creation of gardens follows principles but lacks a fixed formula." Traditional landscape architecture design and construction relied primarily on experience, with design methods emerging from the summary of these experiences, which inherently involved subjectivity, ambiguity, and arbitrariness. In contemporary landscape architecture, both the connotation and extension have undergone significant changes based on traditional foundations, necessitating the accommodation

[1] Wu Liangyong. Introduction to Sciences of Human Settlements [M]. Beijing: China Architecture & Building Press, 2001: 70.

Y. Yuan, *Parametric Landscape Architecture Planning and Design*,
https://doi.org/10.1007/978-981-95-8011-8_1

of multiple objectives. As such, the traditional, experience-based design methods require further development. Technological advancements have led to the emergence of modern landscape architecture. On March 8, 2011, the discipline of "Landscape Architecture" was officially approved as a first-tier subject, presenting both an unprecedented opportunity and a considerable challenge. Landscape architecture is an applied discipline founded on a broad base of natural sciences and humanities, with its core focus on coordinating the relationship between humans and nature. It involves various aspects such as ecology, functionality, space, culture, and construction, and possesses distinctive characteristics. In the early stages of its development in China, landscape architecture was primarily reliant on the fields of architecture, forestry, and horticulture. After being recognized as a first-tier discipline, it now requires further clarification of its scope and boundaries. As one of the three major components of human settlement studies, landscape architecture should not only focus on its own development but also actively draw from the experiences of other disciplines. Modern landscape architecture cannot be separated from art, but it also depends heavily on scientific support, making it a "science of art." Therefore, the discipline of landscape architecture urgently needs to establish its own distinctive methodology and theoretical framework in order to become a rigorous and mature field of study.

1.1 Contemporary Trends in Landscape Architecture Planning and Design

1.1.1 Transformation of Design Concepts

1.1.1.1 Globalization and Regionalism

The development of network technology has significantly reduced the cost and time of information dissemination, enhancing global connectivity. This has led to increasingly widespread and rapid international cultural exchanges, resulting in greater global cultural influence. However, it has also caused a reduction in cultural diversity and led to conflicts between civilizations and values. As Paul Ricoeur stated in Universal Civilization and National Cultures, "While the phenomenon of universalization represents progress for humanity, it also constitutes a subtle form of destruction." As "global culture" tends toward homogenization, "locality" is often regarded as a remedy, frequently re-emphasized and reinterpreted. In the conclusion of Landscape and History since 1500, Professor Ian D. Whyte proposed: "Globalization itself has also created a corresponding demand for localism, as well as for places and landscapes that maintain their uniqueness and continuity."[2] "Locality" is not a

[2] White, Ian D. Landscape and History Since the Sixteenth Century [M]. Translated by Wang Sisi. Beijing: China Architecture & Building Press, 2011: 236.

new concept, and the exploration of "locality" did not originate in the modernist era. As early as ancient Rome, Marcus Vitruvius Pollio in The Ten Books on Architecture suggested that architecture should adapt its form based on the resources and climate variations of different regions. In the early eighteenth century, the "picturesque" garden style in England sparked a trend, as people grew weary of the rigid classical gardens. The English pastoral landscape gave rise to the "Tudor-style" new garden, widely accepted for its distinct regional characteristics. In order to resist the transformation of regionalism into a commercial driving force and a platform for chauvinistic propaganda, Lewis Mumford redefined the principles of regionalism. Alexander Tzonis and Liane Lefaivre referred to it as critical regionalism. This approach is rooted in the long-standing culture and history of a region, drawing inspiration from the geographical environment for design. Since the 1920s, designers represented by Le Corbusier have emphasized industrialized design, focusing on the modularization and standardization of architecture. We are now in an era of non-standardization, an era that values diversity. Critical regionalism interprets a bottom-up design principle, reinterpreting the meaning of "locality" in geographical, social, and cultural contexts, rather than adhering to top-down, self-indulgent design dogmas. This approach preserves the uniqueness and diversity of design. Simon Schama, in his *Landscape and Memory*, extensively discusses the vital role of landscapes in constructing and maintaining national identity. At the same time, critical regionalism also underscores the significance of place in design.[3] Contemporary landscape architecture has long since transcended the pursuit of formal beauty, instead embracing design principles that evolve from the essence of place and reflect regional characteristics.

1.1.1.2 Sustainability and Intensification

The rapid rise of the world economy and the high-speed development of industry after World War II brought about a series of environmental issues, prompting a reflection on past actions. Compared to the fleeting humanistic ideologies such as postmodernism, structuralism, and deconstructionism, the impact of science and technology on landscape architecture has proven to be more lasting. Ecologism has become the most influential and vital ideology in contemporary landscape architecture. The term "sustainability" was introduced in the 1980s and has since become an important global concept, widely promoted and pursued. The Chinese government, in its Agenda 21: China's Population, Environment, and Development White Paper, was the first to incorporate the strategy of sustainable development into the country's long-term economic and social planning, and in 1995, sustainable development was established as a national strategy. "Intensification" is a key means to achieve sustainability, emphasizing the optimal allocation of resources,

[3] Tzonis, Alexander, & Liane Lefaivre. Critical Regionalism: Architecture and Identity in a Globalized World [M]. Translated by Wang Bingchen. Beijing: China Architecture & Building Press, 2007: 23.

comprehensive planning, and the minimization of input-output ratios. In recent years, various national ministries have actively promoted "intensive" policies, striving for "intensification" through societal participation. Accordingly, the concept of intensification and the construction goals of the "two-oriented society"—resource-saving and environmentally-friendly—have placed higher demands on landscape architecture. Guided by the concept of "sustainability," various industries have proposed applicable technological concepts and measures. Human settlement disciplines, primarily consisting of architecture, urban planning, and landscape architecture, have similarly developed appropriate sustainable methods and technologies within their respective fields. The main concept of Life Cycle Management (LCM) focuses on the environmental, social, and economic impacts throughout the entire lifecycle of a product. Its goal is to reduce resource consumption and pollution emissions during the production and consumption processes, while improving its socio-economic benefits.[4] It is evident that, as a design or management method derived from the concept of sustainable development, the ultimate aim of life cycle management is to mitigate the deterioration of global environmental pollution and resource waste. As a "product," landscape architecture projects, from design and construction to post-construction maintenance, also require life cycle management to meet the demands of sustainable and intensive development. As a tool for achieving parametric design in architecture, Building Information Modeling (BIM) contributes to the realization of Building Lifecycle Management (BLM), which encompasses the entire process of a building project from planning and design to construction, operation and maintenance, and ultimately deconstruction. Correspondingly, with the establishment and development of Landscape Information Modeling (LIM), it similarly promotes the advancement of landscape architecture information models and parametric design methods, thereby enabling life cycle management of landscape architecture projects.

1.1.2 The Evolution of Scientific Design

1.1.2.1 Systematization of Design

With the advent of the digital age, both the form and content of design have undergone significant changes, and the design philosophy continues to evolve. Modern design represents an organic integration of science and art, technology and humanity, with no inherent trade-off between these elements, and therefore, it should not be viewed in a one-sided manner. The emphasis on rationality reflects a leap in understanding within the field of design. Advancements in science and technology, as well as the development of scientific theories, have provided a scientific platform and methodology for design, further driving the design industry toward rationality.

[4] REMMEN A, JENSEN A A, FRYDENDAL J. Lifecycle management: a business guide to sustainability [R]. 2007:10–38.

The emphasis on rationality demonstrates that practitioners in the design field are now viewing design with a more mature perspective, seeking the inherent laws and methods within the design process. Approaching things with a rational attitude represents a dynamic process of development, which aligns with the current needs of complexity and systematization in the discipline of human settlement studies. Landscape architecture design, which embodies both artistic and scientific attributes, has always navigated between the realms of intuition and rationality. In response to the contemporary shift from intuition to a blend of "rationality and intuition," the search for new theories and methods in landscape planning and design has become an inevitable development trend.

The emergence of systems theory in the 1930s brought about a profound change in humanity's understanding of the environment and its ways of thinking, providing both theoretical and methodological foundations for the development of modern science, and permeating various research fields. As a scientific discipline, landscape architecture deals with a complex system, and therefore, landscape architecture design research must be guided by systems theory. Contemporary landscape architecture planning and design is considered a composite of environmental science, technology, and art, which gave rise to the concept of "grand design." "Grand design," or systematic design, requires designers to treat the design object as a system, breaking through the limitations of traditional design processes that focus solely on the progression from conceptualization to construction. It encourages viewing the entire lifecycle of landscape architecture planning and design through structural and logical thinking.

1.1.2.2 Scientificization of Design

Before the 1960s, design was regarded as a "black box" process. It was not until the emergence of the "design method" philosophy, which suggested that the design process could be objectively described, that significant progress was made. This intellectual movement directly or indirectly contributed to the publication of Christopher Alexander's *Notes on the Synthesis of Form* in 1964. Horst Rittel, alongside Christopher Alexander, Bruce Archer, and others, initiated the first generation of the design methodology movement. This, in turn, led to the shift from the first generation to the second generation of design methodology research in the early 1970s, transitioning from a purely rational and linear thinking model to one based on bounded rationality and dialogue. The focus of research shifted from establishing "design science" to scientifically describing design, i.e., the "science of design." Herbert Simon defined design research as "a science of design" and elaborated its content as follows: "The difficult intellectual choices and analyses in the design process, partly dependent on form and partly on experience, but all of which can be transmitted as teachable methods" (a body of intellectually tough, analytic, partly formalizable, partly empirical, teachable doctrine about the design process). Similar to architecture, traditional landscape architecture design has emphasized the accumulation of experience and the identification of solutions to problems, often neglecting the study of the underlying principles of those problems. Since the late 1950s, static design methods, characterized by personal experience and a craft-based approach, have gradually been replaced by modern design methods. In the digital age, design theory tends to emphasize the scientific nature of the methods themselves, focusing on rational analysis and scientific induction and deduction. This approach creates environments not merely "designed" in the traditional sense, but rather systems and elements interacting within a diverse, interconnected network to form a logical system. The design outcome is a result of the combined actions of multiple factors and processes. Today, "design research" has become a hot topic in the global design community, and research in this area within the field of landscape architecture, both domestically and internationally, is still in its early stages. Modern landscape architecture planning and design requires the introduction of scientific research methods, along with a scientific articulation of design methods, to establish a systematic theory and methodology for scientific planning and design.

Practice is the ultimate result and the most direct reflection of design; therefore, design research cannot be separated from an exploration of practice. In the first half of the twentieth century, the Bauhaus School in Germany and the Ulm Institute of Design initiated the study of the relationship between science and design, theory and practice. The "Design Methods" movement, led by John Chris Jones and Christopher Alexander, contributed to the development of a scientific methodology for design. In her article *Design Research Paradigms*, Gesche Joost, drawing upon the work of Wolfgang Jonas and Alain Findeli on systematic models for design research, discusses the relationship between theory and practice. She notes, "Research through Design" is not merely a simple connection between research processes guided by scientific methodologies and design practice; the scientific and academic investigations cannot be separated from the context of design practice.[5] The integration of practical research with scientific methods is a significant characteristic of modern landscape architecture design methodology research.

1.1.3 Innovation in Design Tools

1.1.3.1 From the Industrial Era to the Information Age

The invention of the steam engine ushered humanity into the Industrial Age, while the advent of computers broke through the limitations of human intellect and led society into the Information Age. The third scientific and technological revolution, which began in the late 1940s and early 1950s, marked the decline of the Industrial Age and signaled the formal entry of human society into the Information Age, or the Digital Age. On January 31, 1998, former U.S. Vice President Albert Arnold Gore

[5] Just, Gesine. "Design Research Models" [J]. Translated by Cui Qingwei. *Chinese Landscape Architecture*, 2014(7):26–27.

Jr. delivered a famous speech at the California Science Center, titled *The Digital Earth: Understanding Our Planet in the Twenty-First Century*. The concept of "Digital Earth," introduced in this speech, is set to profoundly change people's ways of production and daily life for a significant period in the future. The "3S" technologies, namely Remote Sensing (RS), Geographic Information Systems (GIS), and Global Positioning Systems (GPS), serve as the technological foundation and core of the Digital Earth and have developed rapidly. Alongside these, technologies such as the Internet of Things (IoT), Visualization and Virtual Reality (VR) technologies, and scientific computing have also witnessed rapid advancements.

Ludwig Mies van der Rohe once said, "Architecture symbolizes our era." The establishment of the discipline of Human Settlements Sciences, with architecture, landscape architecture, and urban planning as its "leading disciplines," reflects the demands of the times. As human society enters the Information Age, Earth spatial information science (Geo-Spatial Information Sciences, or Geomatics), as an important branch of information science, has gradually integrated into the academic system of human settlements science, becoming a frontier field of research. The "3S"-based technological methods occupy a crucial position in providing technical support and aiding decision-making within the interdisciplinary system of human settlements science, especially within the field of landscape architecture. In 2011, the Ministry of Housing and Urban-Rural Development issued the *2011–2015 Outline for the Development of Information Technology in the Construction Industry*, which called for the acceleration of the application of new technologies, such as Building Information Modeling (BIM) and web-based collaborative work, in engineering projects, and for the promotion of information standards.[6] This fully advanced the development of information technology within the field of architecture. Additionally, the National Conference on the Informatization of Landscape Architecture was held in June 2013, with the aim of further promoting the application of information technology within the field of landscape architecture.

1.1.3.2 Digital Technology and Construction Technology

The "Digital Technology" born in the Digital Age is inherently tied to computer technology. It refers to the process of converting various types of information into binary numbers that can be recognized by computers and utilizing computers to process this information. By the late 1990s, the discipline of human settlements, with architecture as its representative, entered the Digital Age. From analysis, construction, and expression to actual construction, digital technologies have been deeply and widely applied in every phase of design. In the field of landscape architecture, digital technologies such as data acquisition and measurement/control technologies, data analysis and evaluation techniques, digital generation and construction

[6] Ministry of Housing and Urban-Rural Development. *Notice on Issuing the "Outline for Building Informatization Development (2011-2015)"*. May 10, 2011.

methods, and virtual reality technologies have gradually advanced in their application, becoming essential tools for the future development of the discipline. Digital technologies, based on logical computation systems, have expanded the depth and breadth of human thought, pushing landscape architecture planning and design toward a more scientific and rational approach. Digital platforms provide technical support and platforms for landscape architecture planning, design, and practice. With the assistance of digital technology, landscape designers are able to gain a more comprehensive understanding of the site, objectively evaluate it, and have better control over the design and construction processes. As Professor William J. Mitchell of MIT stated in his speech at the CAAD[7] Futures 2005 conference:"... within the organizational framework of digital planning, design, and construction processes, we are able to precisely quantify every aspect of a project's design and construction content... The architecture that emerges in the Digital Age is characterized by its high complexity. These buildings exhibit a more acute and responsive pursuit of site, function, and the expression of design concepts compared to the industrialized modernist architecture."[8]

It is well known that the Industrial Revolution led to the replacement of traditional handicraft production with mechanized mass production, shifting the focus of construction from "manual" to "mechanical" processes, thereby laying the objective technical foundation for the emergence of the modernist style. Meanwhile, the Information Revolution drove the transition from "mechanical" to "numerically controlled machinery." The development of computer technology, and its application in the design field, has eliminated the divide between art and technology in design. The modern design process has undergone a fundamental transformation, and construction materials have seen major breakthroughs, evolving from "traditional materials" to "multidimensional" and "composite" materials, which strongly support the changes in construction methods. The integration of computer technology with design has shifted from "Computer-Aided Design (CAD)" to "Computer-Intelligent Design (CID)." Technologies such as 3D printing, contour processing, and numerical control machining serve as construction means in design. Their essence lies in the deepening application of computer technology in the design field. Since the beginning of the twenty-first century, the design industry has faced a general trend of moving from subjective intuition to rationality. The emphasis has shifted toward precision, science, and control in design, making the connection between design, construction, and other processes much closer. In landscape architecture construction, the mechanical control technology of 3D Global Navigation Satellite Systems (3D GNSS) has been applied in European and American countries. By inputting the completed 3D terrain model into construction machinery (such as excavators), it is now possible to automate the creation of landscape terrain, saving both labor and resources, while ensuring high precision and speed. The

[7] CAAD:Computer-Aided Architecture Design.

[8] Chen Shouheng, Li Shuyi, Josh Lobel, et al. *Digital Fabrication in Architecture: Design Computation, and Cognitive Theory* [M]. Beijing: China Architecture & Building Press, 2009: 21.

transformation in construction technology has, from another perspective, promoted the development of landscape architecture design methods. This is reflected in the entire design process, from design thinking to design expression, from input to output, and ultimately integrating with implementation.

1.2 International and Domestic Trends in Parametric Research in Landscape Architecture

1.2.1 Exploration of Place Landscape

The study of place theory internationally began with phenomenology and physical geography. In the 1920s, German philosopher Edmund Husserl founded phenomenology, marking the beginning of the era of rationalist epistemology. Husserl emphasized studying the relationship between architectural environments and human perception from a unified perspective of the subject and object of observation. Architectural phenomenology is generally divided into two main research systems: one adopts Martin Heidegger's existential phenomenology, with key representatives such as Christian Norberg-Schulz; the other follows Maurice Merleau-Ponty's phenomenology of perception, with prominent figures like Steven Holl. Heidegger's discussions on "the meaning of dwelling," "poetic dwelling," and "how building belongs to settlement" introduced some of the earliest concepts of place-making, sparking an increased focus on the importance of "place." Starting in the 1960s, Christian Norberg-Schulz established a comprehensive theoretical framework on the concept of place through a series of publications. His work laid the foundation for the theory of place in the context of modern architecture and urban environments.

Scholars in the field of physical geography have gradually expanded the concept of place theory in human geography to include regional environments, architectural design, and landscape planning. Edward Relph, Yi Fu Tuan, and David Seamon all provide relevant discussions in their works. In his 1976 book *Place and Placelessness*, Edward Relph elaborates on the theory of placelessness, arguing that placelessness manifests in environments that fail to recognize the special significance of a place. Such places sever the contextual and value-based connections of a location, replacing diversity with the uniformity of the environment, and replacing sensory order with a conceptualized order. Although scholars in physical geography have discussed issues such as human-environment relations and the connection between architecture and the environment from a phenomenological perspective, the concept of place in cultural geography is broader and encompasses a larger scale of study. While place research in this context offers certain insights for landscape architecture, its applicability may appear somewhat too broad or generalized, especially when considered across different contexts.

Place theory has been widely studied in the fields of architecture and urban planning. Works such as Kevin Lynch's *The Image of the City*, and *Identity by Design* by Georgia Butina Watson and Ian Bentley, explore the relationship between place and architecture, as well as between place and the city, from different perspectives. The study of the relationship between place and landscape architecture also has a long history. As early as the 1770s, a major debate emerged in the field of landscape design: "Is landscape design a process of creating artificial landscapes or a process of managing nature?" French theorist Jean-Marie Morel, in his *Théorie des Jardins*, argued that design is a process of managing natural processes. He applied this concept in practice, especially in his design for Ermenonville near Paris, where he made extensive use of the site's natural hydrology, vegetation, and other processes. In eighteenth-century England, the "picturesque" garden movement emphasized the importance of natural landscapes in garden creation. Lawrence Halprin, a representative figure of the second generation of modern landscape planning in the United States, prioritized understanding the region's landscape before designing. He sought to understand the natural processes that shaped the area, and his designs reflected these processes. His design for the Seacoast Farm Residential Development is a notable example of this philosophy. In the 1960s, Ian L. McHarg's *Design with Nature* laid the theoretical and methodological foundation for the scientific and rational analysis of the relationship between nature, ecological environments, and humans. In this work, McHarg emphasized how natural processes should guide land development.

1.2.2 Development of Digital Landscapes

Since the 1960s, theories such as chaos theory, systems theory, emergence theory, and fractal theory have had a profound impact on modern architectural theory and methods. By the late 1990s, the rapid development of computer and information technologies collectively pushed architecture into the Digital Age. Architects such as Frank Owen Gehry and Zaha Hadid were at the forefront of actively exploring digital design and construction. The academic community responded enthusiastically to the advent of the Digital Age. Many renowned foreign universities began exploring the research, application, and practice of digital technologies in architecture and urban planning, yielding fruitful results. Landscape architecture, being a multidisciplinary field that coordinates the relationship between humans and nature, integrates both artistic and scientific elements. It requires not only scientific analysis and the establishment of all relevant objective factors but also the use of art to narrate and carry the specific semantic meanings of a place. However, the complexity of the objects and goals of landscape architecture has, to some extent, led to a slower adoption of digital technologies in the landscape sector compared to architecture and urban planning.

Digital Landscape refers to the research and application of digital technologies in landscape architecture. In the 1960s, Ian L. McHarg's Design with Nature established a direct link between landscape architecture and ecology. His frequent reference to multi-layered map overlay analysis techniques in the book has served as the blueprint for contemporary digital overlay analysis techniques. This work contributed to the development of Geographic Information Systems (GIS), marking a significant contribution of landscape architecture to the Digital Age. Since 2000, the Department of Applied Sciences at Anhalt University of Applied Sciences in Germany has organized the annual International Conference on Information Technology in Landscape Architecture: Digital Landscape Architecture. This conference has served as a platform for discussing the use and trends of digital technology in landscape design (Table 1.1). Currently, the research on digital landscapes both domestically and internationally focuses on two major areas: landscape digitalization and simulation technology, and landscape analysis, evaluation, and parametric design. The research on landscape digitalization and simulation technologies includes measurement and image processing techniques, landscape environment visualization, and process simulation technologies. Internationally, research on digital landscapes has primarily focused on these areas, with recent shifts in focus towards using digital technology to address practical issues in landscape architecture.

Under the framework of digital technologies, there has been a heated discussion both domestically and internationally regarding Landscape Information Modeling (LIM) and GeoDesign. In the 1970s, Professor Chuck Eastman from Georgia Tech introduced the concept of Building Information Modeling (BIM). In 2009, McGraw Hill Construction released a report titled Building Information Modeling: Transforming Design and Construction to Achieve Greater Industry Productivity,

Table 1.1 International conference on information technology in landscape topics of "Architecture: Digital Landscape Architecture" from 2012 to 2020

Meeting time	Theme
2012	3D-modeling and visualization in landscape architecture
2013	Connectivity and collaboration in planning and design
2014	Developing digital methods in GeoDesign
2015	Systems thinking in landscape planning and design
2016	Representing，evaluating and designing landscapes: Digital approaches
2017	Responsive landscapes
2018	Expanding the boundaries: Landscape architecture in a big data world
2019	Landscape:Informed by science, shaped by design
2020	Cybernetic ground: Information, imagination, impact

which stated: "BIM is the process of designing, constructing, and operating a project using a digital information model." This marked the beginning of the BIM boom in the architecture industry, and discussions on how BIM could be applied in landscape architecture planning and design followed. Some scholars have argued that BIM is not well-suited for large-scale environmental studies and is not particularly effective for site design[9-,10] As a result, they proposed the concept of LIM (Landscape Information Modeling). LIM and BIM share similar concepts and principles but differ in scale, with LIM focusing on a larger scale for landscape architecture projects.

As a practical technology and tool, LIM facilitates collaborative design and full lifecycle management through the creation of digital information models. As a systematic solution, LIM incorporates some parametric ideas and technologies. Currently, LIM includes various platforms and solutions, such as Autodesk Revit, Vectorworks Landmark, and ESRI's ArcGIS and CityEngine.

In the mid-1960s, a significant transformation occurred when Howard Fisher invented the first practical computer graphics software, SYMAP, and established the first computer graphics laboratory at Harvard University. In the late 1960s, Carl Steinitz published A Systems Analysis Model of Urbanization and Change, in which he identified five critical areas worthy of consideration: systems, analysis, models, urbanization, and change. In 1993, Steinitz applied computer simulations to model various landscape processes in a project at Pendleton Camp in California. These simulations included models for soil, hydrology, fire, vegetation, landscape ecology patterns, wildlife habitats, and visual quality. Steinitz pioneered many early ideas regarding the application of Geographic Information Systems (GIS) for landscape analysis and urban planning. He introduced the concept of GeoDesign and, in 2013, published A Framework for GeoDesign: Changing Geography by Design, a book that further elaborated on this innovative approach to integrating geography, design, and technology. The term "GeoDesign" refers to the use of efficient geographic information systems (GIS) that emphasize real-time feedback and evaluation, integrating various factors to assist in broad planning and design processes. From January 6th to 8th, 2010, the first-ever GeoDesign Summit was held at the headquarters of the Environmental Systems Research Institute (ESRI) in Redlands, Southern California. The summit discussed the concept, significance, operation, and prospects of GeoDesign. GeoDesign adopts an interdisciplinary approach, using information technology to provide a unified platform for communication across various fields. This platform is designed to address pressing issues at regional and global scales by optimizing geographic location and orientation. At the GeoDesign Summit held in January 2015, Professor Stanitsz and Hrishi Ballal

[9]Ahmad Mohammad Ahmad, Abdullahi Adamu Aliyu. The Need for Landscape Information Modelling (LIM)in Landscape Architecture[C] //Digital Landscape Architecture Conference 2013. Bernburg, 2012:531–539.

[10]Andrew Nessel. The Place for Information Models in Landscape Architecture, or a Place for Landscape Architects in Information Models [C] //Digital Landscape Architecture Conference 2014. Bernburg, 2013:65–72.

presented the latest development: an online GeoDesign application software, which responded to the "GeoDesign Framework." The standout feature of this software is its ability to rapidly iterate the "GeoDesign Process Change Model." Scholars worldwide have conducted research on GeoDesign, and both domestic and international conferences have been held on this topic. From the current state of research, it is clear that the study of GeoDesign is still in its early stages, primarily focusing on defining its concept, framework, and potential content, as well as predicting future development trends. However, there is a lack of specific empirical research and limited practical applications.

Domestic scholars have also actively engaged in research on digital landscapes and related fields. Zhao Tao published an article titled *"A Brief Analysis of the Application Status of Computer Technology in Landscape Architecture"* in the first issue of *China Landscape Architecture* in 1995. The article analyzed and discussed the application status and issues of computer technology in landscape architecture planning and design. Zhao continued this line of research in a later article, *"Prospects for Computer Technology in Landscape Architecture: A Vision for an Integrated Landscape Architecture Information System"*[11-,12] where he proposed a comprehensive landscape architecture information system guided by systems theory and information theory, with integrated system development as the strategy for future development. In recent years, papers related to digital landscapes have been concentrated in publications, becoming a current research hotspot. The authoritative domestic journal *China Landscape Architecture* featured a "Digital Landscape" theme in its 2015 issue (No. 7) and a "GeoDesign" theme in its 2014 issue (No. 10). The journal also published several articles on "Digital Technology in Landscape Architecture" in the fourth issue of 2014 and "Digital Technology and Landscape Architecture" in the first issue of 2013, exploring the development and current applications of digital technology in landscape architecture, and offering future outlooks. Cai Linghao, in his works *"Digital Strategies for Landscape Architecture Planning and Design"*[13]and *"Conceptual Spectrum and Process Flowchart of Digital Landscape Architecture Planning and Design"*,[14] elaborated on the concepts related to digital strategies in landscape architecture, analyzing key digital design methods and application software platforms, and summarizing the flowchart of the digital planning and design process in landscape architecture. He pointed out that the digital process in landscape architecture planning and design can be broken

[11]Zhao Tao. "Prospects of Computer Technology in Landscape Architecture: The Concept of Comprehensive Landscape Information System (Part 1)" [J]. *Chinese Landscape Architecture*, 1995(4):59–60.

[12]Zhao Tao. "Prospects of Computer Technology in Landscape Architecture: The Concept of Comprehensive Landscape Information System (Part 2)" [J]. *Chinese Landscape Architecture*, 1996(1):60–63.

[13]Cai Linghao. "On Digital Strategies for Landscape Planning and Design" [J]. *Chinese Landscape Architecture*, 2012(1):14–19.

[14]Cai Linghao. "Genealogy of Concepts and Flowchart Representation for Digital Landscape Planning and Design" [J]. *Landscape Architecture*, 2013(1):48–57.

down into four basic stages: environmental cognition, design construction, design evaluation, and design mediation. This includes technologies such as landscape environment space simulation and analysis, plant landscape simulation and prediction, digital ecological analysis and evaluation of landscape architecture, landscape architecture planning and management information systems, digital generation and parameterization in landscape architecture planning and design, visualization and digital simulation technology, virtual reality technology, digital construction technology in landscape architecture, the application of BIM in landscape architecture, digital evaluation methods, numerical control construction systems, innovative digital technologies, interdisciplinary research in digital technology, experimental design and construction, and landscape architecture theories in the digital age. Bao Ruiqing, in his article *"Discussion on Computer-Aided Strategies for Landscape Architecture Planning and Design"*,[15] explored the auxiliary role of computer technology in landscape architecture planning and design from three perspectives: geographic information systems (GIS), ecological-assisted design technologies, and model construction. Since 2013, the Digital Landscape International Symposium, organized every 2 years by the School of Architecture at Southeast University, has further promoted research in this field in China. The symposiums, held in consecutive years, showcased the latest research and developments in digital landscapes both domestically and internationally, systematically discussing current applications and technological developments, and forecasting the future of digital landscape applications. Building on over a decade of research and practice, Southeast University established the Digital Landscape Laboratory in 2012, becoming a pioneering institution in providing technical support for digital landscape research and undertaking systematic research and practical explorations. While much discussion has been devoted to the connotations and application areas of digital landscapes in China, substantial research and practical application are still in the early stages.

1.2.3 The Rise of Parametric Planning and Design

The origins of parametric design methods in landscape architecture can be traced back to the early twentieth century. Around 1910, the widespread use of electricity in the United States led to the invention of the first tracing boards, originally designed to aid in drawing. In 1912, Warren H. Manning was the first to use these boards for map overlays as an analytical tool. Based on this approach, he planned the landscapes of the entire United States, and his results were published in the *Landscape Architecture* journal in June 1923. In 1943, L. B. Escritt's monograph *Regional Planning* provided a detailed discussion on the use of overlay techniques to analyze the landscape environment. Similarly, Ian McHarg frequently referenced

[15] Bao Ruiqing. "Strategies for Computer-Aided Landscape Planning and Design" [J]. *Journal of Beijing Forestry University (Social Sciences Edition)*, 2013, 12(1): 38–44.

multi-layer map overlay analysis in his famous book *Design with Nature*. Overlay techniques, as an early method of quantifying site analysis, had certain limitations, such as treating all factors with equal weight and severing the system's connections between various landscape elements. Therefore, it cannot be considered a true parametric approach. However, its development fostered the advancement of geographic information systems (GIS), contributed to the organic integration of multiple disciplines, and laid the foundation for more scientifically driven analytical methods in the future. Building on McHarg's work, Stanitsz continued to advance the field, conducting extensive research and exploration into the role of computers in landscape analysis and simulation. Stephen M. Ervin, within the framework of GeoDesign, focused on technical research, emphasizing the use of digital computation, computational processes, and communication technologies to promote collaboration and information exchange. The framework, techniques, and methods of GeoDesign have provided valuable reference and insight for the development of parametric landscape architecture planning and design methods. GeoDesign introduces geographic analysis into the design process by using numerous layers of information stored in databases to describe various natural and social elements within the project's spatial scope. This allows the initial design sketch to receive timely suitability evaluations, a concept that shares common ground with parametric landscape planning and design methods. However, GeoDesign's research objects and scope are considerably larger. According to Stanitsz's framework, GeoDesign encompasses "four types of people, six models, and three cycles," bringing together local residents, geoscientists, information technology professionals, and planning design experts into a collaborative management platform. This broader scope makes GeoDesign more expansive than parametric landscape planning and design methods.

In recent years, technological advancements have driven the development of parametric software support platforms, further deepening landscape environmental evaluation methods, analytical techniques, and design methodologies. Articles such as Kuang Wei's *"An Overview and Reflection on the Development of Parametric Planning and Design in Landscape Architecture"*,[16] Cao Kaizhong and Zhu Yufan's *"The Insights of Parametric Diagrams for Landscape Architecture Planning and Design"*,[17] and Chi Zhiwei et al.'s *"Application Progress of Parametric Design and Its Implications for Landscape Design"*[18] have explored the development and application of parametrics in landscape architecture planning and design. However, these studies did not delve into the specific methods and tools for design. Bao Ruiqing, focusing on programming design based on Python, published a series of six books

[16] Kuang Wei. "Overview and Reflections on the Development Status of Parametric Planning and Design in Landscape Architecture" [J]. *Landscape Architecture*, 2013(1):58–64.

[17] Cao, K., & Zhu, Y. (2013). Insights from Parametric Diagrams in Landscape Architecture Planning and Design. *Landscape Architecture*, 2013(1), 65–68.

[18] Chi, Z., Chen, J., & Zhang, D. (2012). Progress in Parametric Design and Its Implications for Landscape Design. *Chinese Landscape Architecture*, 2012(10), 40–45.

in 2015: *"Parametric Logic Construction Process",*[19] *"Parametric Model Construction",*[20] *"Learning Python – Becoming a Designer with Programming Skills",*[21] *"Folded Programs",*[22] *"Programming Landscapes",*[23] and *"Python Programming in ArcGIS".*[24] These books explore the application of Python programming in landscape design.The Southeast University team has actively researched and explored parametric landscape architecture, completing a series of National Natural Science Foundation projects such as *"Research on Parametric Landscape Architecture Design Methods — A Case Study of Vertical Design" (51278115)* (Principal Investigator: Cheng Yuning), *"Research on Parametric Landscape Architecture Spatial Density — A Case Study of Built Environments" (51478104)* (Principal Investigator: Li Zhe), *"Parametric Research on the Scale of Excellent Traditional Landscape Architecture in Jiangsu and Zhejiang" (51578130)* (Principal Investigator: Chen Ye), and *"Research on Parametric Design of Water Environments in Landscape Architecture" (51608108)* (Principal Investigator: Yuan Yangyang). A number of research papers have been published based on these projects.

1.3 The Significance of Parametric Landscape Architecture Planning and Design

Modern landscape architecture planning and design methods face several pressing challenges:

1. Traditional Landscape Architecture Design Relies on Experience: Traditional landscape architecture planning and design methods are often based on accumulated experience. These methods typically emphasize the subjective initiative of the designer, building upon their understanding of a place. As a result, designs are heavily influenced by the designer's intuition, perception, and accumulated experience. However, such methods often lack scientific rigor and teachability. Consequently, the results of these planning and design processes are frequently based on probabilities and assumptions. As a discipline that is both ancient and

[19] Bao, R. (2015). *Parametric Logic Construction Process.* Nanjing: Jiangsu Phoenix Science and Technology Press.

[20] Bao, R. (2015). *Parametric Model Construction.* Nanjing: Jiangsu Phoenix Science and Technology Press.

[21] Bao, R. (2015). *Learn Python: Becoming a Designer with Programming Skills.* Nanjing: Jiangsu Phoenix Science and Technology Press.

[22] Bao, R. (2015). *Folding Programs.* Nanjing: Jiangsu Phoenix Science and Technology Press.

[23] Bao, R. (2015). *Programming Landscapes.* Nanjing: Jiangsu Phoenix Science and Technology Press.

[24] Bao, R. (2015). *Python Programming under ArcGIS.* Nanjing: Jiangsu Phoenix Science and Technology Press.

modern, landscape architecture must recognize that relying solely on a priori methods to address present-day problems is insufficient, especially in the context of continuous social progress.

2. In the past decade, especially in the last 3 years, research on digital landscapes has flourished both domestically and internationally, leading to a surge in the study of parametric landscape architecture. At present, most of the theoretical frameworks for parametric design are borrowed from architecture, with much of the related research being similar to parametric architectural design. These studies are largely confined to *"algorithmic form generation"*, focusing on the study of form and spatial generation under parametric control. However, they have not systematically addressed the specific characteristics of landscape architecture planning and design in terms of parametric design. Not only are parametric practices in landscape architecture rare, but research on the parametric design theory and methods specific to the landscape architecture discipline is also significantly lacking.
3. Under the global trend of internationalization, the boundaries between regions in landscape environments are gradually becoming blurred, and regional characteristics are slowly disappearing. This inevitably leads to the awkward situation of "cities looking alike." The disregard for the unique qualities of places has resulted in a series of designs that are symbolic, patterned, and labeled, which do not align with the character and traits of the local environment. These designs not only consume vast amounts of resources during the creation process but also cause significant waste in terms of human and material resources during subsequent maintenance and management. A key challenge faced by modern landscape architecture planning and design is how to genuinely implement *"contextual design"*—creating design features that reflect the unique character of the place. This would ensure the rational use of environmental resources, meet the demands for sustainable and intensive social development, and avoid the pitfalls of generic, resource-draining approaches.
4. The creation of landscape architecture environments is a complex, systematic project that involves multiple aspects, such as ecology, culture, function, and space. Therefore, it is essential to integrate and coordinate the multiple objectives of planning and design, forming a modern landscape architecture planning and design approach guided by systems theory. The development of digital technologies, through both hardware and software platforms, has provided strong support for the scientific development of landscape architecture. However, there is a noticeable scarcity of related theoretical research and technical applications concerning landscape architecture planning and design mechanisms. Under a systemic framework, constructing a parametric design method that coordinates *"ecology, function, space, and culture"* is a crucial direction and research challenge for the future. This involves scientifically intervening, adjusting, planning, and designing landscape environments through parametric methods. Such an approach is not only an important development direction but also a critical research area for modern landscape architecture planning and design.

In response to the issues mentioned above, the author's research focuses on the following three objectives:

1. Shifting Landscape Architecture Planning and Design from "Feeling" to "Perception";
2. Exploring Modern Landscape Architecture Planning and Design Methods for "Contextual Adaptation" and "Minimalist Design";
3. Creating a Parametric Landscape Architecture Design Mechanism Based on the Coupling Principle.

Herbert Simon, in his 1968 work *The Science of Artificial*, stated: "Design is finding a way to improve the existing situation." Carl Steinitz regarded this as the best definition for the discipline of landscape architecture. The study of a parametric landscape architecture planning and design mechanism based on the coupling principle is an active exploration of design methods grounded in digital technologies. Guided by the current concepts of regionalization, intensification, and sustainable social development, it seeks a landscape architecture planning and design approach that can be effectively applied in practice.

1.3.1 Promoting the Scientific Nature of Landscape Architecture Planning and Design

Landscape architecture is a discipline focused on maintaining and creating a harmonious relationship between humans and the land. Its core lies in the integration of society, ecology, and art, with the primary characteristics of balancing planning and design, landscape ecology, and human society. It involves multiple disciplines and areas of study. Traditional landscape architecture planning and design methods often emphasize the emotional over the rational, remaining at a qualitative level and lacking quantitative support. The author is committed to scientific design, introducing parametric technologies based on digital platforms into landscape architecture planning and design. This approach enriches contemporary landscape architecture design theory and, to some extent, addresses the scientific shortcomings of traditional design methods.

1.3.2 Constructing Parametric Landscape Architecture Planning and Design Theory

Contemporary landscape architecture is characterized by complexity and holism, requiring the simultaneous fulfillment of multiple design objectives, including ecology, culture, function, and space. The transformation of theory and technology proceeds alternately, continuously advancing forward. Traditional design methods struggle to achieve a systematic design for multiple objectives. The development of contemporary digital technologies has provided a solid platform for the scientific advancement of landscape architecture, while also driving progress in related theories. The author, based on landscape architecture planning and design, draws on relevant theories and methods from disciplines such as architecture, urban planning, and ecology, conducting systematic research on the parametric approach from theory to practice within the landscape architecture field. This contributes to the formation of a theoretical and methodological system for landscape architecture planning and design with disciplinary autonomy.

1.3.3 Exploring the Practice and Application of Parametric Landscape Architecture Planning and Design

As a discipline closely connected to practice, theoretical research guides practical application, while the response of practice drives the refinement of theory. Landscape architecture planning and design is a core component of the discipline, closely tied to theoretical research and with a wide scope of practical application. The study of parametric landscape architecture planning and design methods should ultimately be applied in practice. Based on the coupling principle, this approach follows the principle of *Minimization Design*—a design philosophy focused on minimizing resources, in line with current societal demands for sustainability and intensification. The parametric platform aids in achieving full control from planning and design to construction, enhancing the efficiency and accuracy of the construction process in practice, which holds significant practical importance.

The "coupling" refers to the interaction between two systems: *design* and *place*. Addressing the issue of lacking unique landscape features, design rooted in the characteristics of a place helps to create distinctive landscape environments, preserve regional traits, and achieve a sense of place identity. The author aims to explain the internal operating principles of parametric design at the methodological level through research on design mechanisms, and to demonstrate the application of design approaches, strategies, and techniques guided by methodology, using practical case studies. This exploration seeks to build a parametric landscape architecture design system, moving from theory to practice.

1.4 Parametrics and Landscape Architecture Planning and Design

1.4.1 Quantification, Digitization, and Parametrics

1.4.1.1 Quantification

"Quantification" refers to the process of representing objects that are difficult to characterize in scientific research using specific values or ranges. Quantitative research is a relative concept to qualitative research. It is not only a fundamental research paradigm in the social sciences but also one of the important steps and methods in scientific inquiry. From the history of science, we can observe that by the end of the nineteenth century, human knowledge-seeking activities gradually shifted from the idealism of the Enlightenment to empiricism. With the support of advancements in mathematics and logic, the path of quantification and empiricism became the dominant approach. Quantitative research became the main method in modern scientific inquiry, shaping the development of science in the twentieth century. Quantification refers to the method and process of acquiring meaning through numerical or range-based analysis and comparison. The main steps in quantitative research include data collection, classification, organization, and analysis. Compared to qualitative methods, quantitative research emphasizes the objective reality of facts, while qualitative research focuses on the subjective intent of the object. Quantitative research stresses empirical validation, whereas qualitative research focuses on interpretive construction. Quantitative research primarily uses methods such as empirical measurement, statistical analysis, and model building, whereas qualitative research mainly relies on logical reasoning, historical comparison, and other methods. The biggest advantage of quantitative research over qualitative research is its ability to yield more objective and rational conclusions. Since objects are expressed in numerical or range forms, the accuracy of the data directly impacts the alignment between the final analysis results and the objective facts. Indicators set the expected indices, specifications, and standards to be achieved. Using an indicator system for evaluation is an effective way to quantify phenomena. An indicator system composed of various parameters can form an organic evaluation system that describes both the quantity and nature of the subject.

In the 1990s, Mr. Qian Xuesen proposed the methodology of "comprehensive integration from qualitative to quantitative" in response to the issues of open complex systems, offering a philosophical interpretation of the relationship between qualitative and quantitative methods.[25] As a complex system, landscape architecture must adhere to this "qualitative-to-quantitative comprehensive integration" approach. Quantitative research is closely related to scientific experimental research,

[25] Lu, M. (2005). The Formation and Development of the "Qualitative to Quantitative Synthesis and Integration Method": A Tribute to Academician Qian Xuesen on His 93rd Birthday. *China Engineering Science*, 2005(1), 9–16.

and quantitative analysis helps to achieve the scientific nature of research. Today, with the advancement of the scientificization of landscape architecture, various research directions have incorporated quantitative methods, including spatial form quantitative evaluation, landscape resource sustainability quantitative evaluation, landscape structure quantitative evaluation, color quantitative analysis, soundscape quantitative analysis, environmental capacity quantitative evaluation, and suitable technology quantitative evaluation, among others. The application of quantitative methods in landscape architecture primarily involves the collection of data or the construction of evaluation systems to obtain comparable values or ranges. Quantitative analysis and evaluation methods provide an effective pathway for elements of landscape architecture that are difficult to articulate, shifting landscape architecture research from qualitative to quantitative approaches. This has greatly advanced the scientific development of the landscape architecture discipline.

1.4.1.2 Digitization

Digitalization is an extremely broad concept, and its basic process can be described as transforming a vast array of complex and dynamic information into measurable numbers and data, which are then converted into a series of binary codes through digital models. These are introduced into computers for unified processing. The core of digitalization is digital technology, which, as a fundamental tool, is widely applied today and encompasses various derivative technologies based on computer hardware, software, and communication technologies. In the field of architecture, digitalization has not only impacted the way architectural designs are expressed through its powerful digital tools, but has also reshaped the entire construction process of the industry by enabling collaborative work based on information science. Thus, digital architectural design can be briefly defined as the process of utilizing digital technology in architectural design and construction. Since it deals with complex systems, digitalization is not just limited to the use of digital technology tools; it also incorporates mathematical and logical thinking. Early architectural digitalization developed through the creation of drawing software, bringing efficiency to the design process. Today, digital technology is not only synonymous with efficiency, but also helps people to re-understand design, enhancing control over the entire design process. In comparison to architecture, the scope and meaning of digitalization in landscape architecture is "harmonious yet distinct." While the digitalization of landscape architecture still involves the use of digital technologies, due to differences in research objects and principles, its application in the field includes stages such as information acquisition and management, landscape analysis and evaluation, simulation and modeling of landscape environments, deepening of planning and design, and post-construction performance evaluation.

1.4.1.3 Parameterization

"Parametric" originally refers to issues in mathematics, statistics, and computer science. The integration of "parametric" with "design" has become a key innovation in planning and design research in recent years. Based on current developments, it can be broadly divided into two categories: "parametric-assisted design" and "parametric design." The former uses parametric methods as a technical tool to assist in achieving design outcomes, while the latter represents a way of thinking that focuses on associating mathematical logic, geometric logic, algorithmic logic, etc., with design problems, seeking solutions from the perspective of natural sciences. Professor Xu Weiguo believes that parametric design in architecture essentially aims to find a relationship or rule that organizes the main factors influencing the design. In this context, the factors affecting architectural design are considered as parameters (variables), forming a parametric model, which is then described using computer programming languages. Through computer technology, the parameters and variable data are converted into visual representations, and this result forms the preliminary design.[26]

The key stages of the parametric design process include: data-driven environmental evaluation, data-driven design demands, the construction of parameter relationships in the system design, and the establishment of computer software parametric models. The realization of parametric design relies on digital means, covering various digital technologies such as efficient computer computation, computer-aided design (CAD), algorithmic programming, and computer-aided geometric optimization. "Parametric" is essentially a term that describes a process; it defines the relationships between things by establishing connections between variables and outputs, thus forming derivative relationships. Parametric design can be referred to as variable-based design, meaning that the design is controlled by parameters. Each parameter controls or reflects a crucial characteristic of the design result, and changing the value of a parameter will alter the design outcome. It is important to note that parameters are not the same as variables. The difference lies in that parameters are conventions assigned by the designer through analysis and prediction, describing the operational features and relationships of the system. When the system operates, variables may change continuously, but the parameters remain constant. However, this does not mean that parameters are fixed constants. They can still change, and when parameters change, it may lead to changes in the entire system.

Currently, there is a tendency to label parametric design in the fields of architecture and landscape architecture. Parametrics describes a process, but does not represent the outcome. Parametric design is not synonymous with non-linearity, nor is it solely concerned with the pursuit of form. Parametric design emphasizes systemization and process control, and through the systematic integration of information, it provides an interface for digital construction. The complexity of parametrics lies in

[26] Xu, W., & Xu, F. (2010). Parametric Design in Architectural Creation and Thinking in China: An Interview with Professor Xu Weiguo and Mr. Xu Feng from the School of Architecture, Tsinghua University. *Urban Architecture*, 2010(6), 108.

the logical relationships between subsystems, meaning that the subsystems are organized in an orderly manner according to complex logical relationships, rather than being simply listed.

In summary, quantification, digitalization, and parametrization are three interrelated but independent concepts. They are connected to each other, and their relationship can be illustrated by a Venn diagram (Fig. 1.1). Digital design encompasses a very broad scope. As long as a computer is used at any stage of the design process, it can be considered digital design. Therefore, in terms of scope, digital design includes parametric design. It is important to note that quantification is not the same as parametrization. Quantification can be one of the technical methods applied within parametrization. Through quantitative research, values and intervals are generated to describe the specific properties of objects, which can then be transformed into parameters that can be evaluated and used. Thus, quantification plays an important role in parametric landscape architecture research, being integrated into the parametric landscape planning and design process. The characteristic of parametrization lies in understanding and solving problems through a system-based approach. Parametric landscape planning and design not only constructs a quantitative research process, but also establishes a system model for landscape environment evaluation and design that includes both variables and parameters. The "quantities" derived from preliminary research are brought into the model, where calculations are carried out according to defined rules to guide the generation of "planning design outcomes."

1.4.2 Parametrization and Landscape Architecture Planning and Design

Since the twentieth century, the integration of the subjective and objective aspects of design has been a common trend in the design industry. The development of design has moved past its earlier, more laissez-faire stages, including architecture, and is now focused on making design processes controllable and scientific, with closer connections between design, manufacturing, and construction. The rise and

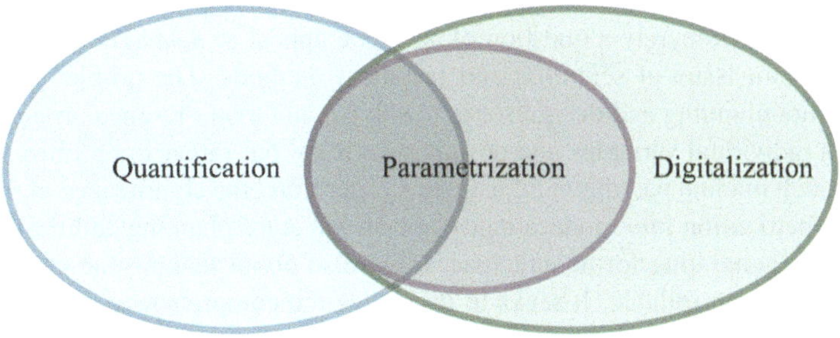

Fig. 1.1 Schematic diagram of the relationship between quantification, parameterization and digitalization

innovation of technology have provided the conditions and possibilities for people to have better control over the entire design process. Parametrization, as one of the most "trendy" design concepts today, has only undergone a few decades of development, but it has already brought significant transformations to design thinking. By controlling parameters and utilizing the computational power of computers, parametric design transforms the traditional design process—previously based on intuitive control—into a scientifically logical process. In the era of big data, digital technologies and methods, including parametrization, are capable of effectively processing and analyzing massive amounts of data related to landscape environments. This greatly enhances work efficiency, particularly with respect to the characteristics of the landscape architecture discipline.

The rise of complexity science has had a significant impact on landscape architecture, introducing systems theory into the study of the discipline. Landscape architecture is a multi-objective system composed of several subsystems, with diverse and complex characteristics. The objects and scales of study are unique, and the application of parametrization in landscape architecture differs from that in architecture design. Parametric landscape architecture planning and design refers to establishing the entire design process on a digital hardware and software platform. It involves rationally constructing logical models according to design requirements, selecting appropriate algorithms, and relying on the computational and graphical capabilities of computers to scientifically analyze and evaluate spatial environments and generate design results. The parametric approach enhances the designer's ability to solve complex environmental and design problems, fosters creativity, and provides a more experiential and authentic design expression. Ultimately, it produces landscape architecture design outcomes that incorporate scientific, artistic, and social values. This approach integrates environmental cognition, comprehensive graphic interpretation, virtual reality expression, digital construction, numerical control manufacturing, evaluation and management, and sustainable development throughout the entire lifecycle of landscape architecture.

Landscape architecture design is multi-objective. Subsystems such as ecology, space, function, and culture each have their own rules and characteristics. These subsystems form an extremely complex "large" system, with numerous variables influencing and constraining one another, such as habitat and plant relationships, engineering volume and cost, etc. The quality of landscape architecture planning and design is not merely a question of aesthetic appeal or spatial construction, nor is it solely an issue of scientific and technical methods. The problem landscape architecture planning and design seeks to solve is not a one-to-one correspondence between individual variables and design objectives, but rather to optimize the system through manual parameter adjustments. Therefore, the significance of introducing parametrization into modern landscape architecture planning and design is not only about generating forms and spaces, but also about making the entire design process fully controllable. It seeks to maximize the comprehensive benefits, minimize environmental disturbances, and optimize design outcomes while ensuring that multiple subsystems work in coordination. The comprehensive benefits apply to the entire lifecycle of landscape architecture, not just a single phase.

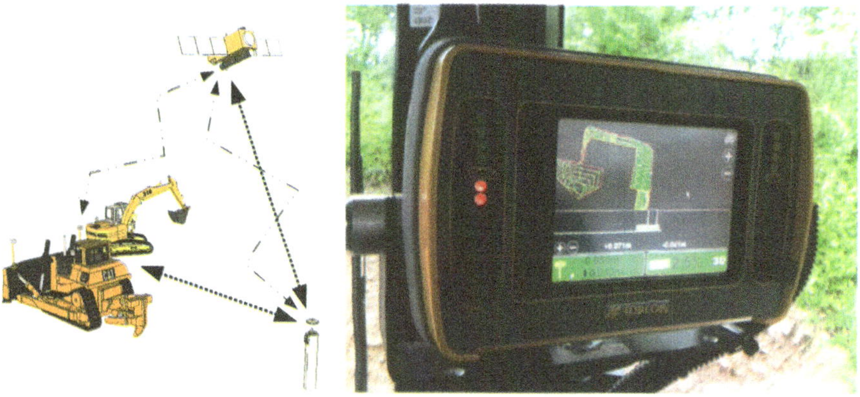

Fig. 1.2 The digital terrain model is loaded into the three-dimensional mechanical control system

The development of modern science, especially modern technology, has bridged various stages of planning, design, and construction. Parametric design plays a crucial role in streamlining the interface for precise construction in the future. This is another important significance of parametric design for landscape architecture planning and design. In the construction and building phase, by introducing digital construction models into machinery, precise construction work can be carried out. In countries such as Switzerland, 3D Global Navigation Satellite System (GNSS) mechanical control technology has become widely used for site construction. By using reference stations (base stations) and mobile receivers (roamers) mounted on shovels for positioning, along with electronic control devices embedded in the machinery, construction can achieve centimeter-level accuracy for natural site terrain (Fig. 1.2).[27] Moreover, digital terrain models (DTM), which serve as foundational data, allow for faster execution compared to traditional piling and line-staking work.

In the future, landscape construction based on parametric design will rely on automated machinery, not only greatly improving efficiency but also enhancing precision and saving labor. For example, in the Danube River restoration project between Ertingen and Binzwangen in Germany, hydraulic digital terrain models (Hydraulic DTM) generated from remote sensing and GIS systems became the basis for editing and analyzing various hydraulic and ecological parameters. These models enabled simulation of the river's post-construction surface morphology, hydrological factors, and water system dynamics, linking the design and construction phases of the water system landscape. By installing 3D mechanical control systems on excavators and using digital terrain models, precise construction was achieved. The automation also significantly reduced labor costs—during the entire construction process, apart from the sheet pile support work, only one excavator operator and one worker were needed to manage the on-site work (Fig. 1.3).

[27] Peter Peiček. (2013). Smart Landscaping. Translated by Guo Yong. *Landscape Architecture*, 2013(2), 33–37.

Fig. 1.3 Digital hydrotopographic model: An excavator equipped with a three-dimensional mechanical control system is working on site. Quoted from: Peter Peiček. (2013). Landscape Water System Optimization Design Based on Digital Hydrological Terrain Model: River Restoration of the Danube River at the Erlangen-Binzwangen Section in Germany. Translated by Li Pan. *Chinese Landscape Architecture*, 2013, 29(08), 30–34

In 1963, Ivan Sutherland from the Massachusetts Institute of Technology (MIT) published his doctoral thesis titled *"Sketchpad: A Man-Machine Graphical Communication System"*, which marked the beginning of interactive computer graphics and the development of computer-aided design (CAD). In its early stages, computers were used solely as a tool to assist with design, making it more convenient and efficient. With the advancement of computer processing power and the development of artificial intelligence technology, parametric design emerged, seamlessly integrating digital analysis with construction. Through the construction of logical models and the addition of algorithms, dynamic control using parameters allows for the rapid generation of diverse design outcomes. Parametrics represents a systematic design process, and compared to parametric-assisted design, it incorporates rigorous logic throughout the entire design process.

Parametric design reflects a bottom-up self-organizing design model. Within the framework of parametricism, the creation of a design work is akin to the self-replication of a living organism, where various parts develop and grow proportionally at the same time, with these proportions being defined by the parameters. Different parameters lead to the vast diversity of living organisms. Similarly, due to the different geographical locations, landscape design sites each have their own distinct characteristics, leading to varying design parameters. The parametric process of landscape planning and design is similar to the "Butterfly Effect," where even the smallest change in one parameter can result in vastly different design outcomes. The parameters obtained from different sites are faithful reflections of the original information set of those locations, bringing the site's unique characteristics into the design results. The parametric landscape planning design mechanism helps preserve and develop the site's distinctive features. The parametric design process also reflects the systemic and open nature of landscape architecture. The establishment of parametric relationships helps discover and understand the underlying logical structures and spatial orders of this complex system of landscape architecture, which are expressed numerically, making them subject to discussion and evaluation. It also

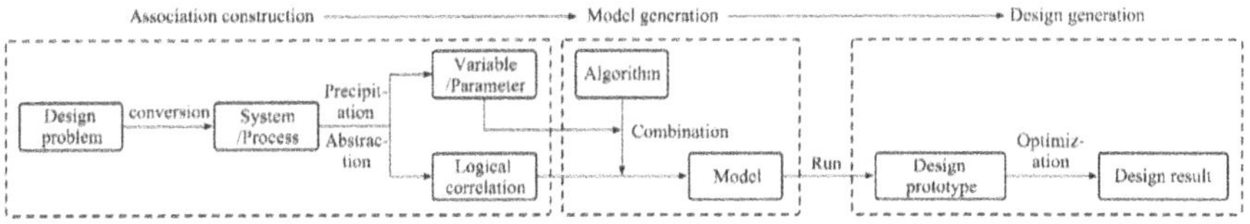

Fig. 1.4 Parametric landscape architecture planning and design process

possesses a certain level of openness, maintaining a close relationship with the environment, allowing for mutual interaction, and continually evolving to better adapt to the environment. Thus, the parametric landscape planning design methodology is characterized by full-cycle, full-scale, systemic, scientific, and controllable features.

The parametric landscape planning and design process involves several key steps: Step 1, Selection of Parameters and Variables, The factors influencing the design are treated as variables, while the design elements selected based on the design goals are defined as parameters; Step 2, One or more suitable algorithms are chosen as the calculation rules for the design. These algorithms will guide the computational process; Step 3, By combining the algorithms, parametric relationships are constructed to describe the design system. This step leads to the generation of a parametric model.

Step 4, The parametric and variable information is input into the computer language environment. Using the algorithmic model as the calculation rule, the final result is produced—this is the preliminary design scheme corresponding to the design goal (Fig. 1.4). The construction of a parametric model makes the design process more flexible, capable of meeting the multi-element and system complexity requirements of landscape planning. By adjusting the parameters and variables based on the algorithmic model, the powerful computational capacity of the computer allows for rapid generation of different results, which facilitates the control and comparison of designs. Furthermore, adjustments to the algorithmic model can optimize the design process, resulting in design outcomes that better meet the objectives. The parametric algorithmic model provides an abstract and logical process model for landscape planning and design, enabling quick feedback—something traditional planning design methods struggle to achieve.

In the field of architecture, parametric research has primarily focused on form generation or the study of formal logic. This involves using algorithms to express the logical relationships between parameters, with the aim of simulating and describing the process of generating buildings and constructed environments, including villages and cities, through artificial intelligence. Such parametric research emphasizes the simulation and description of physical forms and spatial configurations, adjusting parameters to simulate the generative process of buildings or settlements under constraints. This is referred to as "parametric generation." Form and space, being the most direct expressions and external manifestations of design, are essential aspects of any design study. However, form and space are by no means the entire scope of design, especially in landscape architecture planning and design. In my research on parametric landscape planning and design, the focus is not only on form and space generation but also on the regulation and optimization of the system framework. Rather than being confined to the parametric generation of physical forms and spaces, the research interprets the parametric design mechanism itself. The goal is to achieve the coupling of the form, process, and generative rules of landscape architecture planning. This coupling operates on multiple levels and across multiple objectives, leading to the overall optimization of the landscape system under human intervention.

Although landscape planning and design share similar elements, there are certain differences in the specific steps and methods due to differences in problem-solving and scale. The parametric design process in the two fields is also distinct. The former focuses more on the parametric expression of the design process based on analysis and evaluation, aiming for the optimization of the landscape system through reasonable protection and regulation of environmental resources. The latter, on the other hand, deals with the construction of the landscape system, where the parametric design is more focused on the generation of form and space. Therefore, the application of parametric design in landscape architecture varies depending on the research objectives, scale, and subject matter.

The parametric strategy in landscape architecture is not exactly the same as the parametric design proposed in the field of architecture. Therefore, its study must be integrated with the discipline's self-discipline and selectively draw on the research findings of related fields. The development of digital methods and technologies represented by parametric design allows landscape planning and design to transcend traditional design processes, gaining the ability to scientifically, objectively, and rationally understand and analyze the operating and developmental patterns of landscape environments. More importantly, it supports the landscape architecture discipline, providing a platform for dialogue and mutual advancement with related disciplines, promoting the self-organization and development of the field.

1.5 Parametric Software Platform for Landscape Architecture

The parametric landscape architecture platform is the specific implementation platform for parametric landscape planning and design, consisting of software platforms with parametric design capabilities. The purpose of building a parametric landscape platform is to further clarify, interpret, and study the interrelationships between elements and systems, multiple goals, and the balance between the art and science of landscape design. It aims to optimize and coordinate the system through variable adjustments, while also contributing to the consideration of the entire life-cycle of the landscape. Since the elements and systems are described in a parametric manner, the platform must have interconnected, dynamic, and operable features. It uses parameters to control the landscape planning and design process, adjusting the parameters to achieve optimized results, and applying quantitative techniques to compare design outcomes, thus enabling the entire process of parametric landscape planning and design.

With the rapid development of computer and information technologies, new software platforms are emerging continuously, and there is no shortage of software platforms used in digital landscape architecture design. Researchers such as Cai Linghao,[28] Bao Ruiqing,[29] Chi Zhiwei, Chen Jie, and Zhang Deshun[30] have written papers from the perspectives of digital planning and design, computer-aided design, and parametric design, proposing concepts for the application of software platforms in the field of landscape architecture, and providing preliminary summaries. Since digital and parametric approaches are emerging fields within landscape architecture, new connotations and concepts are constantly being introduced. In light of the current software platforms and the development of digital landscape architecture, the author provides a summary and classification of the corresponding landscape architecture design process, and a summary of digital landscape architecture software platforms (Table 1.2).

From Table 1.2, it can be seen that software platforms such as Vectorworks Landmark, Revit, ArcGIS, Civil 3D, and Rhino are capable of controlling the design process. The common characteristics of these software platforms are:

1. Strong data analysis and processing capabilities, capable of performing comprehensive analysis.
2. Ability to quickly present results in a graphical format based on data processing, meeting design requirements.
3. Good compatibility, able to integrate well with other software.
4. Open interfaces, providing the possibility for programming to solve specific problems.

The software platforms listed above are not only powerful in terms of comprehensive performance but also have their own areas of focus. They exhibit good compatibility with various data formats, which facilitates the integration and cooperation of multiple software tools. These platforms have the potential to serve as a foundational platform for parametric design, providing excellent support for the implementation of parametric landscape design processes.

[28] Cai Linghao. (2013). Conceptual Spectrum and Flowchart of Digital Landscape Architecture Planning and Design. *Landscape Architecture*, 2013(1), 50.

[29] Bao Ruiqing. (2013). Discussion on Computer-Aided Landscape Architecture Planning and Design Strategies. *Journal of Beijing Forestry University (Social Sciences Edition)*, 2013(1), 40.

[30] Chi Zhiwei, Chen Jie, Zhang Deshun. (2012). Progress in Parametric Design and Its Implications for Landscape Design. *Chinese Landscape Architecture*, 2012(10), 41.

Table 1.2 Software platforms related to digital design of landscape architecture

Design process	Function	Software name	Function	Remarks
Information collection and analysis	Geographical analysis	ENVI	Remote sensing image processing platform, for processing images and spatial data	It can be connected with GIS
		ERDAS	Remote sensing image processing system, processing images and spatial data	It can be connected with GIS
		ArcGIS	Geographic Information System platform products, processing geographic and spatial information	
		Autodesk123D Catch	3D modeling using digital photos	
	Spatial analysis	Dethmap	Spatial analysis such as spatial structure and field of view can be conducted	
		ArcGIS	It has powerful functions such as spatial data management, spatial analysis, and spatial information integration	
	Habitat analysis	ArcGIS	The relevant modules can be used to analyze vegetation, water bodies, etc	
		Amap	Simulate the plants	
		NetLOGO	It can simulate complex systems that develop over time	
		Phoenics	Conduct fluid simulation	It can be integrated with CAD
		Flow 3D	It can simulate the flow of liquids	
		Vasari	Simulate solar radiation and sunlight trajectories to conduct energy consumption analysis	Connect with Revit
		Urban Wind	Wind environment simulation	
		Xflow	Fluid Dynamics (CFD) simulation	
		Fluent	It is applicable to the simulation of complex flows	
		Ecotect	Sustainability analysis including sunlight and shadow	
		PKPM	It can conduct wind environment calculation and simulation, environmental noise calculation and analysis, three-dimensional sunlight analysis, etc	It can be integrated with CAD
		Fragstats	Analysis of Landscape Pattern	
	Behavioral psychology	Tobii Studio	Eye movement data analysis	
		BSS	Physiological data analysis	

(continued)

Table 1.2 (continued)

Design process	Function	Software name	Function	Remarks
Scheme design	Process control	Vectorworks Landmark	It is capable of conducting two-dimensional and three-dimensional modeling, site analysis, and building landscape information models	It can be connected with GIS
		Revit	Build an information model	
		ArcGIS	Spatial information modeling and analysis, and construction of planning information models	
		Civil 3D	Engineering information model construction	It can be integrated with CAD
	Form generation	SketchUp	3D design and modeling	
		3DMAX	Three-dimensional modeling is possible	
		Maya	Three-dimensional modeling is possible	
		Rhino	It can perform 3D modeling, and the NURBS modeling function is particularly powerful	
	Algorithm programming	Grasshopper	A plugin for generating models using program algorithms running in the Rhino environment	
		Rhinoscript	The parametric editing script program that comes with Rhino	
		Dynamo	Visual parametric plugin based on Revit	
		Processing	Graphic design language	
		Matlab	It has the functions of numerical analysis, matrix calculation, scientific data visualization, as well as modeling and simulation of nonlinear dynamic systems	
		Arcpy	Carry out batch processing of spatial data, develop plugins, and establish geographic processing application models	
	Structural analysis	PKPM	Structural analysis of landscape structures can be carried out	It can be integrated with CAD
		ANSYS	It can calculate loads and other parameters to assist in engineering design	It can be integrated with CAD

(continued)

Table 1.2 (continued)

Design process	Function	Software name	Function	Remarks
Expression of achievements	Design illustration	Photoshop	Image processing and expression	
		Illustrator	Image processing and expression	
		InDesign	Image processing and expression	
		PowerPoint	Image processing and expression	
	Rendering and animation	Vary	Image and animation rendering	
		Vue	Animation production and rendering of 3D natural environments	
		SketchUp	Animation production	
		3DMAX	Animation production	
		Lumion	3D animation production	
	Virtual reality	CityEngine	Three-dimensional urban modeling	
		Smart 3D	Three-dimensional photography modeling	
		VNS	Three-dimensional visualization	
		DVS 3D	Interactive virtual reality software platform	Compatible with multiple format models
		Quest 3D	Real-time 3D visualization	
		Lumion	3D visualization	
		Twinmotion	4D visualization	

Chapter 2
Parametric Landscape Architecture Planning and Design Mechanism

The logic of digital landscapes is about addressing complex landscape architecture planning and design issues through binary digital expressions, digital calculations, and logical derivations. This involves constructing the relationships between different components by translating the logic of identifying, analyzing, and solving problems into algorithms and mathematical formulas. This is the same approach used in all current computer programming for problem-solving. The logic of a digital landscape utilizes digital technology to transform what was once based on experiential judgment into a scientific search for patterns and a rational dependency, thus allowing the complete expression and presentation of these patterns using digital techniques.

The logic of digital landscapes aims to resolve the seamless integration of different stages in landscape planning and design through a digital process, coordinating the quantitative decision-making in land evaluation, proposal generation, construction control, and other stages. As a result, the entire digital flow computation systematically covers the entire traditional planning and design process, minimizing subjective human errors to the greatest extent. This fully reflects the scientific and rigorous nature of planning and design work, thus achieving the fundamental goals of landscape architecture planning and design through the digital flow computation of the entire process.

In the current era of rapid urbanization, elements such as population, resources, capital, information, materials, and energy are constantly interacting and exchanging within social spaces. The diversity of social system elements, the coupling of spatial structures, and the self-organizing behavior characteristics reflect the fundamental properties of the open and complex urban environment. Correspondingly, landscape systems also exhibit complex attributes such as multi-layered, high-dimensional, multi-scale, and non-linear characteristics. Landscape design, therefore, needs to meet the multi-objective demands of four fundamental aspects—ecology, space, function, and culture. As a result, the digital landscape approach, based on advanced digital technologies, has emerged. Digital landscapes seamlessly integrate complex and diverse landscape information into various layers

Y. Yuan, *Parametric Landscape Architecture Planning and Design*,
https://doi.org/10.1007/978-981-95-8011-8_2

of the design process through binary digital flows, transforming traditional experience-based design models into a "human-computer interaction" and quantitatively-assisted qualitative integrated planning design method. Digital landscape is the inevitable product of the development of the digital era.

Design is not only a complex and advanced human cognitive activity, but it also represents an important creative process, where existing knowledge and technology are applied to solve problems or create new things. Since the 1960s, there has been growing attention to the study of the essence of design, design methods, and related topics. The advancement of contemporary science and technology has greatly accelerated the updating of knowledge, theoretical evolution, and technological innovation across various design disciplines, including landscape architecture, generating the key theories, methods, and techniques that underpin modern design practices. The study of "mechanisms" delves into the essence of design and explores its principles at the methodological level. The term "mechanism" originates from Greek, and its English interpretation is "a system of parts working together in a machine; a piece of machinery, a natural or established process by which something takes place or is brought about." As we can see, a mechanism consists of two key components: first, a functional system composed of multiple elements; and second, a natural or artificially established process. In different academic fields, the meaning of a mechanism varies. In engineering, a mechanism refers to the structure and working principle of a machine, or to the structure, function, and interrelationships of an organism. Today, the study of mechanisms is widely explored across both natural and social sciences, such as the feedback mechanisms in physics and engineering science, nonlinear mechanisms, and self-organizing mechanisms in systems theory and chaos theory, or genetic, immune, and cognitive mechanisms in biological research. The core of mechanisms lies in how systems operate. The study of mechanisms involves investigating the internal workings or operational modes within systems, including the interrelationships among their components. The interpretation of a mechanism is a deep understanding of a subject, transitioning from a descriptive phenomenon to an essential explanation of how it functions.

Mechanisms have four characteristics: 1. A mechanism is a method or approach that has been proven effective and stable through practical verification. It represents a relatively stable way in which a system or process operates. 2. A mechanism is a summary and refinement of various effective approaches and methods. It not only possesses interpretive significance for understanding the inner workings of a system but also has a high degree of systematization and theoretical foundation. 3. A mechanism is a combination of multiple methods and approaches. The function of the mechanism depends on the joint operation of these various methods and approaches. 4. A mechanism corresponds to a specific functional need. Mechanisms arise from functional demands, such as evaluation mechanisms, supervision mechanisms, incentive mechanisms, etc.

According to the concept of mechanisms, it is necessary to grasp them from two aspects: 1. System: A thing is a system composed of multiple parts, and there are interrelationships among these parts. This is the prerequisite for the existence of a mechanism. 2. Operational Mode: The mechanism coordinates the relationships between the various parts of a system and the specific ways it functions. Based on these two aspects, the study of mechanisms is divided into two parts: First, what are the components that make up the system, and why are these specific parts involved? Second, how does the mechanism relate the various parts and drive the development of the system? What are its operating principles, methods, and specific paths?

In his classical economics work *The Wealth of Nations*, Adam Smith proposed the idea of the "invisible hand"—a mechanism that governs the value laws behind the production and exchange of goods in a market economy. This metaphor vividly illustrates the basic characteristics of a "mechanism." Similarly, the design mechanism acts as the "invisible hand" guiding the design process, directing design behavior, and leading to the generation of design outcomes. The existence and transformation of things occur due to mutual interactions, and these interactions and relationships can be classified into two types: One is the internal relationships and interactions within a thing. The other is the external influences imposed upon the thing. Mechanisms arise from the internal nature of things and serve as an intrinsic driving force. The design mechanism, in particular, originates within the design process itself. It describes the various elements of the design, their interrelations, and the operational laws that govern them.

Represented by Christopher Alexander's *A Pattern Language: Towns, Buildings, Construction*, the design field has actively sought scientific methods for design. The study of design patterns provides verified, effective solutions to common design problems, enabling the use of past experiences and results to solve design issues while avoiding meaningless repetition. Alexander posits that each pattern is a rule made up of three components, which express the relationship between a specific situation, a problem, and a solution. He believes that the value of architectural design patterns lies in the fact that "each pattern describes a recurring problem in the surrounding environment, followed by the key to solving that problem. This solution can be used hundreds of times without becoming redundant." The study of design patterns began in the 1960s, a period when post-World War II modernist architecture was rapidly developing standardization and modularity. Consequently, design patterns often appeared somewhat dogmatic and rigid. In comparison to design patterns, which aim to offer predefined solutions, the study of design mechanisms places greater emphasis on explaining the operational principles, understanding the underlying laws, and exploring the essence of design. Mechanisms are more flexible and instructive. Moreover, design mechanisms can transform into corresponding design patterns depending on the specific design context, making them more suitable and practical.

In previous concepts, design was seen as a business practice. In recent years, "Design Research" has become a hot topic in the global design community. In the context of the shift in knowledge production models and the changes in scientific knowledge structures, relying solely on the experience and intuition of designers is no longer sufficient to solve the complex and interdisciplinary problems encountered in design. As such, the landscape architecture industry is calling for design and research work guided by scientific methods. Contemporary landscape architecture combines both science and art, and should form an academic research system that integrates scientific and artistic paradigms of thinking. This system should conduct theoretical research based on practice and use the research results to guide and feedback into design practice. Due to its dual nature of art and science, the work and thinking methods of designers differ from those of scientists in engineering and technology, as well as scholars in the humanities. Therefore, the linear thinking process of "problem to conclusion" is not entirely applicable.

Younas and Federi categorize the relationship between design and research into three types: "Research About Design," "Research for Design," and "Research Through Design." "Research About Design" focuses on design as the object of study. The aim is to reveal the inherent characteristics and patterns of design behavior and design thinking. Its research contents mainly include three aspect: Design epistemology research aimed at exploring the designer's cognitive approach; Aimed at exploring design practices and Research on design behavior in the design process Design phenomenology aimed at exploring the form and composition of design products.[1] Research "Research for Design" is research driven by design as the "purpose" and "motivator." This category includes activities such as research, material collection, and other tasks that serve the design process. "Research Through Design" emerges from the specificity of design behavior. It is research that is inseparable from design behavior and uses design itself as both the medium and tool of investigation. There are two levels of goals in "Research Through Design": The first level involves solving design problems through research, starting from concrete design examples. The second level aims at researching and summarizing the design process to form paradigms and models that are both guiding and universal. This provides solutions, approaches, and methods for similar design problems. While "Research Through Design" originates from basic design processes, it goes beyond simply solving design problems. It seeks to elevate and summarize methods at the methodological level. Therefore, the outcomes of "Research Through Design" can exist in a generalized form of knowledge, independent of the specific design process. This kind of research lays the epistemological foundation for the establishment of a design research paradigm. It is a necessary condition for the development of design methodology, emerging from the re-discovery and systematic organization of the inherent characteristics of design behavior and thinking.[2] Research on Design Mechanisms is an example of "Research Through Design," aiming to convey the principles of design, the processes by which these principles work, and their inherent laws to users. This transforms the traditional opaque design process into one that is quantifiable, interpretable, understandable, and controllable. The parameterized landscape architecture planning design mechanism based on coupling principles also aims to establish such connections to clearly express the principles, approaches, and methods of design.

---

[1]Guo, Y. (2011). An overview of methodological approaches in contemporary design research. Journal of Landscape Architecture, *2011*(4), 69.

[2]Guo, Y. (2011). A methodological overview of contemporary design research. *Landscape Architecture*, (4), 69.

Method refers to the means and approaches used to achieve a particular goal. It is an organic integration of goals, approaches, strategies, steps, tools, and technologies used by people to understand and transform the world. According to the Modern Chinese Dictionary, the definition of "method" is: "the way, procedures, etc., for solving problems related to thinking, speaking, acting, and so on."[3] From this, we can see that methods are not monolithic, absolute, or universal. For different purposes and problems, specific methods will vary. Methodology refers to a system or framework aimed at solving problems. It typically involves a discussion of the stages of the problem, tasks, tools, and techniques used to address it. Methodology is the sum of categories, principles, theories, methods, and tools. The English term "methodology" is composed of "Method" and "Ology" (logic), meaning the study of "the principles of methods." Methodology not only explains why to use certain "tools," when to use them, and the effects of using them, but more importantly, it guides the creation of new "tools" when faced with different problems. Design Methodology is the application of systematic procedural theories to design practice.[4] The study of the parameterized planning design mechanism in landscape architecture falls under the category of methodology, as it explores the inherent laws and mechanisms of parameterized design and examines its rationality, effectiveness, and outcomes. Based on design mechanisms, specific design methods can be generated for different problems. The essence of landscape architecture determines that the methodology of landscape architecture planning spans the boundaries of both scientific and artistic methodologies, making it a combination of "science and art." Therefore, research on landscape architecture's parameterized planning design mechanisms necessarily encompasses both qualitative and quantitative research methods, as well as rational and non-rational ways of thinking. This research approach is distinct from pure scientific or artistic research.

The objects of landscape architecture planning and design are ever-changing, with the characteristics of different places varying widely. As a result, design strategies, principles, and methods differ from one project to another. Therefore, the study of mechanisms is of utmost importance. Mechanisms target the underlying principles and represent a study of the essence of design, falling within the realm of methodology. Under the guidance of mechanisms, suitable methods and techniques can be flexibly generated for different design contexts. Starting from the interpretation of the concept of design mechanisms, this paper aims to elaborate on the methodological significance of design mechanisms and explore the generation and operational modes of parameterized landscape architecture planning design mechanisms.

[3] Institute of Linguistics, Chinese Academy of Social Sciences (Ed.). (2017). Contemporary Chinese Dictionary (7th ed.). Beijing: The Commercial Press. p. 366.

[4] Chen, C. (2008). Design cognition: Cognitive science in design. Beijing: China Architecture & Building Press. p. 21.

2.1 Systematic Characteristics of Landscape Architecture Planning and Design

2.1.1 *Complexity of the System*

Complexity is a term frequently used in the fields of natural sciences and technology, originating in the 1940s and 1950s. The theories of systems theory, cybernetics, and information theory were pioneers in the development of complex thinking, breaking away from mechanical, linear viewpoints. These theories established the idea that "the effects of interacting factors are not simply additive," meaning that the whole is greater than the sum of its parts. As research deepened, the concept of complexity was continuously enriched. It no longer merely referred to complexity or chaos, but also to nesting, self-similarity, emergence, and most importantly, the idea of wholeness, order, and progressively increasing levels of hierarchy, organization, and evolution. In addition to the contributions from systems theory, cybernetics, and information theory, the theories of dissipative structures, synergy, catastrophes, and hypercyclic theories explored the environmental conditions, dynamic mechanisms, pathways, and coupling that generate complexity. This led to the emergence of the concept of self-organization in complex systems. In such systems, through interactions, each individual component contributes to the emergence of self-organized behavior, where the system spontaneously adjusts itself. This adjustment is an active process, driven by the system itself.

Although richness and complexity might seem to overlap in their semantic expression, in the context of a landscape, the richness of a site inherently contains complexity, and its complexity is a manifestation of that richness. Richness showcases the vast amount of information embedded in the landscape, while complexity highlights the interrelationships between the various pieces of information within the system. A place records the trajectory of history, preserves the traces of human and nature's development, and accumulates a wealth of information. In terms of the composition of information, the types of information within a landscape are varied, and their external manifestations differ. Humans have five modes of perception—vision, taste, smell, hearing, and touch—that allow us to sense information with material appearances and some immaterial information. For example, we can perceive the shape of plants, the scent of soil, the sound of flowing water, or the force of wind. These types of information are directly perceivable. However, other immaterial information, such as the growth patterns of plants and animals, climate changes, lifestyle habits, or economic forms, cannot be directly perceived. Various types of information exist within a landscape, though their proportions vary. The complexity of this information significantly impacts the planning and design of landscapes. Christopher Alexander, in the introduction of his 1964 book *Notes on the Synthesis of Form*, stated: "More and more design problems have reached a level of complexity that is hard to resolve... With the increasing complexity of problems, the complexity of information and professional experience also grows. This information is vast, dispersed, and disordered, making it hard to grasp... The amount of information has surpassed what a single designer can comprehend."[5] As society has developed, our understanding of the landscape environment has deepened. Designers now have to consider a broader range of factors, and the volume of information they need to process has grown exponentially. Relying solely on experience is no longer enough to manage such complex and diverse design information. For designers with less experience, it becomes even more difficult to handle such a vast pool of information.

2.1.2 Dynamics of Elements

In design, the information that needs to be processed not only increases significantly in quantity and complexity but, as research progresses, it has become apparent that information is not static; rather, it is in a state of dynamic change within the system. The recognition of the dynamic nature of information once again raises the designer's ability to grasp and handle this complexity. Beautiful expressions such as *"the grass grows and the orioles fly," "the waterfall plunges,"* and *"flowers bloom and wither"* are all descriptions of the dynamic phenomena in the landscape environment. The landscape environment encompasses both inanimate materials and living organisms, with plants being a crucial component of the environment. In contrast to architecture and urban planning, landscape architecture deals directly with life and its systems, making it one of the most significant distinctions between landscape design and other design disciplines. The origin of dynamic nature lies in the combined effects of natural forces and time. Under these influences, the natural elements of the landscape—land, water, wind, and plants—interact and continually change. The inclusion of living organisms means that time plays a particularly prominent role in shaping the landscape environment. The growth of plants and seasonal changes are the most obvious natural dynamics in a landscape, and the scenery of a place changes with the seasons, offering new experiences year after year. In ancient Western times, there was a tendency to pursue a relatively eternal and unchanging landscape, most notably represented by classical French gardens such as those at Versailles. In these gardens, the regular shape of patterns was maintained through regular pruning of plants. Although elements like cascading water and fountains were added to introduce dynamic features to enliven the landscape atmosphere, the focus remained on static order. In contrast, the dynamic nature of landscape elements refers to the diverse changes that occur over time and under the influence of natural forces, rather than merely their *"motion state."*

Contemporary landscape architecture planning and design is constantly expanding its scope, no longer limited to small-scale courtyards and parks, but focusing on larger and more expansive natural areas. Unlike built environments, where humans

[5]Alexander, C. (2010). *Notes on the Synthesis of Form* (W. Wang & Y. Zeng, Trans.; Y. Zhang, Ed.). Wuhan: Huazhong University of Science and Technology Press. p. 2.

hold absolute control, in these areas, nature becomes the dominant force. The natural processes within the region play a decisive role, and humans cannot exert absolute control over the landscape. In modern landscape architecture planning and design, the forces of nature are critical for designers to acknowledge, adapt to, and utilize. The dynamic nature of a site, shaped by time and natural forces, should be incorporated into the designer's considerations. Accordingly, a dynamic design process that actively responds to the site's evolving nature must be established. George Hargreaves believes that landscapes are a dynamic process of development and must be viewed from a dynamic perspective. In his designs, he incorporates the natural evolution and development process into an open landscape system, fostering a dynamic ecological view. Through design guidance, he allows nature to manage the details of the landscape, such as in the Guadalupe River Park, where water erosion gradually changed the shape of the grass mounds, and in the Candleholder Point Cultural Park, where years of natural evolution led to the growth of thorn acacia in shallow depressions, gradually altering the surface of the ground. In response to the characteristics of landscape environments, contemporary landscape architecture planning and design should aim to actively and positively engage with natural processes, embrace a process-oriented mindset, and focus on the beauty of the process.

2.1.3 Multi-Objective Design

"Space, ecology, function, and culture" are the four components of a landscape environment. The relationship between these four elements is like the four nodes of a triangular pyramid, where the weight and interrelationship of each node vary in different landscape environments. It is precisely due to the differences between these four elements that the characteristics of the landscape environment display clear tendencies, resulting in a landscape that is ever-changing (Fig. 2.1).[6] Correspondingly, contemporary landscape architecture planning and design are characterized by multiple objectives, with space, ecology, function, and culture being the four fundamental aspects. The landscape architecture planning and design process involves constructing a pathway to achieve multiple objectives based on the conditions of the studied site. Landscape architecture planning and design is not simply the stacking of these four elements, but rather requires balancing and adjusting the four basic components according to the environmental context and demands, ensuring their harmonious coexistence in the same environment.

---

[6]Cheng, Y. (2010). *Modern landscape design theory and methods*. Nanjing: Southeast University Press. p. 58.

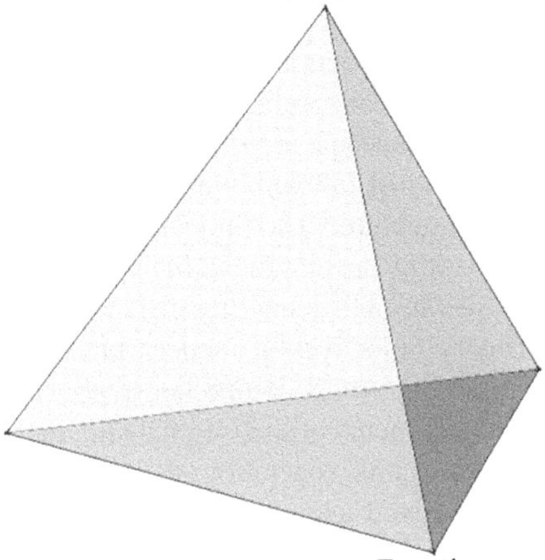

Fig. 2.1 Four objectives of contemporary Landscape Architecture planning and design

2.1.3.1 Ecological Objectives

Landscape architecture design needs to handle not only a vast amount of information with high complexity, but also information that is constantly changing: the information in a landscape environment is in a state of dynamic change within the system. Phrases such as "grass grows, birds fly," "water flows downward," "flowers bloom and wither," and "spring flowers, autumn fruits" are basic characteristics of the dynamic scenes in landscape environments, and they distinguish landscape architecture planning and design from architectural and urban planning design. The fundamental reason for this difference can be seen through the comparison between classical and modern landscape design. The feature of classical garden design lies in maintaining the landscape elements in their "optimal state" according to human will, while modern landscape design emphasizes integrating human needs into the natural process. Unlike the built artificial environment, in the landscape, natural processes play a decisive role, and people cannot exert absolute control over the landscape space. Designers must carefully study and skillfully apply natural forces. Under the combined action of time and natural forces, the dynamic characteristics presented by the site should be included in the designer's thinking process. Therefore, it is necessary to establish an awareness and concept of "process" and construct a dynamic design process that actively responds to the environment's dynamic nature. Parametric design models reflect the relationships between design elements, presenting an open, dynamic system where changes in a single element lead to changes in the design results. As such, the parametric design method can align with the dynamic characteristics of landscape architecture elements, reflecting the evolution and change of nature.

The rapid rise of the world economy and the accelerated industrial development following World War II have brought about a series of environmental issues, prompting a critical reassessment of past actions. Compared to transient humanistic ideologies such as postmodernism, structuralism, and deconstructionism, the impact of science and technology on landscape architecture has proven to be more enduring. Ecologism has become the most influential and vital philosophy in the contemporary field of landscape architecture.[7] The protection of ecological systems is essential for the long-term healthy development of human society. Safeguarding the ecological environment of sites and achieving sustainable development are not only important but also primary objectives in modern landscape planning and design. Landscape planning and design must comprehensively consider two aspects: first, the natural environmental elements of the original site, including geology, geomorphology, climate, hydrology, soil, vegetation, and fauna; second, human activities and their impacts, such as land-use patterns. The native environment possesses inherent stability and sustainability. For natural areas with high-quality habitats and stable ecosystems, it is necessary to maintain the evolutionary capacity of ecosystems by protecting relatively stable ecological communities and spatial forms, based on adherence to natural laws and the intrinsic mechanisms of the environment. Through an integrated approach combining qualitative, quantitative, and spatial analyses and assessments, the ecological suitability and land-use suitability of different site parcels can be determined, leading to the delineation of zones for conservation, utilization, and optimization. Corresponding planning and design strategies can then be implemented accordingly. On the foundation of protecting the ecological environment, natural resources should be utilized selectively, and anthropogenic processes organically integrated into natural processes to ensure the principle of "adapting measures to local conditions" is effectively realized.

2.1.3.2 Space Objectives

Space is the carrier of landscape, where functions unfold, culture generates, and ecosystems are supported. The spatial form and structure generate the basic form of the place and are also one of the key elements in landscape architecture planning and design. As the "skeleton" of the "landscape," space is most easily perceived by people due to its tangible and physical nature, distinguishing it from aspects such as culture and ecology. Therefore, it is most capable of highlighting the characteristics of the garden landscape. It is important to note that the creation of landscape space is not equivalent to "plan design," nor is it limited to "formal design." Thus, one should not blindly pursue the strong sense of place that comes from spatial forms while neglecting the ecological and functional significance of landscape space. Compared with architecture, landscape space is more open and expansive, lacking

[7] Cheng, Y., & Yuan, Y. (2015). Landscape architecture in the context of contemporary science and technology. *Journal of Landscape Architecture*, (7), 15–19.

a fixed "outer shell" and clear boundaries, and exhibits organic and "porous" characteristics. Form, structure, interface, and texture are the four fundamental elements in the construction of landscape space, and they serve as the basis for its generation. Due to the "porous" and open nature, landscape space is closely linked to the surrounding environment. Therefore, in landscape architecture planning and design, the space should be considered within the context of the overall environment, with a coordinated approach to create a harmonious garden landscape environment.

2.1.3.3 Functional Objectives

Place is the setting where behaviors occur, and behavior holds a crucial position within a place. Human behavior within a place is the result of the interaction between the place and the individual. Design ultimately serves human needs, hence it carries a strong human-centered consciousness. The individual in the place—the subject of behavior—is the user of the space and the target of the design. One of the objectives of landscape architecture planning and design is to create a human-centered spatial environment that meets the behavioral needs of diverse groups. Behavior is a complex product of multiple interwoven factors, and there exists a certain correspondence between the environment and human behavior. Well-designed public spaces can facilitate social interaction and enrich outdoor life; specific spatial forms can attract particular groups and trigger corresponding behaviors and activities. John O. Simonds believes that "in landscape design, humans retain their natural instincts, which drive them." To achieve reasonable design, it is necessary to understand and study these instincts. There exists a dynamic relationship between humans and their environment, where human behavior in a place is influenced by the "force" of the place and, in turn, exerts an impact on it. In landscape architecture planning and design, the designer's interpretation of user behavior is a critical element of place cognition. Designers should thoroughly study human behavioral patterns and psychological characteristics within a place, and based on the commonalities and patterns of different behaviors, engage in "adaptive" design, planning functional projects that are suitable for the specific activities of the place.

2.1.3.4 Cultural Objectives

The term "culture" has a broad meaning, and the two indispensable elements for the creation of culture are time and human activity. Culture is manifested as the "imprint" of human activities embedded in a place, and it is also the source of people's identification with that place. The formation of a place originates from the natural environment, which, influenced by human activities, incorporates cultural information. For landscape environments, culture is an essential element present within the place. The natural and cultural environments are closely linked both vertically and horizontally, jointly constituting the landscape's contextual framework. This contextual framework reflects not only the continuity of information over time but also encompasses inheritance and exchange across multiple dimensions such as space and function. Depending on the specific place and the characteristics of the activities within it, the contextual information embedded in the landscape environment varies, possessing uniqueness and particularity. Therefore, contextual information becomes an important source of the place's distinctiveness. These contextual elements within a place include various aspects such as time, space, culture, and nature, presenting a complex and intertwined relationship. The primary goal of emphasizing the cultural significance of a place is to continue the unique human-land relationship embedded in the environment, that is, the "place context." The cultural objective in landscape architecture planning and design is to inherit and develop the culture within the place, respecting the original natural processes, patterns, and the historical cultural accumulation left by human activities. This should form the foundation and background of the design, ensuring the "integrity" of historical context. On this basis, the existing characteristics of the landscape environment should be fully explored. By identifying and selecting human usage patterns from different periods, as well as events with special significance, designers can choose parts that are developable and utilizable. Through techniques such as extension, transformation, grafting, reorganization, and abstraction, these elements can be restructured into a new landscape order.

2.2 Characteristics of the Traditional Landscape Architecture Planning and Design Mechanism

The landscape gardening activities and other landscape-related practices of different historical periods are closely tied to the social development of their respective times. The research outcomes of previous scholars on landscape design stem from years of accumulated experience and the summarization of practical knowledge, representing valuable practical experience and theoretical foundation. However, due to the constraints of the era, these outcomes have certain limitations, which are reflected in four aspects: the subject they serve, the goals they aim to express, the approach they take, and the methods they form. With the development of society and the advancement of science and technology, people's understanding of the environment has continuously expanded, and design thinking has become increasingly profound. In parallel, landscape architecture as a discipline is also undergoing constant development and transformation, with its scope and content continually expanding.

2.2.1 The Singularity of the Subject

Since its inception, the primary function of gardens has always been to serve humanity. In ancient times, limited by the social productivity of the period, people's range of activities and their ability to transform nature were far inferior to what we have today. Garden creation was confined to a relatively small scale, with humans becoming the dominators of the environment, often neglecting the laws of nature and the needs for its development in fulfilling human demands. In traditional Chinese gardens, the concepts of "following nature" and "the unity of heaven and man" were emphasized. The landscape seen by ancient scholars was not pure, unaltered nature, but rather "humanized nature": landscapes refined and abstracted by human intervention. Through these landscapes, human aspirations and ideals were expressed, a concept reflected not only in garden construction but also in closely related artistic works, such as painting. In contrast, while the Western approach to water features focuses more on technical aspects, designers often rely on careful observation of natural water bodies and systems to summarize the various characteristics of water landscapes. They use technical methods and abstract exaggeration to create the scenery, but do not necessarily pursue the expression of artistic conception. Whether it be formal waterscape or geometrically arranged trees, Western classical gardens sought to exert absolute control over nature by humans.

2.2.2 *Imagery of Expression*

Humanity has never ceased its pursuit of "beauty." In eras of underdeveloped productive forces, human-made traces were a manifestation of mankind's conquest of nature, as well as an expression of human power and strength. As a result, artificial, rule-based forms, compared to the organic forms of nature, were more likely to gain people's recognition, becoming a subconscious representation of "beauty." In landscape architecture planning and design, people not only focus on the expression of artistic conception but also never abandon artistic creation, continually seeking the stimulation of the senses through form. This highly subjective pursuit even severed the connection between form and meaning. Landscape design works have thus become "artifacts," reaching the extreme of "beauty." The Earth Art movement, which emerged in the late 1960s, became a representative example. Designers dug pits, reshaped mountains, filled seas, built embankments, or splashed ink and color onto the vast landscape, all in pursuit of the aesthetic beauty of form (Fig. 2.2). Landscape architecture is both a scientific art and an artistic science; however, the overemphasis on artistry has diluted the scientific connotation of landscape architecture as a discipline.

2.2.3 *Unidirectionality of the Process*

In the face of increasingly complex site environments, the volume of information grows exponentially, significantly raising the difficulty of landscape architecture planning and design. This presents a tremendous challenge to the designer's ability to analyze and process information. From the perspective of the tools used,

Fig. 2.2 The Spiral Jetty created by Robert Smithson in 1970. Quoted from the Internet

traditional design methods, which rely on sketches and drawings, depend on the coordination of the brain, eyes, and hands. Due to the limited amount of information that the brain can process at any given time during the design process, and the constraints of the design tools, the typical landscape architecture planning and design process tends to become unidirectional and linear. This makes it difficult to provide real-time feedback and adjustments, preventing timely regulation of the design process.

2.2.4 Empirical of the Method

Traditional architecture, planning, and landscape design, in their conventional sense, rely on inductive reasoning and summarization of methods based on extensive practical experience, which is distinctly empirical in nature. Correspondingly, in traditional teaching models, the paths and methods taught are also derived from such experiential knowledge. "Experience" not only involves probability but also carries a certain level of subjectivity and ambiguity. Experience is directly proportional to time; it is the result of extensive time and practical accumulation, continuously validated over time. It is crucial to note that, due to its foundation in the accumulation of practice and time, the effectiveness of experience transmission is often poor. When detached from time and practice, novice learners often grasp and absorb the methods and paths only superficially. The fundamental reason lies in the fact that they may "know what" but not "know why." At the same time, experience-based design methods are difficult to articulate and are often confined to general descriptions. For instance, in the creation of landscape spaces, theories such as the "Three Views" of height, depth, and horizontality in mountain and water design, or the "Three Far Views" of vastness, seclusion, and mystery, or the "Mysterious and Vast Theory", are examples of such vague expressions. Similarly, in *Yuan Ye* by Ji Cheng, the summation of landscape design practice is elevated to a theoretical level. In his discussions on water management in garden design, for example, he mentions: "Start by investigating the source, then clear the way from the source, and observe the origin of water" and "Build pavilions in elevated areas, dig ponds in depressions." These summaries of design principles, approaches, and methods are qualitative in nature, still remaining quite vague and conceptual for learners. The understanding and acceptance of such concepts often vary from person to person, depending on their different backgrounds and experiences.

2.3 Parametric Planning and Design Mechanism of Landscape Architecture

Landscape architecture has both scientific and artistic attributes. While inheriting and developing landscape culture, it is crucial to acknowledge the role of scientific and technological advancements in advancing the field. Under the dual drivers of science and art, the connotations and extensions of landscape architecture continue to be enriched, necessitating updates to its knowledge system. Scientific ideas and research, exemplified by systems theory, influence contemporary landscape architecture's values and development principles. Similarly, scientific technologies and tools, represented by digital technologies, have altered the paths and methods of landscape architecture planning and design to some extent. The emergence of parametric landscape architecture planning design mechanisms stems from the characteristics and developmental needs of the field, reflecting the evolution and progression of design systems. Researching things from the perspective of their essence and principles undoubtedly holds significant meaning and value. However, due to the targeted nature of methods and the uncertainty of the environment, design mechanisms rely on the characteristics of landscape architecture. Guided by coupling principles, these mechanisms study the system's attributes and the inherent laws of parametric design through system descriptions and interpretations. The focus of design mechanism research lies in understanding the underlying rules, rather than specific methods. The study of parametric landscape architecture planning design mechanisms is an exploration and summary of the principles and modes of parametric applications in landscape architecture planning design at a methodological level. Due to the diversity of landscape environments, the design methods adopted must be flexible and adaptive, ensuring that while attention is paid to the design mechanism, the approach does not lead to dogmatism or uniformity. The research on parametric landscape architecture planning design mechanisms does not merely remain at an abstract level; it is based on the generalization and refinement of specific design processes and operational methods. As an auxiliary method and tool, with the development of information technologies, represented by computer technologies, and the continuous enrichment of design cognition systems, the pathways and tools of parametric design will continue to diversify, providing more possibilities and support for the advancement of landscape architecture planning design methods.

2.3.1 Generation of the Mechanism

2.3.1.1 The Premise of Generation: Adapt to Local Conditions and Conform to Nature

The scope of modern landscape architecture practice has long transcended the traditional "garden" category and broken through conventional disciplinary boundaries. Regional landscapes, scenic environments, rural areas, highways, urban streets, parking lots, building rooftops, and even rivers, rainwater systems, and sponge systems have all become subjects of interest in contemporary landscape architecture. From gardens to parks, and from parks to park systems, encompassing both built and scenic environments, the research scale of landscape architecture continues to expand, leading to an extension of its research boundaries. This has resulted in an ever-widening scope of concerns for landscape architects, who now not only address "point" issues from a small-scale perspective but also consider "surface" issues from a regional level. Moreover, the focus has extended to regional and even national scales, contemplating human settlement systems and structural issues from multiple, layered viewpoints.

With the expansion of scale, nature increasingly assumes a dominant role in the landscape, and natural forces within the scenic environment become the leading elements of the place. It becomes increasingly difficult to apply past design concepts and methods unchanged, as the "beauty" of form can no longer be the central focus of design. Nature's force is infinite, shaping nearly all of the landscapes on Earth, whereas human alterations to the landform represent only a tiny fraction of that. From this perspective, human strength cannot compete with the power of nature. Under current conditions, not only would eliminating the influence of natural forces require enormous human and material resources, but such efforts are often unsustainable. With high maintenance costs, human defenses against nature's forces are often destined to collapse rapidly. Climatic disasters such as mudslides, typhoons, and earthquakes, which occur globally every year, demonstrate this fact in many ways. In reality, "resisting" nature is less effective than "embracing" it. Designers should follow the processes of nature and align with its forces. Nature is the best designer, and harnessing its power not only saves "effort" but also saves "labor."

"Designing according to local conditions" has long become a consensus in the design community, and for landscape architecture planning and design, this principle is particularly suited to the characteristics of landscape environments. Here, "local conditions" refers to the tangible land, representing the site and the natural and cultural processes inherent within it. Unlike "form," which is external, natural processes and laws are the internal driving forces.

In contrast to architecture, landscape environments are always in a dynamic state of development and change, with "form" existing only as a temporary phase. Therefore, understanding the underlying laws is far more important than form itself. Contemporary landscape architecture planning and design should emphasize the trends of future changes and developments within the site. As such, the concept of "Design with Nature" cannot be simply encapsulated by "designing in harmony with nature"; instead, it includes a wide range of complex relationships between design and nature, such as adaptation, collaboration, and coupling, ultimately achieving the fusion of design and nature. The goal of "designing according to local conditions" is to achieve the sustainability of the landscape environment. It is not merely about transforming or imitating the inherent forms within a site but, more importantly, about interpreting and mastering the laws of nature based on understanding its processes. This allows designers to explore the fundamental basis for design generation from the site itself and couple the site's inherent developmental mechanisms, thereby truly achieving "Design with Nature."

2.3.1.2 Background of Generation: The System and Design of Landscape Architecture

Landscape architecture planning and design must be based on the original site to create a new landscape environment that meets multiple objectives. This involves the integration of the designed artificial system with the site's native system, with the path to realization being the coupling of design with the site. The ultimate service object of the landscape environment is humans, so it is inevitable that human needs must be met. The task of contemporary landscape architecture planning and design is to graft and integrate human demands onto the foundation of natural system existence and development laws. For landscape environments dominated by nature, forms are easier to change, but they are not the essential characteristics of the system. The natural processes and laws within the site do not change according to human will. Therefore, the design should focus on the essence, conducting design based on the study of the self-development processes and laws of the landscape architecture system. As a relatively open system, the elements within the landscape architecture system are able to exchange with the external environment, making it possible for the artificial system and the native system to integrate. The landscape architecture planning and design for specific sites consists of various design elements that influence and adjust each other, collectively working together to generate the final design outcome.

The concept of “mechanism” has a profound scientific and philosophical background. The modern scientific revolution has shifted human science from simplicity to complexity, and the introduction of systems theory has promoted the development of mechanism research. Mechanisms reflect the inherent relationships within things, necessarily involving all the components of a system and their connections. Establishing mechanisms within a system involves studying the relationships between its specific components, such as systemic interrelations and causal relationships. Therefore, it can be said that mechanisms are closely tied to systems, and the existence of a system facilitates the generation of mechanisms. Self-organization refers to the process by which a system, driven by its internal mechanisms, continuously increases its complexity and refinement. This is the action of the system’s intrinsic mechanisms. The self-organizing nature of the system corresponds to the spontaneity of the mechanism. The generation of landscape architecture planning design mechanisms arises from the system’s own developmental needs.

2.3.1.3 The Basis of Generation: Nonlinearity and Logic

“Thinking” is the process of cognitive activities such as analysis, synthesis, judgment, and reasoning, based on representations and concepts. It is a uniquely human high-level mental activity. Design is a complex thinking activity that encompasses various types and broad domains, including intuitive thinking, image-based thinking, logical thinking, and creative thinking. The design process is controlled by consciousness, and design thinking influences and constrains the entire creative process. Firstly, from the perspective of Eastern and Western thinking, Eastern thinking, represented by China, often grasps things from a holistic, systemic viewpoint and analyzes the inherent laws from the perspective of universal connections. Western thinking, on the other hand, emphasizes individuality and is adept at isolating things from an overall context, strictly following forms of concepts, judgment, and reasoning to reflect the essential attributes and inherent laws of things. Secondly, Chinese thinking tends to be more intuitive, relying heavily on experience, intuition, and induction, whereas Western thinking is more rational, focusing on analysis, logic, and reasoning during the thought process. These differences in thinking modes result in distinct design philosophies: Eastern design emphasizes emotional appeal and is focused on the design traditions passed down from previous generations, while Western design prioritizes scientific and functional aspects. In landscape design, similar contrasts can be observed. Chinese garden designers focus on conveying the mood or “spirit” of landscapes, emphasizing “resemblance in spirit” rather than “resemblance in form,” and striving for “artistic representation” rather than “realism.” In contrast, Western landscape architects, such as Lawrence Halprin, employ abstract forms to mimic natural landscapes. For example, his design of Lovejoy Plaza (Fig. 2.3) uses irregular terraces to mimic the contour lines of natural mountain ranges, while the water flow trajectory of the fountain follows the mountain streams of California’s Sierra Nevada.

Ways of thinking are rooted in the historical practices and the progression of scientific development, and they evolve alongside the advancement of the times. The development of an era imposes corresponding demands on design thinking, and specific historical periods are marked by certain particular ways of thinking. The scientific evolution of design thinking can effectively reduce the arbitrariness and uncertainty in design work, increasing the determinability, reliability, and rationality of design outcomes. At the same time, it enhances the systematization and orderliness of design work, improving both work efficiency and quality. Similarly, the development of landscape architecture planning and design thinking has followed this evolutionary process. Imagery thinking is the earliest form of human cognition, characterized by subjectivity. In contrast, logical thinking integrates concepts, judgments, and reasoning to handle abstract concepts and achieve scientific judgments and conclusions, making it objective. The complementary relationship between imagery thinking and logical thinking represents the trend in the development of modern cognitive methods. Scientific thinking primarily relies on logical thinking, while artistic thinking predominantly uses imagery thinking. Landscape architecture, which blends both science and art, thus requires a comprehensive thinking approach, encompassing both logical and imagery thinking in its design process. In landscape architecture planning and design, logical and imagery thinking are inseparable and together form a unified cognitive system for the design process. Logic is a rigorous scientific method with two distinctive characteristics: one is the construction of formal language, and the other is the establishment of computational systems. For contemporary landscape architecture planning and design, it is essential to combine rational analysis, induction, and synthesis with sensory analysis. In his 1997 article *Nonlinear Architecture: New Science = New Architecture?*, Charles Jencks noted that the new complex sciences (i.e., nonlinear science) had replaced

Fig. 2.3 Model of Lovejoy Plaza. Quoted from the Internet

the modern linear science originating from Newtonian classical theory. Modern self-organization theory reveals the nonlinear laws that universally exist within systems, particularly in their organizational structures. The self-organization and complexity of landscape environments determine the nonlinear characteristics of design thinking. As the driving force within things, mechanisms reflect nonlinear causal relationships, embodying the complexity of systems. Nonlinearity dictates the non-logical nature of thinking, thereby confirming that landscape architecture planning and design involves both logical and non-logical thinking characteristics. For contemporary landscape architecture design mechanisms, two key requirements emerge: First, non-linear, or non-logical thinking, is needed to handle space and generate forms. Second, logical thinking is essential for grasping the process and the underlying laws. In this way, linear and nonlinear, logic and non-logic interact, couple, and unify organically, with the thinking mode of parametric design mechanisms resulting from this dual-layered integration.

2.3.1.4 Generated Drivers: System Optimization and Minimizing Intervention

"Optimal" refers to the final product that is achieved through the process of comprehensive coordination and mutual compromise to meet a series of requirements in the design process. Systems thinking is a modern scientific approach developed through solving complex system problems, characterized by multidimensional, multi-directional, and dynamic thinking. Systems thinking does not treat the design object as an independent entity, but as a design system, considering the interrelations and internal laws between the design elements, the elements themselves, and the surrounding environment. Systems theory emphasizes studying the object as a whole, where the various components work synergistically, optimizing and maximizing the collaborative effectiveness of each part to achieve the optimal overall benefit. Optimization is the most prominent feature of systems thinking, and the ability to perform system optimization is a key characteristic of design systems. Landscape architecture, as a complex system, retains the original properties of natural materials and follows the laws of natural succession on one hand; on the other hand, the garden space is a human-made environment designed to fulfill human needs and has cultural significance. Therefore, landscape architecture has the characteristic of multi-objectives. How to coordinate different goals within the landscape environment to optimize the system as a whole has become a fundamental requirement for modern landscape planning and design.

Landscape architecture, as a complex system, must adhere to natural laws while also serving human needs, and it possesses a metaphysical dimension, thus addressing the three fundamental values of "truth," "goodness," and "beauty" in landscape design. "Truth" represents scientific rationality, reflecting humanity's understanding of the "regularity" of past experiences; "goodness" embodies human vision and will, with an inherent "purposefulness"; while "beauty" represents the ideal realm, possessing a "spiritual" quality.[8] Compared to the fields of fine arts and architecture, landscape architecture is more complex and diverse. Contemporary landscape architecture, based on the discipline's autonomy, requires transforming disparate clusters of knowledge into a systematic framework, coordinating the relationships between different parts according to cognitive principles in order to achieve optimal system performance. As a system, landscape architecture integrates various elements, generating a "new system" through human intervention that combines both natural and artificial attributes. The goal of system optimization is to maintain the landscape environment's self-renewal capability, sustainable development ability, and to achieve efficiency, low energy consumption, and multi-objective characteristics through design. System optimization is the ultimate design objective, and minimal intervention is the strategy to achieve this goal: maximizing the utilization of system resources and optimizing its effectiveness under the premise of minimal intervention.

2.3.1.5 The Core of Generation: Coupling and Integration

The parametric landscape architecture design mechanism corresponds to the following design process: analysis of the site, the design strategy generated based on the analysis, and the dynamic, feedback process through which the strategy is applied to the site. The design goal is achieved through the coordination mechanism of dynamic feedback. The core of the mechanism generation corresponds to three aspects: first, coordination between elements and the site through coupling; second, the landscape system generated under human intervention has the basic characteristic of optimization, i.e., comprehensive optimization; third, the design integrates with the native environment, possessing self-improvement and renewal capabilities, thus becoming an integrated system for sustainable development. The design strategies generated under the guidance of the mechanism are coupled with the site, and ultimately transformed into specific methods and means that are also coupled with the site. The parametric design mechanism makes the coupling process transparent and controllable, making the design outcomes predictable. Description, control, adjustment, and optimization are all reflected in the parametric process, and parametric design achieves the value and significance of coupling through describing and controlling the design process, and regulating the parameters.

[8] Cheng, Y., & Yuan, Y. (2015). Landscape architecture in the context of contemporary science and technology. *Journal of Landscape Architecture*, (7), 16.

2.3.1.6 Generation Approaches: Quantification and Parameterization

People's cognition and judgment of the external world often follow the process of qualitative analysis first, followed by quantitative analysis. Qualitative research involves discovering problems and understanding events or phenomena to conduct relevant analysis and explanation. In qualitative research, researchers often use methods such as historical review, literature analysis, interviews, and observations to collect data, which are then analyzed and interpreted using non-quantitative means to draw conclusions, emphasizing meaning, experience, and description. The "qualitative" method is commonly used in sociological research, and its advantage lies in its comprehensive expression and perceptual depth. On the other hand, quantitative research is advantageous in its intuitiveness and rationality. It is one of the important steps and methods in scientific research, where problems and phenomena are represented numerically for analysis, testing, and explanation, ultimately deriving meaning. Quantitative research and qualitative research focus on different aspects: one emphasizes "quality," while the other emphasizes "quantity." Their primary methods differ as well: quantitative research employs empirical measurement, statistical analysis, and model-building, whereas qualitative research uses logical reasoning, historical comparison, and other methods. The forms of expression also differ: quantitative research typically uses data, models, and graphics, while qualitative research primarily relies on textual descriptions. In landscape architecture planning and design, both qualitative research methods and quantitative research support are indispensable. Traditional research in landscape architecture has mainly been qualitative, describing phenomena through experience and intuition, leading to judgments. However, contemporary research in landscape architecture requires the organic combination of both qualitative and quantitative methods. Quantification can present the research subject numerically, thus complementing qualitative methods. The value of qualitative research lies in controlling the overall direction, while the value of quantitative research lies in controlling the process. Only the organic integration of process and direction can achieve the expected goals with reasonable "inputs," optimizing and ensuring sustainability.

Quantification allows objects that are difficult to describe quantitatively to be transformed into comparable values or intervals, making it essential for the parametric approach in landscape architecture planning and design. Landscape architecture is a system composed of interrelated and constantly changing elements, characterized by dynamism. These elements play different roles in the system and have different functions at various stages of the design process. Their relative importance or weight varies, but the common feature is that any change in one element will trigger a change in the system, thus influencing the design outcome. It can thus be concluded that landscape architecture planning and design inherently possess parametric characteristics. Parametrics reflect the relationship between the site and the design, serving as an important approach to achieving coupling. Its advantage lies in control and adjustment. For landscape architecture planning and design, parametrics is significant both theoretically and practically. It is not limited to theoretical research that explores the intrinsic relationships and operational mechanisms of the design system, but can also be applied in the actual design process to control both the design process and outcome more accurately and precisely.

2.3.2 Interpretation of the Mechanism

Paradigms and models together form the foundation for interpreting the parametric landscape architecture planning and design mechanism, providing explanations from both theoretical and technical perspectives. Corresponding to the intertwining of the sensorial and the rational, as well as the alternating roles of qualitative and quantitative methods in the entire process of parametric landscape architecture planning and design, paradigms and models progressively reveal the process from interpreting principles to generating forms.

From a metaphysical perspective, paradigms and models represent a manifestation of methodology; they are neither typical nor targeted methods, nor specific paths or strategies. From a practical perspective, however, paradigms and models can guide the methods and approaches for generating designs. Therefore, the research into the parametric landscape architecture planning and design mechanism not only has a methodological height but is also grounded in design practice, ultimately guiding the practice of design (Fig. 2.4).

2.3.2.1 Design Thinking and Paradigm

The American philosopher of science Thomas Kuhn introduced the concept and theory of paradigms, which he systematically elaborated on in his 1970 book *The Structure of Scientific Revolutions*. A paradigm refers to a philosophical and theoretical framework, the theoretical foundation and practical norms upon which normal science operates, as well as the shared worldview and behavioral patterns

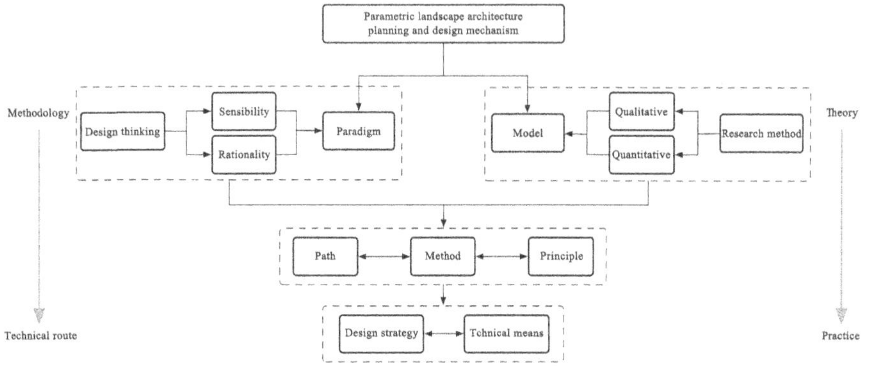

Fig. 2.4 Illustration of the parametric landscape architecture planning and design mechanism

adhered to by a group of researchers in a particular scientific field. Kuhn argued that: "According to the established usage, a paradigm is a recognized model or pattern, and essentially, it is a theoretical system." The study of paradigms enables the induction, definition, and interrelation of various examples, theories, methods, and tools within a particular science, making the formation of a paradigm a sign of the maturity of the academic community. Kuhn pointed out that paradigms have the following characteristics:

1. A paradigm is, to some extent, universally accepted;
2. A paradigm is an integrated whole composed of fundamental laws, theories, applications, and related instruments and equipment, and its existence provides scientists with a research framework;
3. A paradigm also offers scientists a successful precedent that can be imitated in scientific research.

According to Kuhn, a paradigm, as a theoretical system, leads to a scientific revolution when it is broken. Therefore, the essence of a "scientific revolution" is a "paradigm shift." A paradigm establishes a new relationship between applied models and metaphysics, solving the problem of transitioning from general philosophical theories to practical scientific theories. The distinction between paradigms and rules corresponds to different types and categories of research, as paradigms map various rules, each with specific content that varies. For example, in the case of quantum mechanics, only a few physicists engage with its fundamental principles, while most focus on studying how quantum mechanics can be applied to chemistry or solid-state physics.

Sensibility is dominated by perception, while rationality is characterized by logical reasoning. Albert Einstein believed that "the intersection of sensibility and rationality seems to be the essential characteristic of creative thinking." Design, as a creative act, alternates between sensibility and rationality throughout the process, ultimately generating design forms that are perceptible to humans. The design thinking process is a deepening of understanding, reflecting a journey from phenomena to essence, from sensibility to rationality. Modern architectural theory divides design philosophy into the inductive and rigorous Cartesian philosophy and the romantic, divergent Goethe philosophy. Similarly, landscape architecture design thinking is a system where both rationality and sensibility coexist. It is like a large tree, with rational thinking acting as the trunk, and sensibility as the branches and leaves. Without the support of rationality, design thinking becomes chaotic; conversely, without the richness of sensibility, landscape architecture design thinking becomes rigid and dogmatic. Emphasizing either "rationality" or "sensibility" to the exclusion of the other is detrimental to the healthy development of landscape architecture design. Therefore, modern landscape architecture design methods possess both rational and sensorial attributes.[9] Today, various concepts, ideas, and theories prevail, emphasizing design based on environmental cognition, and the significance

[9] Cheng, Y. (2010). *Modern landscape design theory and methods* (p. 59). Southeast University Press.

of this is self-evident. The emphasis on "coupling" affirms the premise of scientific design, which has rational characteristics, but landscape architecture planning and design have always been the result of the intertwining of rationality and sensibility. Emphasizing the rational process does not exclude the importance of sensibility. Of course, from the perspective of the design itself, the subjective initiative of the designer is crucial. However, this subjective initiative is conditional and based on the environment, unlike in painting or calligraphy, where it can exist independently. This is where landscape architecture art differs from other related arts. On the contrary, under a rational framework, fully leveraging the different designers' understanding of the same place, and maximizing the value of the site, is in line with the basic characteristics of landscape architecture art, as well as the creative methods and principles of landscape architecture.

The theory of a discipline is always in a state of continuous development and refinement. When a certain theory can no longer meet current needs, it leads to the emergence of a new paradigm, a "revolution of theory." The advent of the digital age has brought about a transformation in the field of landscape architecture. The traditional paradigms of landscape architecture planning and design can no longer meet the demands of contemporary design, and thus a new transformation is inevitable. Parametric theories, methods, and technologies provide the foundation for the generation of a new paradigm in landscape architecture planning and design. Over the past half-century, design research has embarked on a path of exploring new research paradigms. The study of these new paradigms no longer rigidly relies on rational scientific methodologies or emphasizes humanistic approaches based on sensibility. Instead, it seeks to integrate both scientific and humanistic methods within the realm of design thinking, forming a design methodology that simultaneously guides both design theory and practice.

2.3.2.2 Research Methods and Models

A model is an abstraction or simplification of entities or phenomena in the real world. Through subjective consciousness, it uses tangible or virtual representations to express an objective thing or concept. It is a description of the most important components and their interrelationships within an entity or phenomenon, aimed at describing, explaining, predicting, or designing that entity or phenomenon. Models can be categorized in various ways. For instance, based on the object being modeled, they can be divided into two types: representational models used to describe objects and process models used to describe processes. Based on the mode of expression, models can be classified as physical models or virtual models. In scientific research, models are commonly divided into three categories according to their form: scale models, conceptual models, and mathematical models. A scale model is a representation of the natural material characteristics of the real world, such as Digital Terrain Models (DTM) or hydrological models. A conceptual model uses natural language or flowcharts to express the relationships between system elements. It abstracts the objective objects in the real world into a specific information structure and is an abstract model based on experience and knowledge, constructed logically; it is also a type of virtual model. A mathematical model, in order to solve a specific practical problem, abstracts and simplifies that problem, determines variables and parameters, and applies certain "laws" to establish the relationships between variables and parameters. It uses mathematical structural expressions to describe the characteristics and internal connections of objective things and is also an abstract model (Fig. 2.5).

In the introduction titled "The Necessity of Rationality" from *The Timeless Way of Building*, Christopher Alexander presents the challenges contemporary design faces: "The number, complexity, and difficulty of problems in design have increased, and the changes in social forms and cultures are occurring faster than ever before." He further notes that, in such circumstances, due to the limited nature of human cognition and creativity, relying solely on intuition is insufficient for addressing increasingly complex and changing problems. Alexander strives to base design problems on a logical and scientific foundation in *The Timeless Way of Building* to overcome the subjectivity of experience and the uncertainty of intuition. His philosophy of grounding design in rationality still holds relevance today. Nigel Cross points out that, in terms of the objects of study, science targets the natural world, the humanities focus on human experience, and design addresses the artificial world. In terms of methods, science uses controlled experiments for classification and analysis; the humanities employ analogy, metaphor, and evaluation; while design utilizes modeling, diagramming, patterning, and synthesis.[10] In the field of landscape architecture, various forms of models are commonly used, including process models, conceptual models, and mathematical models. The parametric landscape architecture planning and design mechanism describes the relationships among various elements within a landscape design system. It relies on models to express these relationships and guide specific practical operations. For parametric landscape architecture design, the model primarily focuses on describing relationships and processes. It is a logical conceptual model that incorporates a combination and integration of scale models and mathematical models in the parametric design process. For example, in the road alignment model mentioned earlier, the model is based on the relationships between various elements in the landscape environment, determining parameters, and describing the entire alignment process. In the application of the model, a scale model of the design site is built using ArcGIS software, specifically the Digital Elevation Model (DEM). Path distance algorithms, shortest path algorithms, and other mathematical models are then used for parametric calculations. The model studied in this research differs slightly from the models generated in parametric modeling in architecture. The similarity lies in the fact that both use parameters and variables to control the generation of design solutions. However, due to differences in object scale, design goals, and output forms between landscape architecture and architecture, the parametric model for landscape architecture

[10] Guo, Y. (2011). An overview of methodological approaches in contemporary design research. *Journal of Landscape Architecture*, *2011*(4), 69.

<table>
<tr><td rowspan="2">Scale model</td><td>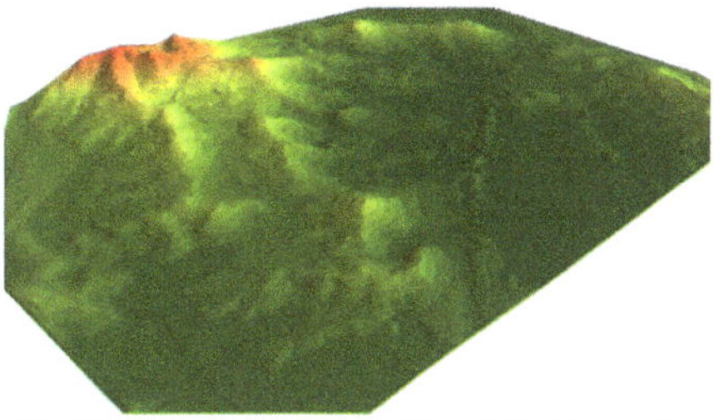</td></tr>
<tr><td>Digital topographic model of the northern area of Niushou Mountain Scenic Area</td></tr>
<tr><td rowspan="2">Conceptual model</td><td>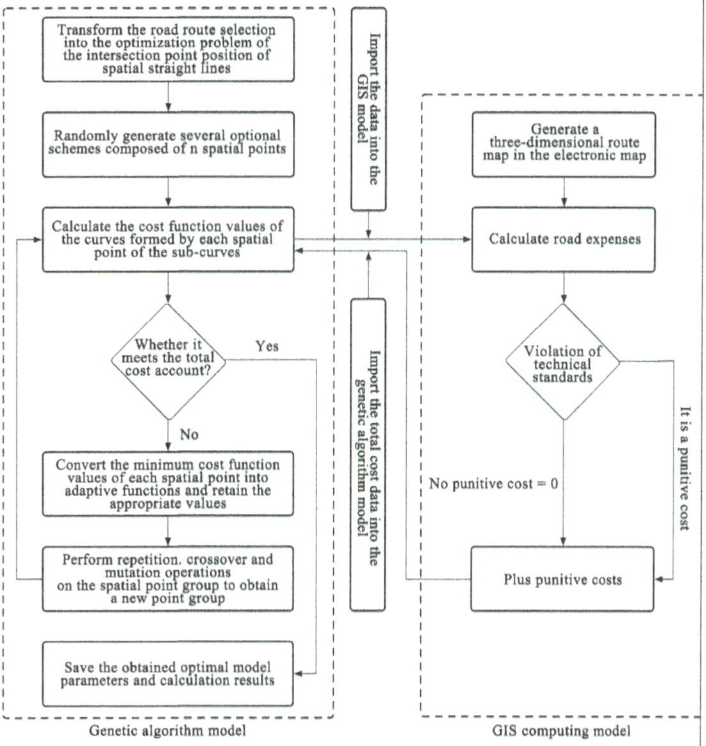
</td></tr>
<tr><td>Road route selection model combining GIS and genetic algorithm[40]</td></tr>
<tr><td rowspan="2">Mathematica l model</td><td>$$\begin{cases} \min \sum_{(v_i,v_j)\in E} a_{ij}x_{ij} \\ x_{ij} \geq 0 \\ \sum_{(v_i,v_j)\in E} x_{ij} - \sum_{(v_j,v_i)\in E} x_{ji} = \begin{cases} 0, i = 1 \\ 0, 2 \leq i \leq n-1 \\ -1, i = n \end{cases} \end{cases}$$</td></tr>
<tr><td>Mathematical model of the shortest path</td></tr>
</table>

Fig. 2.5 Scale model, conceptual model and mathematical model

focuses more on the description of the system and process, while parametric modeling in architecture emphasizes the integration of multiple factors in the generation and optimization of forms (Fig. 2.6).

The construction and application of parametric models optimize the traditional design process, presenting a systematic design thinking and dynamic characteristics. In the traditional design process, designers primarily rely on their experience and intuition: first, they conceive a general plan and design strategy in their minds, then they create hand-drawn sketches, followed by modeling the preliminary design in 3D software. Throughout this process, the design is repeatedly refined, adjusted, and modified until the final plan is confirmed, after which the drawings are produced. This design process is divided into several stages, and each stage can only proceed after the completion of the previous one, with clear boundaries between the stages. If modifications are necessary, the work becomes cumbersome and often involves a great deal of repetitive labor. Moreover, since the design is entirely based on the designer's experience, it often reflects a strong personal will, making it difficult to comprehensively reflect the site's conditions. The design process tends to be ambiguous, and the design outcomes are subjective. The application of parametric models changes this situation. With parametric models, designers identify the elements that influence the design as parameters, and through the construction of the model, they relate the various elements of the design system. The process is then described, and ultimately, through a computer platform, the parameters are adjusted to control the generation of the final design results. This design approach is bottom-up and based on process logic, considering both the actual site conditions and the roles played by various system elements in the design. It is more rational and precise. Design cannot be strictly derived from calculations alone. Parametric landscape architecture planning and design models are based on logical reasoning and describe the design process, rather than providing calculative formulas.

The positivist paradigm emphasizes quantitative research methods, while the constructivist paradigm advocates for qualitative research methods. These two

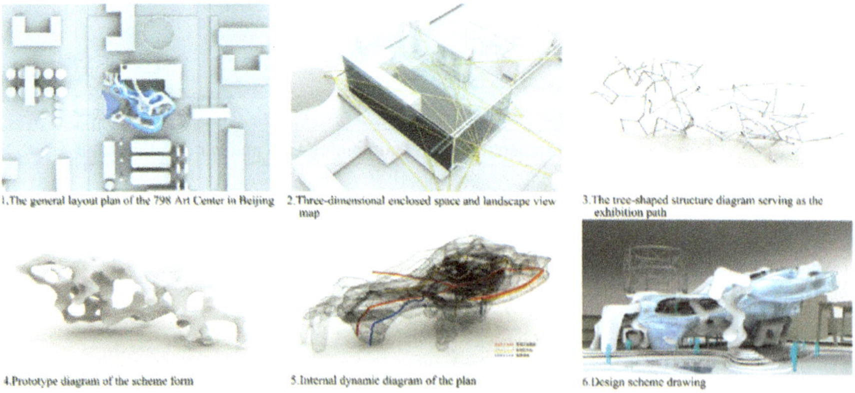

Fig. 2.6 Parametric Modeling of Architectural Design. Quoted from: Xu, W. (2011). Parametric design and algorithmic form-finding. *World Architecture*, (6)

paradigms have sparked numerous debates and "battles," often referred to as the "qualitative-quantitative debate," with quantitative proponents (QUANs) and qualitative proponents (QUALs) insisting that the two paradigms are "incompatible." In the early 1990s, some scholars proposed compatibility theories to address the ongoing conflict between these two paradigms in the social and behavioral sciences. During this period, researchers began adopting a philosophical and methodological approach tailored to specific issues within their studies. Pragmatic theorists and researchers introduced the concept of "mixed methods," incorporating both "quantitative pathways" and "qualitative pathways." The unique nature of landscape architecture, which combines both art and science, allows it to integrate the characteristics of both humanities and engineering disciplines in its research methodologies. Mixed methods, combining both qualitative and quantitative approaches, are equally applicable to landscape architecture studies. The research on models in landscape architecture is closely aligned with the need for both qualitative and quantitative research. Within the combined qualitative and quantitative research methods, there are two key relationships: first, the temporal relationship, referring to the sequence in which the two methods appear in the research; second, the hierarchical relationship, referring to the primary and secondary roles that each method plays in the research.

Thus, the mixed research methods include three types: The first is the sequential type, where qualitative research and quantitative research are on an equal footing, including the transition from qualitative to quantitative, from quantitative to qualitative, and the concurrent use of both; The second type is the primary and secondary type, including qualitative analysis as the main approach and quantitative analysis as the secondary one, and vice versa. The third type is the comprehensive type, that is, the mixed and multi-level use of the above two types. The parametric landscape architecture planning design model exemplifies the integrative connection between qualitative and quantitative methods. In the operation of the model, qualitative and quantitative methods alternate and complement each other, creating a synergistic relationship between the two.

Chapter 3
Construction of a Parametric Landscape Architecture Planning and Design Framework

Design is not "a tree without roots or water without a source," and this is especially true for the creation of landscape environments, which are inherently tied to their context. Systems science has had a profound influence on contemporary landscape planning and design. As a design system with multiple objectives, it must achieve sustainable development through an approach that is adapted to local conditions. Only designs rooted in site interpretation can fully utilize site resources and generate results that align with the characteristics of the place. Designs that appear to "grow naturally" from the site are more likely to blend into their surroundings and endure over time. The concept of "coupling" involves the relationship between "design" and "site." A coupling-based design approach seeks design rationale from the site itself, emphasizing "site appropriateness" as the core principle. This is an effective means to achieve the integration of design and site, and reflects a form of design intelligence. "Coupling" provides a methodological explanation of the essence of landscape planning and design, and serves as a guiding philosophy throughout the entire design process. It is a contemporary interpretation of the principle of "designing in accordance with local conditions," highlighting the close connection and interaction between the two systems of "design" and "site" throughout the full lifecycle of landscape architecture. The principle of coupling not only helps generate site-specific features in the landscape environment, but also promotes reduction, efficiency, and sustainability in landscape planning and design. It facilitates design activities that minimize environmental disturbance and maximize resource utilization. Introducing the concept of "coupling" into landscape architecture has significant meaning and value for contemporary landscape planning and design. Based on the interpretation of "coupling," this study explores its implications for landscape planning and design from the perspectives of ecology, morphology, and methodology. It further explains the goals, core principles, and criteria of "coupling" in relation to the unique characteristics of landscape planning and design.

The concepts of "system" and "process" signify humanity's deepening understanding of the world's continuous development and universal interconnectivity. "System" reveals the relationships between elements and between parts and the

Y. Yuan, *Parametric Landscape Architecture Planning and Design*,
https://doi.org/10.1007/978-981-95-8011-8_3

whole, while "process" illustrates the dynamic changes within systems. The study of systems and processes aims to explore the laws of development in specific temporal and spatial contexts, and to understand how interconnected elements evolve within systems. Research on parameterized mechanisms in landscape planning and design must be grounded in an understanding of "systems" and "processes."

Landscape architecture is a discipline focused on the relationship between humans and nature. The organic integration of natural and artificial systems is the ultimate goal of planning and design. "Coupling" offers a pathway for achieving harmony between the natural and the artificial. A scientific and comprehensive analysis and evaluation of the site is essential to enhance the degree of coupling between design and site, and parametric methods provide an effective means to do so. The principle of coupling forms the conceptual basis of parametric design, while the systematic nature of landscape planning and design serves as the foundation for implementing parameterization. The pursuit of comprehensive multi-objective optimization is a condition for generating parametric mechanisms. The goal of parametric design is to establish a dynamic and controllable design mechanism to achieve system optimization (see Fig. 3.1). With this mechanism in place, landscape planning and design is no longer a "black box," but is transformed from sensation to perception, from intuition to reason, and from vagueness to logic. Based on the study of process and guided by the principle of coupling, the author reconstructs the landscape planning and design process, creates a parametric landscape design system, interprets its characteristics, explores the formation of a parametric platform for landscape planning and design, and preliminarily constructs and discusses the parametric design process for landscape architecture.

3.1 The Definition of "Coupling" in the Context of Landscape Architecture Discipline

The term "coupling" is widely used across various disciplines such as probability theory, electronics, physics, and computer science. In physics, "coupling" is a fundamental concept that refers to the phenomenon in which two or more systems or modes of motion influence and integrate with each other through various interactions. It represents a dynamic relationship characterized by mutual dependence,

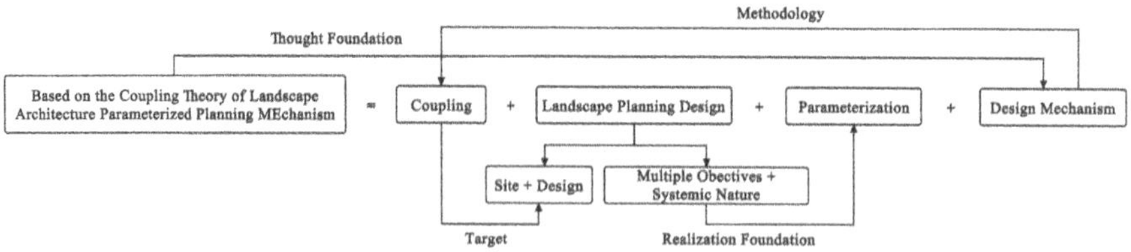

Fig. 3.1 Parameterization planning and design mechanism of landscape architecture based on coupling theory

coordination, and promotion among subsystems through positive interaction. Therefore, the concept of "coupling" encompasses three key aspects: system, relationship, and dynamics. Complexity science considers the interrelations among system components as coupling relationships. As a complex system, modern landscape planning and design has evolved far beyond mere construction-oriented planning and beautification. Instead, it is now rooted in the essential characteristics of the discipline, moving toward a systematic design approach. It organizes various environmental elements in a comprehensive, coordinated, and scientific manner, exhibiting dynamic, diverse, and composite effects. Moreover, modern landscape planning and design is inherently multi-objective, needing to balance four fundamental dimensions: ecology, space, culture, and function. These dimensions may appear distinct but are highly integrated—they are independent yet interdependent, forming a symbiotic relationship that influences the landscape as a whole without compromising their individual identities. Thus, landscape planning and design is not a simple overlay of these four dimensions, but a unified organic whole with complex characteristics. Balancing and coordinating the four basic dimensions of landscape planning and design to achieve harmonious coexistence within a new environmental system is a key quality of excellent design work. Extending the core idea of "coupling" into landscape planning and design emphasizes respect for the site and the use of "natural forces," linking multi-objective design with the inherent order and elements of the site. The goal is to enhance overall environmental quality while utilizing environmental resources. As a fundamental strategy, "coupling" is also the most universally applicable method in landscape planning and design. Its basic principle involves reorganizing and reprocessing the original site elements along with design elements and objectives to form a new spatial order that meets multiple design goals. The focus of coupling is to minimize alterations to the site's intrinsic attributes and to reconcile the conflict between "external" design elements and the "intrinsic" characteristics of the site. This embodies the design principle of generating a harmonious whole and represents the core of the coupling principle. The scope of the coupling principle extends from ontology to form, function to technology, and design to construction, covering the entire lifecycle, including post-construction management.

"Coupling" emphasizes beginning with the site and responding to it accordingly. The methods and appropriate technologies employed for coupling vary across different environments, scales, and levels. At medium and large scales, coupling—as a methodology—offers greater operability and systematization compared to "regiondialism." At smaller scales, the application of coupling principles still promotes dialogue with the environment, resembling the concept of "organic architecture." However, while organic architecture seeks internal harmony within a building and between architecture and its environment, often with a unidirectional focus, the coupling principle reflects a dynamic mutual adaptation. It also incorporates ecological, spatial, functional, cultural, and even technical and mechanical systems. The aesthetic quality of landscape environments is merely an external manifestation—their true essence lies in internal harmony. Coupling concerns more than formal harmony; it strives for true "seamless integration" of new design elements with the environment, from form to process.

"Coupling" possesses the attributes of dynamism and process. In this process, the design goals and the site influence each other, ultimately achieving the goal of harmonious symbiosis. Therefore, the principle of mutual adaptation between design objectives and the site lies at the heart of coupling. Although landscape planning and design projects vary greatly in nature, scale, and complexity, they share fundamental theories and principles. Chief among these is the principle of mutual adaptation, which runs through the entire process of modern landscape planning and design.

"Coupling" is a fundamental rule, while the principle of "mutual adaptation" is its concrete method of application. Literally, "mutual adaptation" means being mutually suitable or adaptive. This suitability is bidirectional: one aspect involves the active adaptation of design objectives to the site—selecting appropriate projects, design methods, etc., based on the environment. The other aspect involves moderate transformation of the environment based on thorough analysis to achieve integration between the project and its setting. This is similar to the approach taken toward architectural heritage: instead of blind preservation, it advocates a balanced strategy combining conservation and adaptive reuse based on careful evaluation. As a bidirectional process, coupling emphasizes appropriate human intervention that fully utilizes site resources and natural forces.

3.2 The Significance of the Coupling Principle in Landscape Architecture Planning and Design

Adapting measures to local conditions "is the crystallization of China's long-standing agricultural civilization wisdom, guiding the Chinese dialectical approach to resolving the "human-land relationship," with the unity of heaven and humanity as its highest ideal. Today, with China's urbanization accelerating, we face massive transformations of living environments that require modern scientific technologies, concepts, methods, and tools to meet practical demands. By comparison, the traditional notion of "adapting measures to local conditions" often can only be intuitively grasped rather than explicitly conveyed. As a wise approach to land stewardship, this concept has become a consensus among contemporary designers. There is a need for clear, operable ways to reinterpret the principles of site-responsive planning and design for today's context. "Coupling" not only responds to and elaborates on the idea of "adapting to local conditions" but also seeks to establish an intelligent design philosophy. In landscape architecture planning and design, this significance mainly manifests in three aspects: ecology, morphology, and methodology.

3.2.1 The Ecological Significance of Coupling

Ecology is the science that studies the interactions between organisms and their environment. In landscape architecture, ecology focuses on the relationship between humans and the environment, including adaptation, utilization, and optimization based on an understanding of nature. The most direct expressions of this are practices such as planting appropriate trees according to site conditions or adapting to the natural terrain—reflecting respect for and alignment with climatic conditions, landforms, and the habits of plant species. For landscape planning and design, specific elements such as road construction, water feature creation, and site selection for built projects must not only follow the principle of adapting to local conditions and conforming to natural laws, but also respond to varying environmental requirements. This reflects a process of coordination and mutual adaptation between design and the environment.

3.2.1.1 Coupling and Adapting to Local Conditions

Ji Cheng once wrote in The Craft of Gardens (Yuan Ye): "Naturally interesting in itself, without laborious human effort." The traditional Chinese landscape design method of "site adaptation" (xiang di shu) has long been widely applied in site selection and environmental construction. It emphasizes starting from the site itself and seeking the basis for design within the characteristics of the place. Lewis Mumford also once stated: "Make the fullest use of the potential nature offers." Clearly, "adapting measures to local conditions" has been a shared design philosophy in both Chinese and Western landscape architecture throughout history, highlighting the importance of place cognition in order to reveal the site's characteristics. In landscape architecture planning and design, "yin" means "according to" or "based on," and "di" refers to "site." Thus, "yin di" stresses initiating the design from the site and tailoring the design to suit it. "Zhi yi" refers to the appropriate selection and optimization within the design. The concept of "adapting to local conditions" encompasses two fundamental levels: on one hand, it refers to the maximization of site use—that is, making full use of the site's existing resources and development trends to maximize its inherent advantages and resource benefits; on the other hand, it refers to the minimization of site disturbance, meaning that during the landscape construction process, design goals should be achieved while minimizing interference with the original site elements, avoiding redundant design and excessive consumption or disruption of resources. Therefore, a design based on "adapting to local conditions" can realize intensive utilization of the site, becoming an important pathway to creating efficient, compact garden landscapes. The implementation of "adapting to local conditions" includes two active processes—cognition and selection, where cognition is the prerequisite and foundation for selection. The core of the landscape planning and design approach based on "adapting to local conditions" lies in site cognition, that is, the process of interpreting the site.

Landscape design is dedicated to creating environments with distinct characteristics. The method of "adapting to local conditions" emphasizes that landscape features originate from the site itself, starting from the site's physical and contextual qualities, excavating its inherent characteristics, and maximizing the use of its resources. This approach also echoes current values of intensive and sustainable landscape design.

The implementation of landscape planning and design based on "adapting to local conditions" needs to follow the principle of "coupling." This means promoting positive mutual adaptation between the design and the elements of the site to achieve minimal environmental disturbance and maximal resource utilization. This is the essential meaning behind the promotion of coupling principles, which can be interpreted on three levels: first, it maximizes the expression of inherent site characteristics. Under the premise of ecological priority, the site's intrinsic elements are highlighted in the project, aiding in the protection and enhancement of ecological quality, as well as the inheritance of spatial and cultural site features, thereby forming the basic channel and strategy for achieving personalized and distinctive design; second, "coupling" enables design reduction, achieving the reorganization of site elements with minimal intervention, truly realizing reduced or minimalistic design; third, aided by scientific and technological methods, the cognition of the site becomes more objective and accurate, helping shift design from intuition to perception.

3.2.1.2 Coupling and Scale

Scale is an ancient and enduring concept, and in contemporary scientific research, its importance is more prominent than ever. Traditional notions of scale have gradually faded or become less defined, while the connotation and extension of the term have deepened and expanded. Literally, scale refers to measuring with a ruler, but it can also be extended to mean a standard or perspective for understanding things. As a concept commonly used in multiple disciplines, scale has different meanings in different environments and academic contexts. In the study of human settlements, scale has two primary interpretations: one originates from the field of architecture, where it refers to proportional relationships and derives a quantifiable comparison of objects; the other comes from ecology, where scale examines the temporal and spatial scope of the characteristics and changes of things (or phenomena). In the discipline of landscape architecture, scale encompasses both of these meanings.

In the 1970s, American filmmakers Charles Eames and Ray Eames produced the documentary Powers of Ten, which presented the universe from a micro to macro perspective, using powers of ten meters as increments to sequentially reveal spatial scales of the world, Earth, and cosmos. With the advancement of social productivity, the scope of human activity has expanded vastly. Correspondingly, the scale of modern landscape planning and design has also broadened, moving beyond small-scale courtyards, pocket parks, and urban green spaces to regional and national territory. Contemporary landscape architecture research encompasses everything from the built environments of cities and villages to studies of natural systems (Fig. 3.2). This can also be referred to as the study of "landscape systems."

Chinese gardens originated over 3000 years ago during the late Shang and early Zhou dynasties as gardens that took the natural environment as their absolute subject, seamlessly integrated into the vast native landscape. With social progress and development, gardens gradually separated from their original natural settings, instead depicting mountains and waters within artificially created environments where human control supplanted nature as the dominant force. Today, ecological design philosophies guide public attention, and designers focus their vision on areas dominated by nature. This indicates that, as society has advanced, the focus of landscape attention has undergone a cyclical process: from nature to artificial and back to nature. The development of Western gardens follows a similar trajectory. As the cradle of European civilization, whether in the Nile or Mesopotamian basins, vast natural environments gave rise to practical plantations and hunting reserves. From the Roman Empire through the Renaissance, Western garden-making gradually detached from the natural environment, constructing numerous artificial gardens attached to architecture that exhibited an overtly artificial character, with nature no longer the garden's controlling subject. The onset of the Industrial Revolution marked a return to nature in Western garden design: the rise of the English landscape garden brought gardens back to nature's embrace. Subsequently, under the advocacy and development by figures like Frederick Law Olmsted, modern landscape architecture emerged, heightening awareness of natural scenery and fostering movements for the protection and utilization of natural landscapes. The ecological movement, which focuses on human impacts on nature, began in the 1950s and

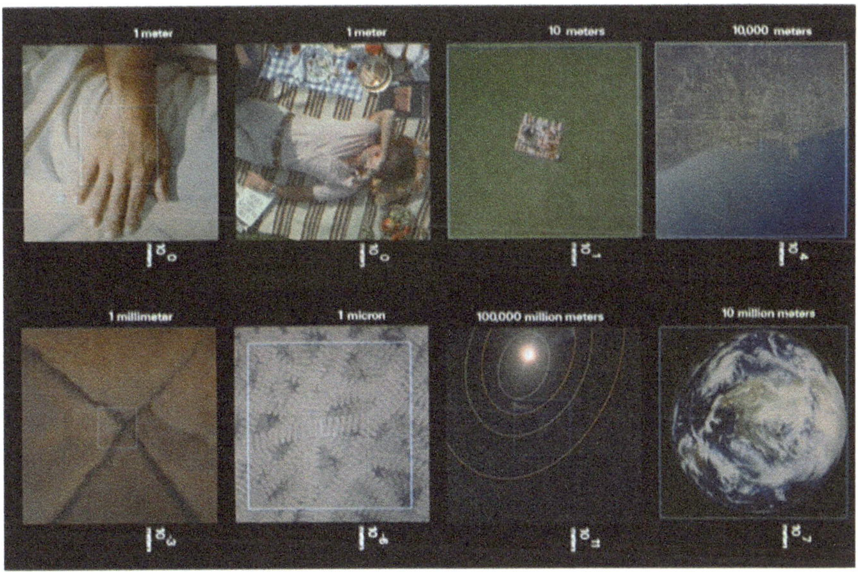

Fig. 3.2 Documentary "Powers of Ten" showcasing different spatial scales. Source: Internet

1960s and remains the most active current trend, signifying renewed attention to the natural subject in landscape environments. Regarding the alternation of natural and artificial dominance in landscape architecture, Mr. Feng Jizhong humorously remarked, “Ultimately, it’s because people have seen too much and want to try something new.”[1] The underlying reason lies in continuous reflection and exploration of landscape architecture, alongside ongoing optimization and refinement of landscape planning and design. Research on landscape systems and built systems together constitute the contemporary research horizon in landscape architecture, broader than any previous historical period.[2] This research must transcend the concept of the “garden” and embrace the notion of the “landscape.” Attention to scale differences reflects varying complexity; thinking, perspectives, design strategies, and procedures also differ accordingly—these are critical issues for deep reflection in contemporary landscape architecture. Ecological issues under different scales call for systematic classification and management. Landscape architecture addresses vast natural regions, focusing on protecting and conserving human habitats and natural systems. Traditionally, ecology has concentrated on complexes formed by plants and their habitats. However, from a disciplinary perspective, the scope of traditional landscape ecology is limited. At the scale of territorial and regional landscape systems, traditional landscape ecology is insufficient to fully cover the domain. Therefore, the concept and vision of “landscape ecology” should expand into broader fields. Attention should shift from garden habitats to human ecological environment systems—this represents the developmental trend of landscape architecture. From near-surface climate to vegetation systems, from natural zoning to land patterns, the “large” ecological system including human habitats ought to be the contemporary landscape architecture research domain.

3.2.2 *Morphological Significance of Coupling*

The term morphology derives from the Greek morphe (form) and logos (logic). “Morphology” thus refers to the study of the logic behind form composition. Originally rooted in biological research, by the mid-twentieth century it had been applied across disciplines such as literature, mathematics, sociology, and others, primarily exploring the “form” of entities. In urban studies, urban morphology (also known as urban form or urban landscape) seeks to view the city as a living organism: by observing and studying its growth mechanisms, one aims to establish theoretical analyses of urban development. The American geographer Carl Ortwin Sauer, arguing that Ellen C. Semple’s environmental determinism overly

[1] Feng J Z. The Realm and Space: On Planning and Design [M]. Shanghai: Oriental Publishing House, 2010:52.

[2] The discussion on “landscape environment” and “built environment” can be found in detail in: Cheng Y N. The Theory and Methodof Modern Landscape Design[M]. Nanjing: Southeast University Press, 2010:139.

emphasized the impact of environment on humanity, in 1925 published Landscape Morphology[3], criticizing the determinist view and asserting that human culture determines the characteristics of a place's landscape. From an aesthetic perspective, Wu Jiahua studied landscape morphology as a "language of design," exploring fundamental issues of form, logic, and emotion and the structure of design language.[4] From the above, whether in urban planning, geography, or landscape aesthetics, scholars have investigated the logic behind shape generation by examining mechanisms, drivers, and other factors. In landscape-garden planning and design, the pursuit of "beauty" is an enduring theme, and form is both its vehicle and expression. One may ask: what is the logic behind form? Is form "implanted" artificially? The attributes of morphology reveal that it first encompasses a holistic, top-down approach (from philosophy to practice), revealing an ecological concept; second, it follows the principle of "Tao follows nature" rather than searching for replicable patterns; third, while acknowledging the impossibility of an ideal form, it seeks methods to optimize the environment, achieving some balance across different spaces and systems[5]. The logic of form generation arises from place, and "coupling" provides the route and guarantee for that logic.

As both a response to and realization of "adapting to local conditions", "coupling" bears a special relationship with "place" (di). Here "di" refers to the setting faced by landscape-garden planning and design. Place signifies the locus of activity. In European languages expressing "to happen"—in English "take place," German "stattfinden," Italian "avere luogo"—there is a shared association with "place" (place, statt, luogo). This indicates that all human activity must find a suitable "place" in which to occur. Thus, "place" is in fact an indispensable element of human activity. Studies of place began with phenomenology and physical geography; in the 1970s, Norberg-Schulz published a series of articles and books that gradually formed the theoretical system of architectural phenomenology. In Genius Loci: Towards a Phenomenology of Architecture, he declared that "the most concrete expression of environment is place." Generally speaking, it is the locus of actions and events. Here "environment" is the bearer of all phenomena in the world. For him, place is not merely an abstract location, but a distinct space with clear characteristics: not only a whole composed of tangible objects, but also where life happens. Territorial scales—national land, regions, landscapes, settlements, buildings (and sub-places within buildings)—gradually shrink in scale to form a "hierarchy of environments" that includes "natural places" at the apex and "built places" at the base. Steven Holl states that place is a third kind of existence that arises from the fusion of specific external and internal perceptual orders in architecture; it is a product of both nature and human consciousness.

---

[3]C. Sauer. The Morphology of Landscape[M]. University of California Publications in Geography, 1925:19–54.

[4]Wu J H. Landscape morphology: Comparative study of landscape aesthetics[M]. Beijing: China Architecture & Building Press, 1999:365.

[5]Wu J H. Landscape morphology: Comparative study of landscape aesthetics[M]. Beijing: China Architecture & Building Press, 1999:365.

Place and site are two distinct concepts and should not be confused or equated. The Hanyu Da Cidian (1997) defines "site" (changdi) as an open space suited to a particular need—for sports, construction, storage, etc.—generally conceived as a carrier surface for activities. Compared with "site," place has richer meaning: it comprises both the physical site and the activities. In landscape and garden environments, place is the basis for activity; it is not limited to a two-dimensional plane but covers the entire three-dimensional region in which activities—natural forces or human—may occur. In a sense, a place is a "site of human activity"; a location devoid of human action cannot be called a place. In landscape-garden planning and design, most places involve human processes and differ from pure natural environments. Pure natural environments—"first nature"—are original landscapes undisturbed by humans, and are rare in contemporary landscape practice. Professor Cheng Yuning, in Theory and Methods of Modern Landscape Design, argues that place comprises objective environmental elements inherent in the site—ecology, climate, soil, hydrology, topography, morphology, fauna, flora—a crucial thread in design thinking. Beyond space, context, and habitat, as a "site of human activity", place also contains human "behavior." These four fundamental aspects together constitute the essential meaning of place, corresponding exactly to the four basic elements of landscape-garden planning and design. Environmental psychologist David Canter believes the concept of "place" can be applied at all environmental scales. Thus, from its basic definition, place is a concept encompassing all scales and covers most sites in landscape-garden planning and design.[6] Places are dynamic and constantly changing; designers must, on one hand, protect and inherit the existing order of a place, and on the other, imbue new order according to its characteristics—allowing new and existing orders to coexist harmoniously while meeting new functions.

The genius loci or spirit of place originates in ancient Rome. According to Roman belief, every independent entity has its own soul (genius), a guardian spirit. This soul gives life to both humans and places, persisting with them from birth until death and determining their characteristics and essence. Norberg-Schulz points out: "Ancient people experienced the environment as having clear characteristics—what they called 'genius loci'." The spirit of place is a general atmosphere, a concentrated and abstracted reflection of environmental features. While places may change, this does not necessarily imply a loss or alteration of spirit of place, because within a certain period, a place gives settled groups a sense of orientation and identity. These feelings arise from memories of place-based information. Maintaining an understanding of genius loci during change prevents confusion or loss. In landscape-garden environments, since place comprises both site and activity, behavior and extent are its two basic features. People first perceive places through the senses, then cognitively understand environmental characteristics and determine their relation to the environment. Ultimately, they ascertain their belonging in that space, forming specific cognition and establishing a connection with place. That is the basic

[6] Ndubisi F. Ecological planning: a historical and comparative synthesis[M]. JHU Press, 2002.

process of place cognition. Here, place forms the foundation of cognition, and all actions occur within a place. Thus, people, through cognizing place, form a sense of belonging—a sense of place—and thereby the spirit of place arises. The specificity of a place corresponds to people's recognition of its uniqueness.

With modern technological advances, information can be transmitted more rapidly and conveniently. Traditional concepts of region are gradually fading and national or ethnic consciousness is dissolving, contributing to a loss of landscape character and identity. Alexander Tzonis asserts that design should not merely focus on universal aesthetic principles. Unique territorial characteristics—landform, plant distribution, etc.—should be emphasized to evoke awareness of the past, of place, and of ancestors. Landscape character is not rootless; its emergence is rooted in the unique personality of place. Whether according to Sauer's view of "human dominance in landscape" or environmental determinism, place includes dual attributes of "people" and "environment," and is the source of distinctive landscape-garden design. As for form generated from place, emotional factors and the emergence of place identity are inseparable. Hence, the creation of landscape-garden environments relies on place, and constructing distinctive landscapes requires full utilization of place and amplification of its inherent features. Landscape-garden planning and design that adhere to the principle of coupling express precisely this process: seeking design rationale from place and thus evoking—through design—place-rooted memory.

3.2.3 The Methodological Significance of Coupling

Every theory or method has its own scope and applicable scale. In the face of the expanding scope and scale of contemporary landscape architecture planning and design research, "coupling," as a principle, corresponds to the internal mechanisms of landscape architecture planning and design and is a concept applicable across all scales. "Coupling" is a kind of design wisdom; as a guiding rule, it can be flexibly transformed into operational principles, methods, and pathways depending on the scale, scope, level, and type of the design object. Vertically, at the macro level, it functions as a methodology; at the meso level, it is a synthesis of methodology and landscape philosophy; and at the micro level, it can be transformed into a landscape philosophy. "Coupling" is both a concept and a method that can be applied in practice.

From a methodological perspective, the first significance of coupling lies in making full use of existing site resources on the basis of preservation. Secondly, coupling helps realize design minimization. "Design minimization" refers to controlling human intervention in the environment within a limited scale, achieving design goals while maximizing the expression of the inherent meaning and characteristics of the environment. This can be likened to using "a small force to move a great weight." Furthermore, coupling is a guarantee of sustainability. A design that conforms to the laws of nature and ecological succession is inherently sustainable, whereas one that goes against these laws requires considerable human, material, and financial resources to maintain. Therefore, coupling reduces maintenance "costs" by aligning with natural laws, thereby enabling sustainable development.

3.2.3.1 Coupling and Minimization

"Coupling" is an effective strategy and method for achieving design minimization. The term "minimization" corresponds to the English word "minimization," meaning to reduce to the smallest amount. The significance of design minimization lies in achieving design goals through subtlety—using "skillful means" to reach design objectives. It emphasizes starting from the environment, relying on the existing conditions and trends, minimizing disturbances to the site's ecological system, and using natural forces in the design process. Literally, "design minimization" stands in opposition to "overdesign." "Overdesign" refers to design practices that detach from the site's original conditions, characterized by appropriation, formalism, and excessive construction. These approaches lead to the destruction of the original environment, resource waste, and the loss of site character and human-centered qualities. Design minimization includes two core meanings: the minimization of artificial disturbance and the maximization of resource utilization.

Landscape architecture planning and design strive for appropriate intervention and optimization based on the natural environment. Ji Cheng, in The Craft of Gardens (Yuan Ye), expressed this idea as "ingenious in borrowing, precise in suitability," reflecting the traditional Chinese attitude toward landscape creation—emphasizing clever use of nature, with the brilliance of design residing in appropriateness of scale. With the advancement of material civilization, the concepts of sustainability and ecological consciousness have gained widespread acceptance in contemporary society. Since modern times, the development of ecological planning has helped people recognize natural processes in landscape environments, greatly influencing modern landscape architecture planning and design. In Ecological Design, authors Sim Van der Ryn and Stuart Cowan define ecological design as "any form of design that minimizes environmentally destructive impacts by integrating itself with living processes." The shift of the site's subject from humans to nature, as well as increasing attention to scale, has led landscape architecture planning and design to focus more on natural laws and processes within a site. Before ecological awareness became a societal consensus, the activities of landscape architects were mostly confined to small or medium scales, with a focus on meticulously shaping the environment to fit aesthetic ideals—sloped lawns following the topography, rigid linear retaining walls, and geometric flower beds and hedges. These often introduced excessive disturbances to the site unconsciously, resulting in unnecessary, redundant design. In recent decades, people have gradually realized the energy and effects of natural forces. There is a need to guide natural processes reasonably to achieve proactive integration between design and environment. Respecting the natural process does not mean passive submission; rather, it requires comprehensive investigation and understanding, and the use of natural forces to let nature "do the work." Landscape architecture planning and design should conform to nature and its processes, allowing natural forces to fully play their role in the creation of place. Under the guidance of natural forces, designers should optimize the site as much as possible, rather than relying solely on human effort to alter the existing state. This calls for a path of improvement that borrows from and is compatible with natural forces—precisely what the principle of "coupling" provides. "Coupling" is a method and pathway for landscape architects to seek harmony and symbiosis with the site in the process of landscape creation, striving to minimize negative impacts and disturbances to achieve minimized design. This aligns with traditional Chinese landscape design concepts such as "adaptation to local conditions," "harmony between heaven and man," and "creating a natural interest without artificial effort." To achieve "adaptation to local conditions" through minimized design, there must be "coupling" between design and site. This coupling allows for a scientific and optimal determination of design content and construction intensity, enabling the symbiosis between design and site.

The Oosterschelde Weir project in the Netherlands, designed by West 8 landscape architects under Adriaan Geuze, is an exemplary case of minimized design. The construction of the Oosterschelde Weir caused certain disruptions to the local ecosystem and ecological balance, leading to a sharp decline in fish populations and, consequently, a decrease in bird populations that fed on those fish. In response, West 8 simply leveled the docks and construction sites left after the dam's completion into a gravel plateau, then artistically repurposed discarded shells from nearby aquaculture farms. Clam and mussel shells were arranged in black-and-white checkerboard patterns with rhythmic variation. This shell layer became not only a large land art piece but also an excellent camouflage environment for endangered seabirds, providing critical habitat. Over time, under the effects of wind, sun, and rain, the shell layer will gradually fade, and a naturally restored dune landscape will emerge (Fig. 3.3).

Fig. 3.3 Oosterschelde Weir landscape design. Source: Internet (http://www.west8.nl)

Minimization and intensification are interrelated propositions. Landscape architecture, by nature, involves multiple objectives. Through systemic principles, it can potentially meet multiple demands simultaneously. Unlike approaches that merely emphasize saving resources, intensification seeks to maximize the comprehensive efficiency of input and output across the lifecycle of a landscape system, based on systems thinking. This is aligned with the philosophy of design minimization. Intensive landscape architecture aims to reduce resource and energy consumption throughout the landscape's life cycle (planning, design, construction, operation, reuse), effectively decrease waste generation, and maximize improvements to the ecological environment. By optimizing the use of land and other resources and improving ecological functions, intensive landscape design ultimately realizes the goal of harmonious coexistence between humans and nature.

3.2.3.2 Coupling and Sustainability

Concern for the environment ultimately aims to enable better human development. Since the concept of "Sustainable Development" was proposed in the 1980s, it has gone through more than three decades of evolution and has become a consensus of all humankind and a guiding principle for long-term global development. Sustainable technologies are the specific pathways to achieve sustainable development. Under the guidance of the concept of "sustainability," various industries have proposed applicable technical ideas and measures. The disciplines of the human living environment—comprising architecture, urban planning, and landscape architecture—have likewise developed suitable sustainable methods and technologies within their academic scope. Since the 1990s, China's Ministry of Housing and Urban-Rural Development has issued the "National Garden City Standards," requiring urban built-up areas to achieve a green coverage rate of 36% or more and a green space ratio of 31% or more. This portion of land covered by vegetation has significant effects on the urban environment. Renewable materials, clean energy, water ecology, and low-maintenance technologies—components of sustainable technologies—have become essential topics in the development of contemporary landscape architecture. For a rapidly urbanizing society like China, there is a greater need to strengthen understanding and application of sustainable technologies. Whether through the introduction of the LEED (Leadership in Energy and Environmental Design) rating system or the emphasis on LID (Low Impact Development) concepts, these all represent responses to sustainable technologies from different perspectives.

There is a saying in the landscape architecture field: "Three parts planting, seven parts maintenance." In order to ensure landscape effects and the fulfillment of functional roles, maintenance and management require significant input of manpower, materials, and finances. Compared to the design and construction processes, landscape maintenance is a long-term and continuous effort. In a context of intensive development, sustainability principles should permeate the entire life cycle of landscape—from design to construction and maintenance—thus giving rise to the concept of "low maintenance." Low-maintenance technology refers to the adoption of specific technical means and methods in the design phase, such that the cost of maintenance after construction is minimal or even unnecessary. Landscape environments contain many living elements that change over time; therefore, landscapes should be viewed as dynamic, evolving processes. The concept of "coupling," which upholds the principle of "adapting measures to local conditions," is an important pathway and method for achieving sustainability. For instance, extensive use of native plant species and the adoption of quasi-natural configuration methods for planting and community construction are key strategies. Native plants refer to those species naturally distributed in a given region, well-adapted to local climatic conditions, and characterized by strong regional identity. The use of native plants reflects the principle of "right plant for the right place," which not only enhances survivability but also achieves low or even no maintenance. The quasi-natural configuration emphasizes the clever use of natural "forces" by simulating species compositions and forms of natural plant communities to create habitats that closely resemble natural states, allowing the landscape environment to sustain itself and maintain a relatively stable condition.

The Qiántáng River fish-scale stone embankment was first built over 1700 years ago to guard against tidal disasters and has withstood millennia of tidal impact. It is praised as the "Great Wall of Sea Defense," and, along with the Great Wall and the Grand Canal, is one of China's three great ancient engineering feats. Its construction demonstrates ancient wisdom in adapting to nature. From the Eastern Jin to early Ming Dynasty, the riverbank experienced nearly a thousand years of continuous erosion. According to legend, during the Song and Yuan periods, the riverbank extended beyond the Zhan'ao Pagoda, but repeated destruction by the surging river forced continual retreat of the embankment. In the 15th year of Jiading in the Southern Song Dynasty (1222), official Liu Hou of western Zhejiang stated: "With land eroded and no means left, we retreat with stakes and stones, not fighting the tide." During the Chenghua period of the Ming Dynasty, Hangzhou magistrate Chen Rang proposed: "Withdraw a mile from the city, support the inner embankment of the canal, extend ten miles, and manage force with leniency, not contending with the sea for profit." By the Qianlong period of the Qing Dynasty, the embankment had retreated to the foot of the Zhan'ao Pagoda (Fig. 3.4). People no longer sought to contest nature but instead built stone embankments along the natural flow patterns of the Qiántáng River. In high tide conditions, people would scatter rice husks along the shore, and after the water receded, the residual marks helped determine the base position of the seawall. Therefore, unlike conventional seawalls, the fish-scale stone embankment has a shoreline that forms an S-shaped curve when viewed from the side. After centuries, the embankment remains robust, still enduring twice-daily tidal impacts, each exerting approximately six tons of pressure per square meter. The fish-scale stone embankment is a product of human–nature negotiation. By following the rising and retreating patterns of the tides, the embankment embodies the principle of "coupling." A design aligned with natural laws is undoubtedly rational and sustainable. The Dujiangyan irrigation system, one of China's great pre-Qin water engineering feats, has benefitted the Sichuan Basin for over two millennia. In

Fig. 3.4 Zhan'ao tower and fish scale stone ponds. Source: Internet

Fig. 3.5 Aerial view of Dujiangyan. Source: Internet

contrast, similar ancient irrigation systems such as those in Egypt, Babylon, and the Zhengguo and Ling canals in Shaanxi and Guangxi have disappeared over time. The guiding philosophy of Dujiangyan is "responding to the terrain and guiding by advantage," which reflects not only adherence to natural laws but also the incorporation of ancient expertise such as "shaping the fish mouth, erecting spillways, deep dredging shoals, and low dams" for water diversion, sand prevention, and flood discharge. This exemplifies the high degree of coupling and unity between humans, land, and water. Dujiangyan's construction respected natural resources and utilized them to serve human needs, making it a sustainable and harmonious example of human–nature coexistence (Fig. 3.5).

3.3 Coupling Principles and Landscape Planning and Design

3.3.1 *The Objective of Coupling: Systematic Design*

The goal of coupling is to establish a relationship between the two major systems of site and design. "Coupling" occurs within the system and describes the relationships and interactions among elements within a system. Landscape planning and design are essentially processes of establishing integrated relationships between newly introduced environmental elements and the original site. Therefore, for landscape architecture, "coupling" operates within the design process to establish a coupling relationship between site and design. The study of "coupling" is thus about constructing this type of relationship, designing within the framework of systems, seeking justification for multi-objective designs, generating site character, and ultimately achieving sustainable site development. The key to systematic design is to regard the site and its elements as a complex system, emphasizing the connections among system components, and focusing on multiple variables rather than single factors.

3.3.2 *The Core of Coupling: Site Suitability*

In ecological planning and design, Forster Ndubisi employed the concept of "place-making" to "determine whether a consistent fit exists among human experiences, natural processes, and the physical spatial environment" (Fig. 3.6). In his book Ecological Planning: a Historical and Comparative Synthesis, he states: "Amos Rapoport and Kevin Lynch believe that only when such a fit exists does the environment (landscape) become meaningful to residents and users. Therefore, it is the task of us planners to seek or maintain such suitability and alignment, ensuring the integrity of place, preserving natural and cultural processes as well as spatial-temporal

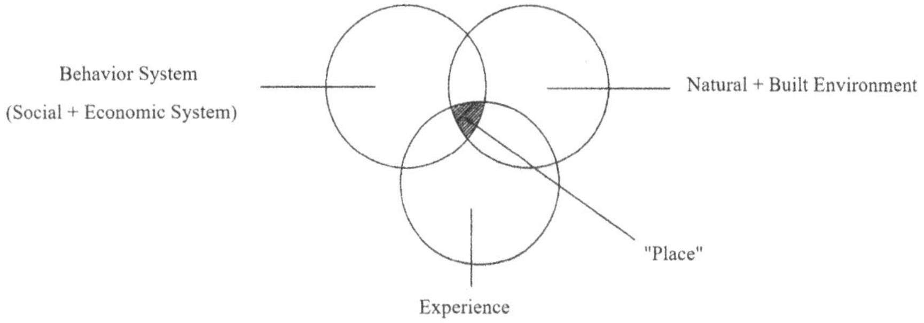

Fig. 3.6 Place as the intersection of natural processes, behavior systems, and experience. Source: Ndubisi F. Ecological planning: a historical and comparative synthesis[M]. JHU Press, 2002

connections, and instilling a sense of identity and belonging in users and residents."[7] Ndubisi pointed out that only through the suitability of functional needs, natural processes, and the original material space of the site can a sense of place be maintained. Landscape planning and design based on "coupling" precisely pursue this suitability, emphasizing the integration between newly introduced designs and the native environment. The core of "coupling" is the analysis of the "mutual suitability" between design elements and the original environment, aiming to utilize the inherent connections within things to generate new relationships. Through human intervention, site optimization is achieved during the design process, which requires that design objectives and environmental conditions become an organic whole. The design created under the guidance of coupling theory is a system—an integrated entity capable of self-improvement and possessing a self-regulating structure.

"Suitability" means appropriateness, correctness, or propriety; site suitability refers to the attributes that the design, under comprehensive consideration of multiple design objectives, should possess in order to be most appropriate and well-matched to the site. Beyond indicating the significance of design to a site, "suitability" also implies the "capability" of the site—that is, its ability to support and sustain a given land use. This capability can continue to support such specific uses without causing degradation of the site's natural and cultural characteristics. The pursuit of site suitability demonstrates an effective pathway for achieving sustainable development in landscape planning and design. For landscape planning and design, from project conception to the formation of a master plan, and down to the siting and design of landscape nodes and individual elements, three fundamental stages are involved, with different design scales encompassing different stages. "Coupling" refers to selecting appropriate projects based on environmental conditions and then introducing them into the site—this corresponds to the initial project conception stage in contemporary planning and design. The second stage corresponds to the master planning phase—that is, selecting suitable environments to support different projects. The third stage is the control of individual projects, including regulatory elements such as building scale, coverage ratio, massing, height, and construction intensity—this pertains to the level of regulatory detailed planning in design. Therefore, the three stages of "coupling" respectively correspond to the stages of "project conception – master planning – detailed planning" in landscape planning and design, progressively covering the entire landscape planning and design process.

[7] Ndubisi F. Ecological planning: a historical and comparative synthesis[M]. JHU Press, 2002.

3.3.3 The Standard of Coupling: Coupling Degree

Traditional design methods mostly emphasize the exertion of human initiative based on cognition of the site, often relying heavily on the designer's intuition, resulting in outcomes that tend to be subjective and uncertain. Contemporary landscape planning and design, while affirming artistic value, are imbued with a more rational spirit. The objective characteristics of a site do not change with differences in the designer's perception. Landscape designers, based on their own judgment and design tendencies, may choose to emphasize different features of a site, resulting in diverse design outcomes. Almost no design has a single "correct" solution; designers' differing perceptions of the same site, varying personal intentions, and disparities in design abilities all influence the design process and the generation and evolution of proposals. Traditionally, people are accustomed to comparing and selecting among various design schemes. Today, this practice corresponds to the prevalent design bidding process in the market, which seeks the scheme that best fits the site's characteristics. However, due to a common lack of understanding about the site, such scheme comparisons often devolve into mere "beauty contests." The so-called best scheme is often determined by the decision-maker, thus losing the true value of comparison. First, it must be acknowledged that sites possess objective conditions that do not change based on human will. These conditions merely provide the basic elements of design, not its entirety. Because different schemes vary in their understanding of the site, this determines the quality and merit of those schemes. "Coupling degree" refers to the level of correlation between a design proposal and the site; it can also express the range of values for the degree of adaptation between the design and the site. "Coupling degree" describes the responsiveness of the design to the site and is an important indicator for judging the quality and merit of design proposals. Selecting a scheme based on the principle of coupling means choosing the idea that best matches the site from among multiple schemes. Emphasizing the "coupling" between the design and the site is the prerequisite for evaluating the quality of a proposal. "Coupling degree" provides the evaluation basis for comparing and judging the merits of different schemes and is a fundamental indicator for verifying the feasibility of a design scheme.

Ontology is the origin of all things. Ecology, space, culture, and function are four fundamental dimensions that reflect the multi-objective nature of landscape design and form the foundation of landscape planning and design. By using GIS software analysis, digital overlay, and other quantitative techniques to analyze site data, and by quantitatively evaluating target systems of space, ecology, function, and culture within the design, an evaluation system for the coupling degree between landscape planning and design proposals and the site can be constructed. The generation of multi-objective landscape planning and design proposal comparison and optimization methods undoubtedly provides a scientific approach for comparing and selecting among multiple design schemes.

3.4 Landscape Planning and Design Process Based on the Coupling Principle

"Process" refers to the procedures and stages through which things develop. In the field of philosophy, "process" is defined as the manifestation of the continuity of development and change of things in time and space. From the perspective of dialectical materialism, process is the dialectical unity of matter, motion, and time-space. In Ludwig Feuerbach and The End of Classical German Philosophy, Engels pointed out: "The world is not a collection of ready-made things, but a collection of processes." Alfred North Whitehead, founder of "process philosophy," advocated that "the world is process." He believed that all things are driven forward by their internal contradictions, constantly developing and changing, manifesting as sequential progression in time and continuous succession in space. Everything has its corresponding process, which consists of cause, course, and result. Since things are constantly developing, the result is only relative; some results may become causes, others may become parts of the course. The design process is also a kind of "process," reflected in both time and space. From the temporal dimension, the design process corresponds only to a specific stage of a thing's development. The landscape planning and design process studied by the author is a continuous developmental process, taking the generation of the design proposal as the result of the process. This process includes three interrelated stages: site cognition, proposal generation, and proposal optimization. The landscape planning and design process unfolds within the site; beyond serving as the object of design, the site also acts as the spatial carrier in which the design process occurs.

Academician Wang Jianguo summarized the urban design process in Modern Urban Design Theory and Method as two types with different value orientations and approaches: Bottom-Up and Top-Down. The "Bottom-Up" approach refers mainly to the design and construction of cities (or towns) through the cumulative overlay of individual intentions over many years, following the principles of organic growth under the influence of "natural forces" or "objective forces." The "Top-Down" approach mainly refers to the method of designing and constructing towns under the influence of human power, in accordance with the will or ideal models of certain classes or even individuals. Typically, this approach is implemented through statutory planning and design regulations.[8] The process of landscape architecture planning and design is undoubtedly a "Top-Down" process: guided by laws and policies, integrating collective consciousness to generate design proposals, and ultimately implementing them in the site. In order to avoid deviating from the "natural" characteristics of the site, design must be based on an in-depth understanding and analysis of the site. Therefore, the "cognitive process of the site" is a manifestation of the "Bottom-Up" approach within the design process and is a crucial phase of the entire

[8]Wang J G. Modern Urban Design Theory and Method[M]. Nanjing: Southeast University Press, 2001:25–32.

process. Unlike the urban design process, the landscape architecture planning and design process embeds the "Bottom-Up" within the "Top-Down" framework.

Design activities in disciplines such as urban planning, architecture, and landscape architecture generally follow a fundamentally similar process. Traditionally, the design process was often conceptualized as a linear sequence. With the advent of systems theory and the development of contemporary design, it has gradually become apparent that the object of design is a complex system rather than a mere collection of individual elements. John Zeisel, in Inquiry by Design: Tools for Environment-Behaviour Research, described the design process as a "design spiral," a cyclical and repetitive process in which the designer continually repeats a series of activities, each solving certain problems, all oriented toward the same goal (Fig. 3.7). Design must meet a variety of needs, making it a creative endeavor under constraints, realized through a series of specific design activities. The design process is a complex systems project—an effective integration and coupling of diverse knowledge domains. Accordingly, the design process can be defined as: a creative, systematic method for solving design problems, one that is planned, step-by-step, and progressively refined, requiring multiple iterations to arrive at a satisfactory outcome. The iterative and cyclical nature of the design process can be described as a "design–analysis–evaluation–redesign" sequence. Patrick Geddes advocated the use of scientific methods to understand cities and summarized the modern urban planning process as "survey–analysis–plan."[9] In landscape architecture, he focused on the relationship and interaction between landscape features and social processes, using natural ecological regions as the basic units of planning, paying attention to natural climatic conditions, vegetation coverage, and animal distribution, and creatively applied the "survey–analysis–plan" logical framework. Modern landscape architecture planning and design must use scientific methods to understand the site. Its planning and design process aligns with that of urban planning, similarly proceeding from investigation to analysis to planning. In addition, it must fulfill the characteristics of systematization and iteration, where each subdivided process is interconnected and mutually reinforcing.

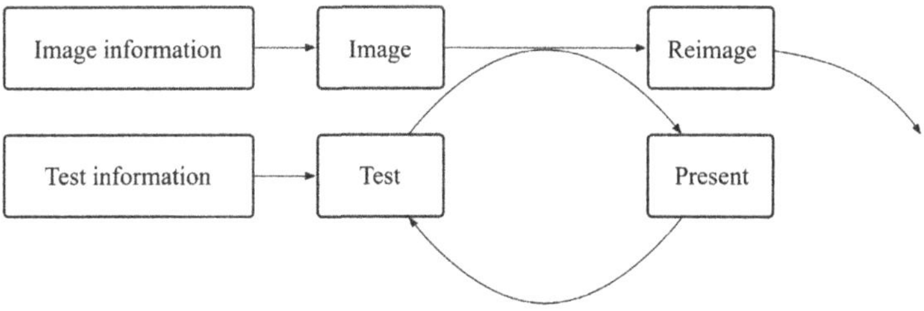

Fig. 3.7 The spiral repetitive design process. Redrawn from: Chen C C. Design cognition: Cognitive science in design [M]. Beijing: China Architecture & Building Press, 2008:33

[9] Hall P, Tewdwr-Jones M. Urban and regional planning[M]. Routledge, 2019.

From an information theory perspective, the "information" of the site flows into the design activity at its inception, and through filtering, extraction, and recombination, ultimately generates an "information-rich" design proposal. The design process is an aggregation of design activities, which interact and interconnect, with "information" flowing and interacting within these activities. In Process and Reality, Alfred North Whitehead argued that each entity is an opportunity for experience or consists of a nexus of experiential occasions. There are no material substances; rather, what exists are systems of interrelated events.[10] According to his view, relationships between things are more important than the entities themselves. Steven D. Eppinger summarized the relationships between two design activities in three forms[11] (Fig. 3.8):

1. Sequential. There is a unidirectional relationship between two activities, in which the output of the preceding activity is required as input for the subsequent one. The information flow follows a unidirectional sequence.
2. Parallel. There is no precedence relationship between the two activities; they can proceed simultaneously, with information flowing separately into both activities.
3. Coupled. The two activities are closely interrelated, with information flowing interactively between them, often repeatedly. Coupling is commonly found in design activities and reflects the iterative mechanism inherent in the design process.

In addition to the iterative nature, systematization and nonlinearity are also key characteristics of the modern landscape architecture planning and design process. J. Christopher Jones argued that since the 1950s, under the influence of computers, automation, and systematization, design should rely more on logical and systematic thinking than on traditional intuition and experience.[12] Systematization requires that landscape architecture planning and design emphasize the roles of system analysis

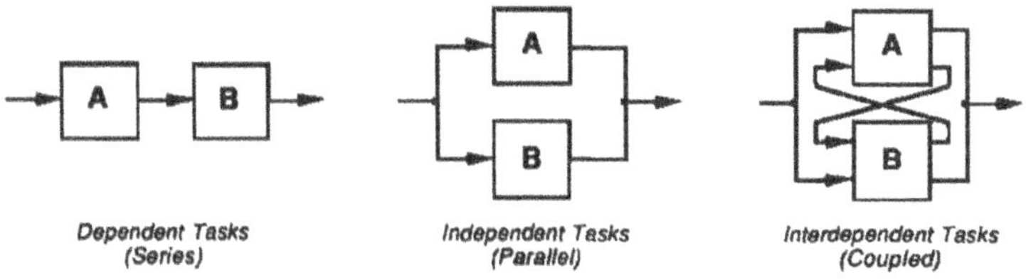

Fig. 3.8 Three types of relationships between two design activities. Source: Steven D. Eppinger. "Model-based Approaches to Managing Concurrent Engineering" [J]. *Journal of Engineering Design*, 1991, 2(3): 284

[10] Li S Y, Qu Y H. About Process Philosophy[J]. Journal of Tsinghua University, 2004(2):24–28.

[11] Steven D E. Model-based Approaches to Managing Concurrent Engineering[J]. Journal of Engineering Design, 1991(3):284.

[12] Chen C C. Design cognition: Cognitive science in design [M]. Beijing: China Architecture & Building Press, 2008:33.

and comprehensive judgment. System analysis treats all elements within the landscape site as interconnected parts of a whole, analyzing not only each element independently but also their interrelationships. Comprehensive judgment synthesizes the results of system analysis to weigh various elements and make design decisions. These two activities form the interactive relationship mentioned above. The discussion of linear versus nonlinear thinking originated in the natural sciences. For a long time, studies of nature were confined to linear methodologies. However, the achievements of the natural sciences indicate that humanity faces a complex nonlinear world that cannot be fully understood through linear means alone. The discourse on nonlinearity has quickly extended into various fields, including design. Landscape architecture planning and design deals with a complex, nonlinear world and therefore requires nonlinear thinking to address design problems. Nonlinear thinking introduces complexity, nonlinearity, and uncertainty into the design process. The openness and uncertainty of design results lead to multiple possible solutions, yet the implemented solution is singular. Designers cannot merely rely on subjective evaluations to determine the quality of a design; thus, a regulated mechanism of selection and evaluation within the design process becomes both inevitable and necessary.

From the above discussion and analysis, the design process of landscape architecture planning can be initially outlined (Fig. 3.9). The process includes three main design activities: site cognition, proposal generation, and proposal optimization. This process not only reflects the classic "survey–analysis–plan" paradigm of Geddes's urban planning model, but also incorporates an iterative feedback mechanism aligned with contemporary design principles. From the initial site survey to the final proposal optimization, the inflow and outflow of information are integrated through both sequential and coupled relationships, presenting a design process that is characteristic of landscape architecture planning and design.

3.4.1 Site Cognition Process

Cognition is a term from modern cognitive psychology, referring to the process through which humans perceive and understand external objects—in other words, the process by which humans, as subjects, process stimuli acting on their sensory

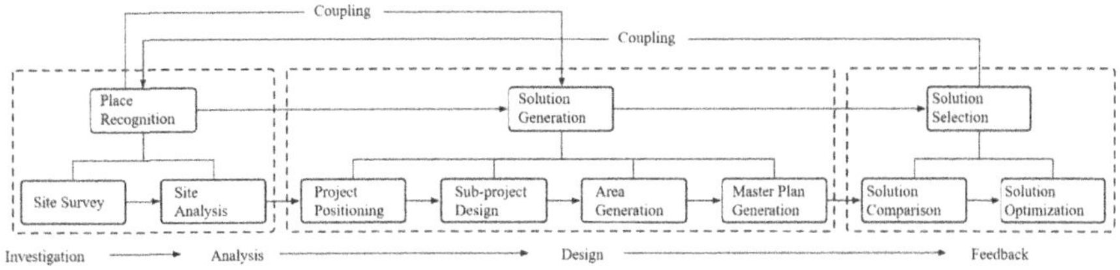

Fig. 3.9 Landscape planning and design process based on the coupling principle

organs. There are various interpretations of "cognition." One definition sees it as "the process of information processing," describing it as "a computer-like information processing operation carried out by the human brain, that is, the entire psychological process and internal mechanism through which individuals receive, store, retrieve, and apply information (or knowledge)."[13] Ulric Neisser, a pioneer of modern cognitive psychology based on the theory of information processing, defined the cognitive process in his book Cognitive Psychology as the full process through which information is transformed, simplified, processed, stored, retrieved, and used after being input through the senses. This viewpoint emphasizes the flow of information in the human brain, from the perception of information to the final behavioral response.[14] In terms of process, "cognition" goes a step beyond "perception" or "understanding"; it not only involves the input and processing of information but also includes the transformation and output of that information by individuals.

Any site is an aggregation of information, containing a variety of both natural and human attributes. Site cognition refers to the process by which people perceive and understand the information embedded within a site. Different sites contain different types of information, which together form the unique characteristics of each site. Therefore, the site cognition process is a targeted understanding of a specific site. Since landscape architecture planning and design involve numerous elements within a site, comprehensive understanding of site information is required. Thus, site cognition is the process of acquiring and analyzing the information required for landscape architecture planning and design, and it is a fundamental and essential step. The cognition of a site by humans is an active process and manifests in various forms. Recognition of the apparent factors of a site constitutes the first level of the cognition process. The site information within a landscape environment is the object of perception for the designer (as the perceiving subject). Interpretation of these apparent factors includes two aspects: one concerning natural elements, and the other concerning human-made elements. Recognition of ontological factors represents the second level of site cognition. In the context of site cognition, "ontology" refers to laws, requirements, and principles. Understanding these ontological factors involves delving beyond appearances to uncover underlying patterns. First, decoding these ontological factors allows designers to initiate creative activities based on a solid grasp of a site's essential rules, thereby promoting divergent thinking. Second, a scientific understanding of the essence beyond surface phenomena helps designers identify latent patterns and relationships within a site, enabling them to grasp the regularities and inevitabilities in the development of landscape environments. This is also a fundamental approach to achieving sustainable and efficient design concepts. Corresponding to the two layers of site cognition in landscape environments, the site cognition process in landscape architecture planning and design includes two progressive stages: site investigation and site analysis. Site

[13] Che W B. A New Dictionary of Contemporary Western Psychology [M]. Changchun: Jilin People's Publishing House, 2001:298.

[14] Liang N J. Contemporary Cognitive Psychology [M]. Shanghai: Shanghai Education Press, 2003:4.

investigation involves the collection of site information through various methods and tools—essentially a process of information gathering. Site analysis, as the core part of site cognition, builds upon the data collected during site investigation. It involves classifying, organizing, and analyzing the collected information to summarize rules and patterns. During this stage, the site information is processed and restructured based on different usage purposes, forming new information sets that serve as the foundation for subsequent design phases.

Given that cognitive subjects vary individually, the "external forces" acting upon their cognition processes—such as cultural background, experience level, and value systems—differ, leading to variations in cognition. In design, these differences are particularly evident in the site cognition process. For example, compared with novice design students, experienced designers often demonstrate a deeper subjective understanding of a site. Scientific and technological tools can help designers perceive sites more accurately, comprehensively, and efficiently. With the advancement of relevant supporting technologies, disciplines related to landscape architecture (such as architecture and urban planning) have widely and deeply adopted digital technologies in all stages from preliminary analysis, logical construction, and design expression to implementation. New technologies like BIM, digital fabrication, and digital management continue to emerge. Similarly, digital technologies have provided new tools and methods for the site cognition process in landscape architecture planning and design. This is mainly reflected in two aspects:

First, in the preliminary site investigation stage corresponding to the perception of apparent site factors, new equipment and technological tools assist in the digitization and visualization of information. Through the collection, classification, and organization of data, site information databases can be established. Second, based on the data collected, the site is analyzed and evaluated. This involves secondary processing of the data to obtain quantifiable indicators, establish parametric relationships, and construct an information model library of the site.

3.4.1.1 Site Investigation

Site investigation typically relies on two means to obtain basic information: data provided by third parties and field surveys. The contents of investigation include the natural and cultural attributes inherent to the site. Information related to natural attributes includes geographic location, landforms, climatic conditions, water bodies, soil, vegetation, and other inherent site features. Cultural attributes—literally referring to information left by human activities—cover the current built environment, historical and cultural resources, and human behavior and psychology within the site. Among these, the current built environment refers to the present human use of the site, including existing buildings and structures, roads and plazas, as well as installed utility networks.

Among conventional investigation methods, field surveys are one of the most important approaches, relying on designers' hands-on exploration. As the movement trajectory is linear, the information tends to expand along linear paths, which limits the investigation's dimensionality. Most of the collected information is presented in the form of paper or electronic documents. The development of science and technology has provided new methods and tools for site investigation, extending the scale from "line" to "surface" and opening new possibilities for parametric landscape architecture planning and design.Take site topographic maps—a must-have in landscape architecture planning and design—as an example. Typically provided by clients or acquired from planning bureaus, these maps are usually measured by certified agencies and are often three-dimensional digital terrain maps. They use a regular grid and elevation points to express landforms. Traditional surveying techniques include electronic total stations, digital levels, total station surveying, and fully automated total station systems. The majority of topographic maps in use today are produced using these methods. In recent years, the rapid development of satellite navigation and laser scanning technologies has significantly advanced surveying methods, especially as laser measurement has evolved from static point measurements to dynamic real-time tracking and ultimately to three-dimensional volumetric scanning. Supported by laser measurement technologies, digital surveying instruments such as 3D laser total stations and 3D laser scanners can generate highly accurate, high-density 3D point clouds, providing highly detailed and digital 3D terrain maps for parametric landscape architecture planning and design (see Fig. 3.10).The advancement of remote sensing technology enables designers to

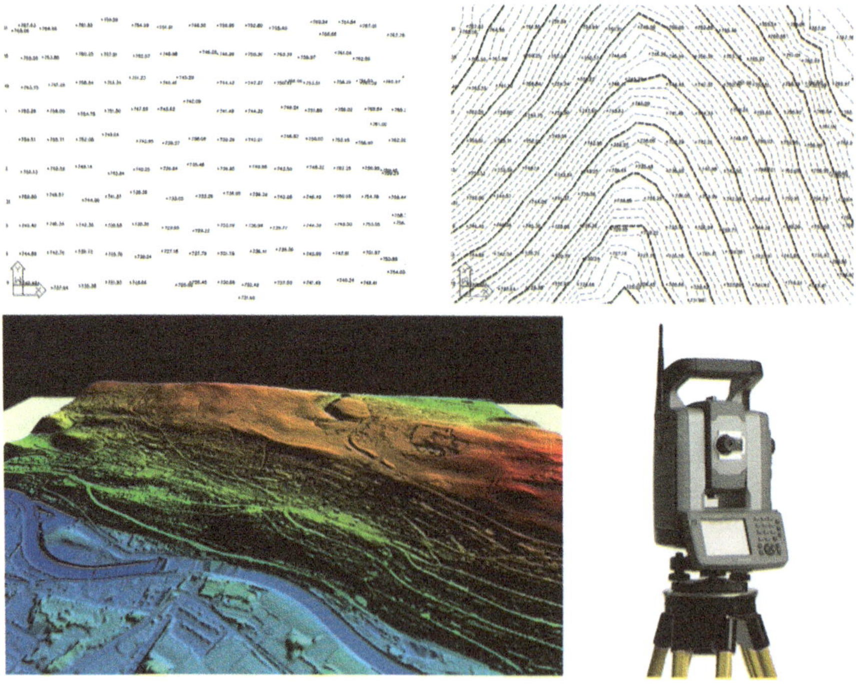

Fig. 3.10 3D laser total station and digitized terrain point cloud generated by scanning

access high-resolution images of the earth's surface that comprehensively present both natural and cultural phenomena and reflect the forms and distributions of various surface features at a macro scale. Remote sensing maps contain a wealth of data about geological features, topography, soil, vegetation, hydrology, and artificial structures, making large-scale landscape architecture planning and analysis possible. For investigating human behavior and psychology, traditional methods include questionnaires, cognitive maps, and behavioral maps. However, developments in hardware (e.g., eye trackers and portable galvanic skin response devices) and related software have provided new technological means and digital, quantitative evaluation platforms for traditional investigative methods. These advances have not only diversified the types of data collected but also enhanced their precision and efficiency, significantly promoting the scientific nature of landscape architecture planning and design methods and providing essential material for parametric implementation.

3.4.1.2 Site Analysis

Site analysis is the subsequent phase following the completion of site information collection and corresponds to the processes of classification, extraction, organization, and summarization within site cognition. Site analysis consists of two components: analysis of site elements and site suitability analysis. The two are hierarchically related, representing an in-depth study of site information from superficial to essential levels. Designers' understanding of a site is gradually revised, enriched, and improved through the site analysis process based on intuitive perception.

Site Element Analysis

Corresponding to the four levels of landscape architecture planning and design, site elements can be categorized into four types: spatial, ecological, functional, and cultural. In the parametric landscape planning and design process, datafication of information is the first step. The analysis of site elements refers to the process of converting directly observable and qualitative information into numerical and quantitative forms. Through analysis and refinement, this information is visually represented using a series of diagrams or textual forms, enabling designers to access and understand it conveniently.

In traditional site investigation and analysis, landscape architects manually record site information and further study it by drawing analytical diagrams. For instance, Lawrence Halprin used the Ecoscore method to document and interpret traces of natural process evolution left in the environment, describing the complex ecological systems therein. In the environmental survey and analysis of the Sea Ranch coastal farmhouses, he documented five relevant threads using the Ecoscore method: climate, plants, animals, site culture, and human history (Fig. 3.11).

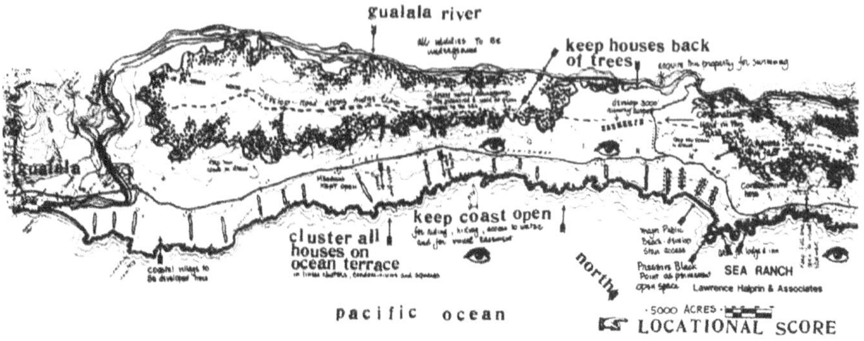

Fig. 3.11 Plan of Lawrence Halprin's ecological survey of the sea ranch coast, USA. Source: Walker P, Simo M L. Invisible gardens: the search for modernism in the American landscape[M]. Mit Press, 1996

With the advent of the big data era, digital analysis software such as ArcGIS and Ecotect has become widely used, making manually calculated and drawn analytical charts more intuitive and precise, while also allowing for three-dimensional virtual and dynamic presentation. More importantly, the development of computer and software platforms not only changes the mode of expression but also provides new ideas and ways of thinking for designers. Parametric thinking is also reflected in the modern technology-supported site element analysis. Designers can adjust input data and select algorithmic expressions as needed, enabling design diagrams to respond and provide feedback quickly.

Spatial Analysis

Spatial analysis includes regional location, topography and landforms, and visual field analysis. Regional and locational analysis is an essential component of every design project. The term "location" implies positioning. On one hand, it reflects the geographic position of the site; on the other, it embodies the site's relationship with other elements in geographic space. Geographic location forms the broader environmental context for a site, and changes in location bring about differences in climate, society, economy, and culture, thereby exerting varied influences on site development. Thus, the uniqueness of geographic location is the root of a site's distinctive character. Locational analysis entails dissecting and presenting the above-mentioned series of information.

Topography and landform constitute the spatial and physical foundation upon which a site exists and represent fundamental data inherent in every site. Since landscape planning and design sites are often located in areas dominated by natural resources, analyzing topography and landform helps in grasping the spatial form characteristics of the site. Utilizing ArcGIS software to analyze elevation, slope, aspect, and other components of topography and landform is already widely applied. Moreover, ArcGIS can also analyze geographical indicators such as surface

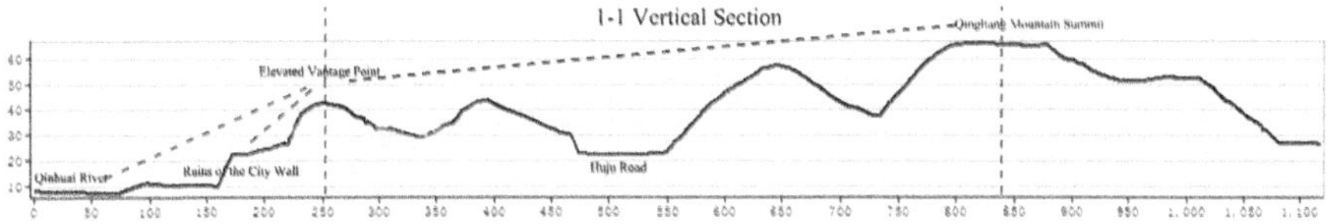

Fig. 3.12 Visual analysis of landscape space using ArcGIS software. Source: Information about Cheng Yuning Studio

undulation, surface roughness, and object density, thereby further datafying and quantifying site information. Visual field analysis reflects the visibility conditions within a site. By analyzing the visual connections of key nodes, designers can obtain spatial information that is difficult to gather directly during field surveys and preserve it in digital form for use in subsequent design stages (Fig. 3.12).

Ecological Analysis

Ecology refers to the state of living organisms and their relationship with the environment. In landscape planning and design, ecological analysis mainly focuses on habitat conditions. Habitat refers to the environmental conditions necessary for biological survival. In landscape architecture, habitat analysis primarily includes three aspects: hydrology, soil, and microclimate. Hydrological analysis involves studying the variations and movement of water in the site, including runoff, catchment, water quality, and depth. As a component of the underlying surface, soil significantly affects water movement and the growth of flora and fauna. In landscape planning and design, soil analysis mainly focuses on soil type, physicochemical indicators (pH, porosity, aggregate structure), and organic matter content.

In addition to wind direction, precipitation, temperature, and sunlight, microclimates should also be studied. Microclimate refers to climate conditions within the site that differ from the surrounding environment, and certain sites' microclimate characteristics often represent key site features. Among biotic resources, plant resources are a vital part of landscape creation, thus warranting detailed analysis, including investigation of vegetation types, forest structure, canopy density, plant growth conditions, and forest age. Information on rare flora and fauna under national or provincial protection, including the scope of protection, is a necessary consideration in design and an important part of ecological site analysis.

Functional Analysis

Functional analysis refers to the study and evaluation of human activity within the site. A site is referred to as such (rather than merely a "plot") precisely because it includes human activity. Since landscape planning and design serves human users, analyzing human activity patterns in the site helps achieve both human-centered and ecologically sound designs. This includes:restoring negatively impacted areas subject to excessive human disturbance; preserving behavioral imprints significant to the site; addressing human behavioral needs. Human activity in a site is primarily reflected in current land use, transportation conditions, and infrastructure status. The nature of current land use indicates how land resources are presently utilized, indirectly showing how people interact with different areas, and serves as pre-information for carrying capacity analysis. In landscape planning and design, land-use planning is a necessary design component and a direct tool for managing the allocation of land resources. Site transportation analysis is usually divided into external and internal parts. External transportation covers the channels through which people and goods access the site and plays a crucial role in defining the site's overall planning position. Internal transportation reflects the current accessibility within the site. As the road network forms the framework of a landscape site, extracting and organizing internal transportation data is equally essential. Infrastructure conditions refer to existing data concerning water, electricity, gas, heat, environmental protection, sanitation, disaster prevention, and other fundamental systems.

Cultural Resource Analysis

"Culture" is a term with an exceedingly broad definition. Philosophers, sociologists, anthropologists, historians, and linguists have all attempted to define it from their respective disciplinary perspectives, yet no single, strict, and precise definition has been universally agreed upon. Culture is born from human activities and represents a form of human ideology.

As a "place for human activity," the site is closely tied to culture, making culture particularly significant in landscape contexts. Except for areas untouched by humans—the so-called "first nature"—any site bearing traces of human activity is marked by culture, constituting what is known as "second nature" and "third nature."[15] As an ideology, culture requires a material carrier for its expression. These material carriers, as externalized "symbols," evoke collective memories within a group and become tools for ideological communication and transmission. Cultural resource information is the most difficult type of site data to quantify. Its collection and analysis require reliance on physical carriers such as ruins, relics, books, and paintings. Cultural resource analysis mainly involves examining the type, level, and distribution of cultural resources.

Site Suitability Analysis

Suitability Analysis originated in the field of ecological planning. After Charles Eliot, in the 1960s, Warren Manning, George Angus Hill, Philip Lewis, and Ian McHarg gradually quantified the processes of site survey and analysis. They

[15] For the discussion on "first nature," "second nature," and "third nature," see: Wang X T, Lin Q. The Meaning of Nature [J]. Chinese Landscape Architecture, 2007 (1),6–17.

published seminal works such as The Ecological Basis of Land Use Planning (1961), Regional Design for Human Impact (1969), and Design with Nature (1969), each developing their own methods of landscape system analysis and evaluation. These works have become critical components of ecological factor evaluation and land suitability analysis. In contemporary practice, suitability analysis is widely used across various fields including agriculture, urban planning, and ecology as an important method for studying land use. "Suitability" denotes appropriateness or compatibility and thus expresses a relationship between two entities. Frederick Steiner stated: "Suitability analysis is viewed as the process of determining the appropriateness of a specific parcel of land for a specific use. Land suitability can be defined as the degree to which a particular land area is appropriate for a certain use."[16] Whether discussing "land suitability" or "ecological suitability," the focus is on the appropriateness between "land" or "ecology" and "specific land use." For site suitability in landscape architecture, it approaches this relationship from the perspective of landscape planning objectives, prioritizing ecological considerations while examining the compatibility between "site" and "use demands," reflecting the degree of fit between the site and design. The goal of landscape planning and design is to meet human usage demands; any site that undergoes artificial intervention will inevitably change, but such changes must be predicated on compatibility with the site. Judgments regarding suitability should adhere to nature-based value standards, upon which human demands are matched and satisfied.

Site suitability analysis can be divided into two progressive parts: ecological sensitivity analysis and land-use suitability analysis, each conducted from the perspectives of site ecology and site development respectively. Ecological sensitivity analysis examines elements such as slope, aspect, vegetation, hydrology, soil, and current land use to produce ecological sensitivity zoning. The significance of this analysis lies in grading the ecological sensitivity of the site, clearly identifying areas that are highly sensitive, moderately sensitive, and weakly sensitive. Areas of extreme ecological sensitivity must be strictly protected and human activities should be limited therein. Ecological sensitivity analysis provides a scientific basis for the designation of ecologically sensitive zones in landscape planning and design. Land-use suitability analysis builds upon ecological sensitivity grading and re-evaluates areas outside those of extreme sensitivity, delineating zones as suitable, moderately suitable, or unsuitable for development. The term "development" here is not confined to the construction of buildings but refers to the full spectrum of human alterations aimed at enabling recreational and aesthetic activities in landscaped environments, including vegetation adjustment and optimization, water feature construction, terrain shaping, and other landscape creation activities, as well as the building of roads, bridges, structures, and buildings. According to the General Planning Standards for Scenic Areas (GB/T 50298-2018), ecological zoning should be primarily determined based on comprehensive evaluations of ecological value,

[16] Frederick Steiner. The living landscape:an ecological approach to landscape planning[M]. McGraw-Hill Inc., 1991:186.

ecosystem sensitivity, and ecological conditions.[17] Detailed planning for scenic areas must also include land-use suitability analysis within the planning boundaries.[18] Both ecological sensitivity and land-use suitability analyses provide scientific and reliable methods for defining protection categories and assigning protection levels.

The interrelationships among internal site elements are difficult to fully capture through surface-level data collection, synthesis, and expression alone. Therefore, it is necessary to analyze site information with different objectives in mind, leading to deeper insights and summaries regarding the interrelations among site elements. Ecological sensitivity and land-use suitability analyses convey information that cannot be directly or intuitively obtained by designers, enriching the site investigation content. They provide not only materials for site cognition but also a reliable basis for subsequent design decisions. In the process of site analysis, each site element can be digitized and quantified under unified analytical and algorithmic rules. The resulting digital data, shaped by the requirements of landscape planning and design, vary by site and can comprehensively reflect the characteristics of the site. As the first stage of parametric planning and design, scientifically surveying and evaluating the site lays a solid foundation for realizing the principle of "designing according to local conditions."

3.4.2 Design Generation Process

The design generation process is the second stage in landscape planning and design. Using a culinary metaphor: site cognition is akin to a chef preparing ingredients—site investigation resembles the gathering and categorization of ingredients, while site suitability analysis equates to washing and cutting the ingredients for further processing. Once all preparations are completed, the actual cooking begins—this is the design generation stage. The design generation process is the most critical part of the design phase, wherein the designer proposes solutions to design problems based on design objectives and within defined constraints. The design objects in landscape planning are often large- to medium-scale environments such as scenic areas, resorts, and parks. This stage of design generation comprises two progressively detailed parts: project planning and positioning, and the generation of the design scheme.

[17] National Quality and Technical Supervision Bureau, Ministry of Housing and Urban-Rural Development of the People's Republic of China. Scenic and Historic Interest Area General Planning Standard (GB/T50298-2018) [S]. Beijing: China Architecture & Building Press, 2018.

[18] National Quality and Technical Supervision Bureau, Ministry of Housing and Urban-Rural Development of the People's Republic of China. Scenic and Historic Interest Area Detailed Planning Standard (GB/T51294-2018) [S]. Beijing: China Architecture & Building Press, 2018:4.

3.4.2.1 Project Planning and Positioning

The essence of project planning lies in seeking a coupled relationship between the project and its site. "Planning" is typically understood as the process of developing detailed textual and visual strategies to accomplish a specific task or achieve a desired goal through careful and logical consideration of methods, approaches, and procedures.[19] The concept of planning has deep roots in ancient China, as reflected in idioms such as "strategizing in the command tent," "Tian Ji's horse racing strategy," and "those who plan ahead succeed, those who do not plan fail." The meaning of "planning" is broad. In contemporary times, "planning studies" has evolved into an independent academic discipline with widespread application across fields such as politics, economics, design, tourism, and marketing, becoming an indispensable component of these domains. Modern project planning originated from project investment evaluation and developed rapidly around the world after World War II. Its primary purpose is to analyze and assess all potential influencing factors, synthesize them into a coherent strategy, and provide guidance and control for future development. It is also a rigorous, constructive, and logical thought process.

Currently, in China, before initiating any engineering or product project, the investor must submit a project proposal (project initiation application) to the supervising authority, presenting an overall vision for the proposed project and discussing its necessity and feasibility. Landscape planning is conducted after a project has been approved and feasibility studies completed. The "project planning" discussed here does not refer to the project planning stage involved in submitting a project proposal as described above. If project initiation is akin to deciding where to buy what kind of house, then "project planning" is like determining what type and style of furniture to place inside the house. The term "project" in "project planning" refers to attraction units aimed at achieving economic, social, and environmental benefits through resource integration and targeting tourists and local residents. One of the primary functions of landscaped environments is to provide venues for sightseeing, recreation, and leisure. The "project" therefore constitutes the "content" of human activities within a site. Planning is a goal-oriented, anticipatory activity for human endeavors. As a component of planning, "project planning" involves analyzing information to pre-evaluate possible human activities within the site. In landscape planning and design, the project planning process reflects the coupling relationship between site and project: generating projects through site analysis, and locating appropriate sites through project analysis. Compared to tourism planning, which emphasizes creativity, project planning in landscape design focuses more on functional organization and arrangement. In practice, landscape planning places greater emphasis on practicality and implementation of "projects," bringing them closer to real-world application rather than treating them as mere formalities. Project planning aids in proposing appropriate construction types, scales, and methods

[19] Zhuang W M. Introduction to architectural planning [M]. Beijing: China Water & Power Press, 2000:8.

within the design area, thereby providing a basis for subsequent design and making the planning process more purposeful and targeted. Project planning begins with positioning, proceeds to determining the content—including the theme, type, and form of the project—and finally makes preliminary estimations of the project's scale based on functional requirements.

Analysis of Project Positioning

Project positioning is the first step in project planning. It requires the analysis and synthesis of a series of information, including tourists, policies, and site resources, and involves research into three key aspects: market, function, and characteristics, to determine possible projects suited to the site. Different types of scenic areas have varying tourist source regions due to differences in their characteristics, functions, and classification levels, including distinctions between major, important, and potential tourist source regions. The purpose of tourist market analysis is to more accurately select and define the development direction and target of the tourist market, and thereby predict, select, and determine the scale and structure of future visitors.[20] As a space designed to serve "people," the planning of landscape environments requires research into potential visitor groups, the primary target audience, their age composition and spending patterns, and subsequently, their environmental needs and preferences in modes of touring. This allows for more targeted project selection and ultimately the establishment of a clear market position. As a prerequisite and guideline for landscape planning and design, it is necessary to conduct studies on higher-level plans such as the "Urban Master Plan," "Master Plan for Scenic Areas," and "Tourism Development Plan," as well as relevant government policies and regulations, in order to analyze and derive an appropriate functional positioning. For example, based on the government's development plan for the entire urban area, if a certain planning area is designated as a tourism service zone themed around Buddhist culture, then the functional positioning of the area should be Buddhist cultural tourism, and all planning activities should proceed on this basis. To avoid resource waste, special attention should be paid to shaping the distinctive features of the planning area. Take the planning of Nanjing Ziqing Lake Tourism Resort as an example: located in Tangshan, Nanjing, this area is rich in hot spring resources, and several hot spring-themed tourism zones already exist in the vicinity. Therefore, the planning must not only leverage the advantage of hot spring resources but also ensure differentiated operations from nearby tourism offerings to avoid redundant "hot spring theme" developments, which would otherwise lead to resource waste and negatively affect the future development of the scenic area.

[20] National Quality and Technical Supervision Bureau, Ministry of Housing and Urban-Rural Development of the People's Republic of China. Scenic and Historic Interest Area General Planning Standard (GB/T50298-2018) [S]. Beijing: China Architecture & Building Press, 2018.

Determination of Project Content

Based on the project positioning, the content of the project—its specific components and forms—can be preliminarily determined. In addition to reflecting the thematic orientation, the project content must strictly comply with relevant national regulations and standards. For example, the "Regulations on Scenic Areas" (2016) lists advocated, restricted, and prohibited construction activities (Table 3.1); the "General Planning Standards for Scenic Areas" (GB/T 50298-2018), in Sect. 4.2 "Scenic Tour Planning," outlines 8 categories and 59 types of touring activities (Table 3.2). Based on the degree of association between project content and site resources, projects can be classified into three types: original, extended, and derived projects. Original projects refer to activity types inherently existing in the site, representing primary, native functions. Extended projects derive from material or cultural resource information present at the site and are interpretive extensions of such information. Derived projects have the weakest connection to the site's original resources and are created by extrapolating or grafting new content based on existing information. For instance, if there is a renowned temple in a landscape environment, activities such as incense offering, Buddhist rituals, and vegetarian dining are original activities inherent to the site. The Buddhist resources can further be interpreted into extended themes such as Buddhist culture or Zen culture, which in turn generate extended projects like meditation practices and cultural exhibitions. Related

Table 3.1 Contents of encouraged, restricted, and prohibited Items in the "Scenic Area Ordinance" (2016)

Category	Article	Content
Encouraged items	Article 32	According to the characteristics of scenic areas, promote the protection of traditional culture and ethnic customs, develop health-oriented sightseeing and cultural recreational activities, and promote historical, cultural, and scientific knowledge
	Article 33	Reasonably utilize scenic area resources, improve facilities, and enhance services and tourism experience
Restricted items	Article 29	(1) Large-scale amusement activities; (2) Changes to terrain and vegetation that are not in accordance with planning; (3) Activities causing landscape visual pollution; (4) Activities that harm the ecological environment and landscape features
Prohibited items	Article 36	(1) Mining, quarrying, logging, land reclamation, and activities that alter the natural landscape, vegetation, or landform; (2) Construction of facilities with strong commercial, recreational, or harmful properties such as hotels, amusement parks, or animal performance venues
	Article 37	Construction of buildings or facilities that do not comply with the planning of scenic areas and their buffer zones, such as hotels, shopping centers, nursing homes, or real estate developments in conflict with landscape protection

Source: The Central People's Government of the People's Republic of China. Scenic and Historic Interest Areas Ordinance [EB/OL]. [2025-6-16]. https://www.gov.cn/gongbao/content/2016/content_5139422.htm

Table 3.2 Classification of tourism activities

No.	Tourism type	Tourism activities
1	Aesthetic appreciation	① Viewing; ② Photography; ③ Sketching; ④ Painting; ⑤ Visiting; ⑥ Sentiment; ⑦ Appraisal; ⑧ Commentary; ⑨ Writing; ⑩ Creative production
2	Nature exploration	① Hiking; ② Swimming; ③ Trekking; ④ Mountaineering; ⑤ Camping; ⑥ Wilderness survival; ⑦ Cave exploration; ⑧ Stargazing; ⑨ Rock climbing
3	Scientific education	① Study tours; ② Observation & research; ③ Science education; ④ Learning activities; ⑤ Collecting; ⑥ Seeking roots; ⑦ Museum Tours; ⑧ Memorials; ⑨ Publicity
4	Cultural experience	① Folk customs; ② Traditional culture; ③ Ethnic customs; ④ Religious culture; ⑤ Labor experience; ⑥ Community gatherings
5	Entertainment & Leisure	① Amusement parks; ② Training camps; ③ Gardening; ④ Water activities; ⑤ Boating; ⑥ Fishing; ⑦ Recreational sand activities; ⑧ Grassland recreation
6	Outdoor sports	① Fitness; ② Competitive sports; ③ Themed competitions; ④ Other outdoor skill sports
7	Wellness & Recovery	① Hot springs; ② Spa; ③ Sand therapy; ④ Steam baths; ⑤ Medicinal baths; ⑥ Mud baths; ⑦ Sunbathing
8	Others	① Role-playing; ② Interactive experiences; ③ Animal interaction

Source: Ministry of Housing and Urban-Rural Development of the People's Republic of China. General Planning Standard for Scenic and Historic Interest Areas (GB/T 50298-2018) [S]. Beijing: China Architecture & Building Press, 2018

wellness and leisure activities, such as yoga or spa services, represent derived cultural forms based on the extended themes of meditation and relaxation.

Forecasting the Scale of the Project

Forecasting project scale is a crucial part of project planning. It is not only closely related to the project content previously defined, but also marks the beginning of the subsequent master plan development. This forecasting is a key differentiating aspect of landscape planning compared to project planning within the field of tourism studies. Forecasting project scale involves a preliminary judgment of the land area, plot ratio, and potential building area, height, and architectural style required to meet the functional demands of the project. Article 15 of the "Regulations on Scenic Areas" (2016) clearly states: "Detailed planning for scenic areas shall be formulated according to the different requirements of core and other scenic zones, determining the site selection, layout, and scale of infrastructure, tourism facilities, cultural facilities, and clearly defining the scope of construction land and design conditions." Similarly, the "General Planning Standards for Scenic Areas" (GB/T 50298-2018) call for the organization of scenic spot capacity and scope in the sections "Scenic Tour Planning" and "Typical Landscape Planning," and set out detailed hierarchical control measures regarding the nature and function, content and scale, standards and grade, location and height, volume and form, color and style of various types of buildings.[21] It is thus evident that forecasting the scale of a project is not only necessary but mandatory. This step helps guide the implementation of the project, facilitates the realization of the planning vision, and prevents the plan from remaining a mere formality.

3.4.2.2 Scheme Formulation

Project Site Selection

Site selection refers to the process of choosing the most suitable location. This issue arises in many aspects of life, including transportation, logistics, retail, and public facilities. The German economist Alfred Weber's 1909 work Theory of the Location of Industries is considered the origin of modern site selection studies. In contemporary times, operations research, management science, engineering, and computer science have all paid significant attention to site selection problems. In ancient China, there was the concept of "Xiangdi," referring specifically to the process of surveying environmental and natural conditions to select a location for garden construction, which can be seen as the ancient Chinese approach to site selection. Today, to develop a construction project on state-owned land allocated through administrative means, it is necessary to apply for and obtain a "Construction Project Site Selection Opinion." Site selection has great significance not only for project development but also for resource protection. To preserve resources within scenic areas, the "Regulations on Scenic Areas" (2016) require that major construction projects within such areas undergo site selection approval. This regulation concretizes the requirements of scenic area master planning and ensures implementation. In landscape planning and design, site selection corresponds to the planning stage and serves as a preliminary phase of site selection for construction projects. It involves choosing appropriate locations within the planning area for the proposed projects outlined during planning. Similar to "Xiangdi," project site selection entails comparing predicted project scale, environmental requirements, and other conditions with the site, in order to find a land parcel that meets the project requirements. This is a concrete manifestation of the "coupling" between project and site.

[21] National Quality and Technical Supervision Bureau, Ministry of Housing and Urban-Rural Development of the People's Republic of China. Scenic and Historic Interest Area General Planning Standard (GB/T50298-2018) [S]. Beijing: China Architecture & Building Press, 2018:20–22.

Route Selection for Roads

Route selection refers to determining a rational path based on the road's function, nature, classification, start and end points, and control points, through comprehensive consideration of terrain and landforms, geological conditions, climate, hydrology, soil conditions, and other site conditions, supported by technical and economic analyses and comparisons. The road network in a landscape environment serves the function of transporting people and goods and is an important part of the infrastructure. In medium- and large-scale landscape planning and design, route selection is inevitably involved. While the route selection of landscape roads shares similarities with highway alignment, it also has its unique characteristics. Highway alignment emphasizes the shortest and most economical path under safe conditions, whereas landscape roads do not necessarily pursue the shortest route or overly prioritize economy. In a landscape setting, undulating and winding roads can enhance the enjoyment of visitors, creating a spatial effect of "changing scenes with every step," thus enriching the touring experience. Although safety and a certain degree of economy must still be considered, unstable geological areas and steep slopes pose high construction costs and significant safety hazards, making them unsuitable for route alignment. Roads should efficiently and conveniently transport visitors to various scenic spots. The road network in a landscape environment should connect all landscape nodes within the area to meet accessibility requirements. Thus, scenic projects become control points and directional constraints for route selection. In landscape planning and design, route selection closely follows site selection in the design process.

Water Feature Design

"A scene without water lacks vitality"; it is the presence of water that highlights the stability of mountains and showcases the dynamism of the landscape. Throughout history and across cultures, people's fondness for water features is beyond doubt. In medium- and large-scale landscape environments, water features are significant components and key subjects in landscape planning and design. Design originating from the principle of "coupling" emphasizes deriving from the site, i.e., seeking design rationale from the place itself, and water feature design is no exception. Beyond natural water bodies that exist within the site, a reasonable analysis of site runoff conditions can reveal areas with the potential for water feature creation. Through moderate human intervention, surface water can be accumulated, satisfying aesthetic needs and cultural expressions while offering positive ecological benefits. The design and construction methods of quasi-natural water features in such environments—based on the analysis of runoff and rainwater conditions—embody the principle of reductive design and continue the philosophy of "coupling."

Vertical Optimization

Vertical design in landscape planning includes four components: road grading, water system elevation, site grading, and earthwork balance. In medium- and large-scale landscape environments, disturbance to the original topography should be minimized while meeting drainage and planting requirements. Road grading, water systems, and site grading are key elements of landscape vertical design. In parametric landscape planning and design, corresponding to an interactive design process, vertical design is closely related to route selection and water feature creation, proceeding upon the completion of these stages and feeding back into them. Once the road alignment is determined, road design can begin, and estimates of earthwork can be made through longitudinal and cross-sectional designs. Similarly, for water features, once the basic form is established, excavation and fill estimates can be made to assess earthwork balance, providing feedback to the water feature design process. If a balance is not achieved, it becomes necessary to revisit the previous stage and adjust the water system's form. By integrating the estimated earthwork volumes for roads, water systems, and site construction, designers can understand the earthwork balance during the design phase and manage quantities of earth and rock from the outset.

Zoning Formation

With the finalization of project siting, the organization of different elements within the site can be generally determined. Projects of similar types often have similar site requirements, allowing for clustering. By coordinating and adjusting among various projects, the final project locations can be determined. On this basis, functional zones can be generated according to the characteristics and types of clustered projects. Projects identified in landscape planning and design are not isolated but are interrelated, forming a comprehensive project system. In addition to functional zoning, the road network, water features, and vertical designs are also finalized at this stage, thereby culminating in the formation of the general layout for landscape planning and design.

3.4.3 Scheme Optimization Process

The particularity of design thinking prevents designers from arriving at a single correct answer through rules, calculations, deductions, or experiments as researchers in other scientific fields might do. In nearly all design scenarios, there is no unique "solution"; the designer's perception of the site, subjective intent, and design capabilities all influence the process and development of the design scheme. In reality, only one scheme will be implemented, and it should be a comprehensively optimal

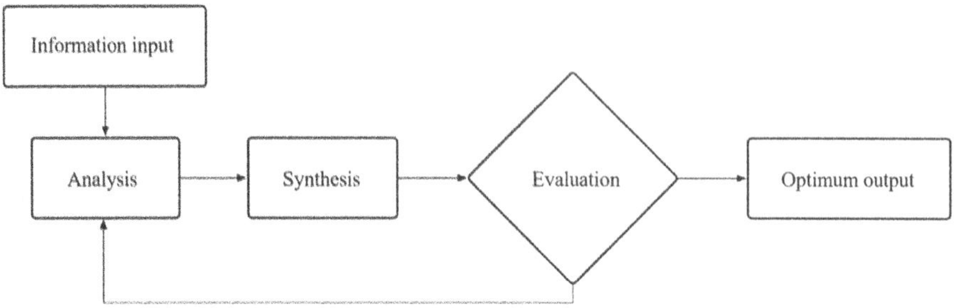

Fig. 3.13 Cyclical design process. Redrawn from: Chen C C. Design cognition: Cognitive science in design [M]. Beijing: China Architecture & Building Press, 2008:33

scheme—one that is the most "appropriate" under multi-objective conditions. It must be noted that this "optimality" is theoretical and relative; given the complexity of design and real-world problems, it is nearly impossible to define perfect criteria for an "optimal" solution, let alone determine an absolute "best" choice. The process of selecting the optimal scheme represents the final stage of the design process in landscape planning and design based on the coupling principle. It consists of two parts: scheme comparison and scheme optimization. The first generation of design methodology, founded on Cartesian epistemology and rational analysis, viewed the design process as one of analysis and logical reasoning. J. Christopher Jones divided the design process into three stages: "Analysis," "Synthesis," and "Evaluation." He believed that the design scheme is generated based on analysis and synthesis, followed by an evaluation to determine feasibility. If the scheme fails to meet standards, the design process restarts until the final scheme is produced (Fig. 3.13).

In this first-generation method, design is described as a process of problem identification through analysis, problem-solving through action, and final scheme formation. The second generation of design methodology posits that the three stages are cyclical and spiral, with mutual integration. Both the first and second generations of design methodology follow a general sequence from analysis to evaluation, but the second generation is nonlinear, emphasizing feedback and optimization throughout the process. Scheme optimization corresponds to the evaluation phase of design, assessing whether the previous design solution meets the intended goals. The optimization process aligns with the systems theory principle of optimization—using various tools and methods to seek the most optimal scheme under given conditions. The process involves analysis, integration, and comparison, requiring scientific evaluation and decision-making tools. The decision-making process in landscape planning and design aims to achieve specific design goals by analyzing, calculating, judging, and selecting the best scheme from feasible options, which is then implemented. Since landscape planning deals with complex systems with tightly interrelated issues and frequent multi-objective conflicts, determining superiority can be difficult. Hence, scheme decision-making involves selecting the most appropriate option within set constraints, resulting from comprehensive balance and compromise. The result depends on the comparison criteria—i.e., the scope of

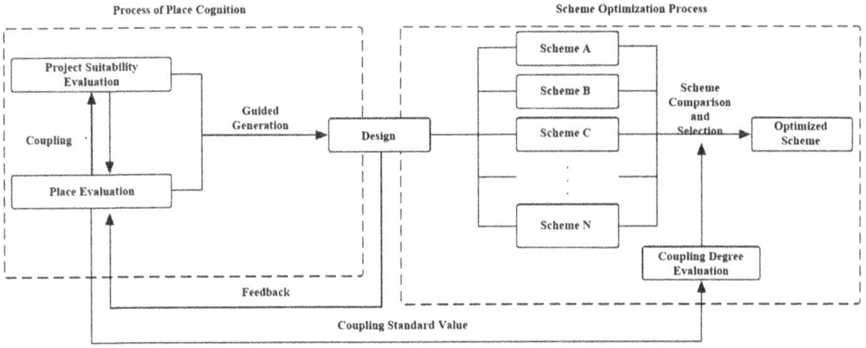

Fig. 3.14 Cyclical design process. Redrawn from: Cheng Y N, Yuan Y Y. Landscape Architecture Design Teaching Based on Place cognition [C]//Proceedings of the 2012 Landscape Architecture Education Conference. Nanjing: Southeast University Press, 2012:165

selection, evaluation methods, and rules. Whether a scheme is truly "appropriate" hinges on these criteria. Reasonable formulation of comparison rules is a prerequisite for scientifically optimizing the design scheme. The "appropriateness" of a scheme refers to its compatibility with the site and its ability to satisfy multi-objective design requirements. A site possesses objective conditions independent of human will, but it only provides the foundational context for design, not the entirety of it. The varied understanding of a site by different schemes determines their quality and effectiveness. Scheme comparison based on the coupling principle is essentially the selection of the concept most compatible with the site. In the process of landscape planning and design based on the coupling principle, the fundamental "coupling degree" between design and site is the prerequisite for evaluating the quality of a scheme. "Coupling degree" refers to the degree of correlation between the design scheme and the site, representing a numerical range that expresses the adaptability of the design to the site. Scheme optimization requires establishing an evaluation system based on ecological, spatial, cultural, and functional objectives to assess the coupling degree between the design scheme and the site. This evaluation provides feedback to the initial design schemes, leading to the development of optimization strategies (Fig. 3.14).

3.5 Parametric Design in Landscape Planning and Design

Landscape planning and design confronts an integrated and complex system. Although the system itself possesses uniqueness, its constituent components are variable—changes in one variable may trigger a series of transformations that affect the entire system. Furthermore, the landscape system is an open system, constantly engaging in the exchange of information and matter with its external environment, thereby endowing each element in the system with dynamic properties. Parametric design in the landscape context differs from that in architecture. While architectural parametric design emphasizes using parameters to express constraints on form or spatial generation, ultimately producing the form or space of a building or building complex, landscape parametric design is not limited to form and spatial generation under parameter control. Instead, it views design as a system, wherein parametric design not only describes systems and processes but also optimizes planning and design by adjusting parameter variables within the system.

3.5.1 Characteristics of the Parametric Design System

3.5.1.1 Systematic Nature

The idea of the "system" has a long-standing origin. As a scientific discipline, systems theory was established in the twentieth century by theoretical biologist Ludwig Von Bertalanffy, becoming an important component of modern scientific research and significantly influencing numerous disciplines. A system is generally defined as: an organic whole composed of several elements linked in a certain structural form and possessing specific functions. According to this definition, a system includes four parts: elements, structure, function, and environment. The core concept of systems theory is the holistic view that any system is an organic whole—not a mechanical assemblage or simple sum of its parts. The overall functionality of a system is an emergent property absent in the isolated elements. Thus, "the whole is greater than the sum of its parts," and each element exists within the system in a specific position, playing a particular role. The interrelationship between elements forms an inseparable whole.

In the 1920s and 1930s, methods of landscape planning and design underwent major transformation. British scholars G. E. Hutchings and C. C. Fagg recognized that a landscape is a system composed of many interrelated and complex elements. Major changes to one part of the system would inevitably impact other elements. A "system" is "an organic whole composed of several interacting and interdependent components with specific functions," or, in brief, "an integrated body with specific functions." Landscape is a complex and comprehensive information integrator. Based on digital technology, the parametric platform enhances human capability to perceive the environment and rationally integrate, process, summarize, and manage complex information. The parametric landscape planning and design system constructed from a design mechanism perspective, due to the multi-objective nature of design, possesses systematic attributes—achieving integration and optimization of parameter variables within the system. It is important to note that the results of system simulation are not the design itself but serve to open up diverse possibilities for planning and design thinking.

3.5.1.2 Dynamic Nature

Every system undergoes a process of formation and evolution and is in a state of continuous dynamic change. Research into a system's dynamic nature is essential; by interpreting the interaction mechanisms among internal elements and between the system and its environment, the operational and evolutionary rules of the system can be grasped. The dynamic nature of a system originates from the flow and exchange of information between the system and its environment as well as among internal components, and the system's internal structure also evolves over time. System evolution is a directional dynamic process. Objectively, as a system, the landscape is in constant development and transformation—partly driven by human factors, and partly governed by the laws of nature. Over 70% of the elements involved in landscape planning and design relate to natural materials, and nature itself has dynamic attributes that are not subject to human will. Thus, natural changes bring about dynamic variations in landscape planning and design. In addition, time is an important factor that must be considered in landscape planning and design—particularly in spatial environmental research. For example, the seasonal cycles of trees, and the transformation from young to mature to old trees, are continuous transformation processes, with landscape spatial forms constantly undergoing changes and repetition. In confronting the dynamic nature of the landscape environment, design must both ensure short-term effectiveness and account for long-term development, adopting a view of dynamic evolution. In parametric landscape planning and design, the dynamic characteristics of the overall system result from both the dynamic nature of the individual elements and their interactivity.

3.5.2 Construction of the Parametric Platform

The parametric platform for landscape planning and design refers to the operational platform composed of two parts. The first part consists of the technical methods that the design relies on, including evaluation methods, algorithms, and models built from those algorithms. The second part consists of software platforms such as ArcGIS, Civil 3D, and other collaborative tools. In parametric landscape planning and design, the extensive technical operations require software support to assist in realization, and the software platform is used to visually present both analytical and design results. Therefore, the implementation of parametric landscape planning and design necessitates the close collaboration between technical and software platforms.

3.5.2.1 Technical Platform

Evaluation Methods

Typically, qualitative analysis is considered subjective, whereas quantitative analysis is seen as objective. In the 1990s, Mr. Qian Xuesen proposed the methodology of "integrated combination from qualitative to quantitative" while addressing problems related to open and complex systems. He argued that remaining solely at the level of qualitative description without quantitative research limits a deep understanding of system behaviors and characteristics, thus making it difficult to reach accurate conclusions. Therefore, it is necessary to combine qualitative and quantitative approaches, where qualitative conclusions serve as the basis for quantitative demonstration, and quantitative analysis guides and corrects qualitative understanding.[22] As scientific research methods, qualitative and quantitative approaches are not contradictory or mutually exclusive, but rather interdependent, reciprocal, and permeable. Qualitative methods have fuzziness and subjectivity, and without quantitative support, their results may tend toward rough estimations. As a discipline that blends science and art, landscape architecture inevitably requires support from objective materials; however, the analysis and synthesis of such materials rely on human subjective judgment. Therefore, the integrated research method combining qualitative and quantitative approaches is equally applicable to the study of landscape architecture.

Marx once stated: "Science only becomes truly perfected when it successfully applies mathematics." Contemporary scientific research is inseparable from mathematical support, particularly with the advancement of computer technology, which has greatly improved the ability to analyze, synthesize, and summarize data, thereby promoting progress in related disciplines. The role and significance of quantitative analysis and evaluation have become increasingly important. For the field of landscape architecture, the value of quantitative analysis lies in its ability to process phenomena that are difficult to numerically quantify through qualitative descriptions and logical reasoning, thus producing comparable values and intervals. This not only enables direct evaluation and analysis of values and intervals but also lays a foundation for parametric design. Given that the landscape environment is a complex, multifactorial, and integrated system, with interrelated and interdependent factors, evaluation cannot rely on a single index; instead, it must begin with multiple factors and establish a comprehensive evaluation index system. In summary, the integration of quantitative and qualitative analysis, using a multi-index system for comprehensive and systematic evaluation, is an objective requirement and inevitable trend in the scientific evaluation of landscape architecture. Generally, the evaluation process in landscape architecture includes the following four stages:

---

[22]Yu L. Combining qualitative and quantitative: meta-synthesis digital urban planning[J]. Urban Development Studies, 2014, 21(2): 83–90.

1. Composition and selection of evaluation factors. This stage involves selecting evaluation factors based on different evaluation objectives.
2. Selection of evaluation techniques. Depending on the characteristics and final goals of the evaluation, appropriate techniques are chosen, procedures are arranged, and implementation details are designed—this is the core of the evaluation process.
3. Data collection. Based on the selected evaluation factors, reliable site information is gathered, forming the foundation for the reliability and authenticity of evaluation conclusions.
4. Data analysis. Qualitative and quantitative analysis techniques are determined. This stage equally impacts the reliability and validity of the conclusions.

To a certain extent, the essence of scientific evaluation is an information management process, encompassing the collection, organization, analysis, and utilization of information—this runs throughout the entire evaluation process. The theoretical framework of scientific evaluation begins with the definition of the evaluation problem and includes eight parts: evaluation program design, information collection, process control, method application, method support, methodological guidance, and theoretical support (Fig. 3.15). The evaluation system of landscape architecture resembles this framework. Due to the various attributes and characteristics of elements within the landscape environment, classification for comparative purposes must be carried out based on evaluation objectives. Therefore, scientific and accurate categorization of evaluation objects is especially important during the evaluation process. Only by ensuring the comparability of evaluation factors can they be incorporated into a unified evaluation standard and evaluated using the same procedures and methods, thus maximizing the scientific nature of evaluation results.

One of the most important parts of landscape evaluation is the quantification of qualitative indicators—transforming qualitative descriptions into comparable values and intervals. Generally, the quantification of qualitative indicators can be divided into two types based on specific objects: "direct quantification" and "indirect quantification." Direct quantification involves assigning a specific numeric

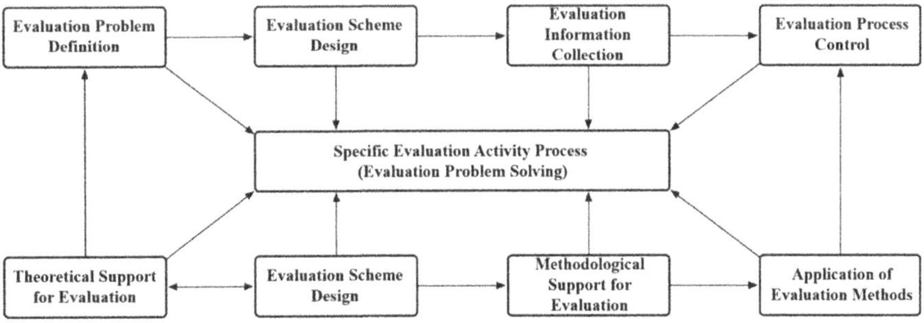

Fig. 3.15 Scientific evaluation theoretical system based on specific evaluation activities. Source: Qiu J P, Wen T X. Evaluation science: Theory, Method and Practice [M]. Beijing: Science Press, 2010:114

value directly to the object to be quantified, such as the scoring method. Indirect quantification first lists all possible values of the qualitative indicators, records each evaluated unit's specific values for the variable, and then quantifies the elements in the "qualitative value set," converting each unit's qualitative values into numeric quantities—for example, by using grading methods, interval scoring methods, or fuzzy evaluation methods.[23] One key step in the quantification process is the dimensionless treatment of evaluation indicators, normalizing values with different properties and characteristics into a unified system to allow for direct comparison. The Analytic Hierarchy Process (AHP) is a commonly used method in landscape evaluation, capable of transforming qualitatively described environmental elements into values and intervals. The AHP evaluation process includes the following steps:

1. Establishing a hierarchical structure model (Fig. 3.16);
2. Constructing the judgment matrix;
3. Calculating the weight of each level's indicators;
4. Conducting consistency checks.

There are numerous classifications of scientific evaluation methods. Table 3.3a, b present commonly used evaluation methods based on different classification approaches. According to their nature, these methods can be divided into three categories: qualitative, quantitative, and comprehensive evaluations. Landscape architecture evaluation primarily relies on comprehensive methods, involving both qualitative and quantitative approaches. The Delphi method, which is qualitative, and the Analytic Hierarchy Process (AHP), a comprehensive method, are among the most frequently used. When addressing complex evaluation problems, it is often necessary to combine multiple methods of varying types and characteristics.

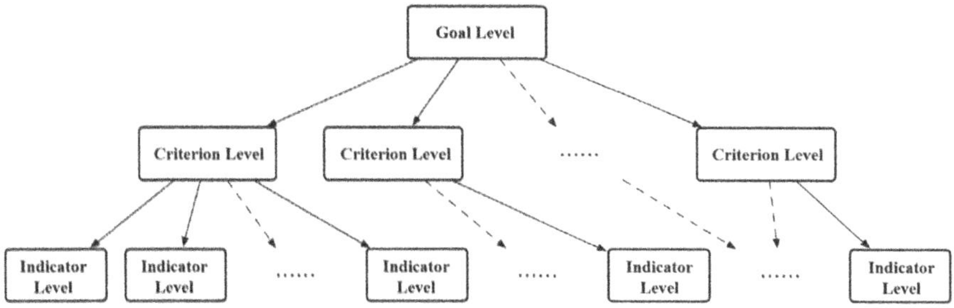

Fig. 3.16 Hierarchical structure model

[23] Qiu J P, Wen T X. Evaluation science: Theory, Method and Practice [M]. Beijing: Science Press, 2010:114.

Table 3.3 Main methods of scientific evaluation

(a)

Method classification	**Method nature**	**Representative methods**
Subjective evaluation based on expert knowledge	Qualitative evaluation	Consensus evaluation, expert evaluation, Delphi method, case study method, qualitative analysis, set pair analysis
Objective evaluation based on statistical data	Quantitative evaluation	Literature analysis, scientific calculation, economic analysis
Comprehensive evaluation based on system models	Comprehensive evaluation	Analytic Hierarchy Process (AHP), fuzzy mathematics, operational research, statistical analysis, systems engineering, and intelligent evaluation methods

Source: Qiu J P, Wen T X. Evaluation science: Theory, Method and Practice [M]. Beijing: Science Press, 2010:114

(b)

Method category	**Representative methods**	
Multi-index comprehensive evaluation methods	Comprehensive scoring, visual charts, compromise methods, optimization, linear programming, logical choice methods, Analytic Hierarchy Process (AHP), goal programming methods	
Index and economic analysis methods	Index methods, cost-benefit analysis, input-output analysis, value engineering	
Mathematical methods	Operational methods (mathematical)	Mathematical programming (linear programming, dynamic programming), data envelopment analysis, ranking theory
	Statistical methods	Multivariate statistical analysis (classification analysis, discriminant analysis, principal component analysis, factor analysis, Bayesian analysis), regression, clustering, dimensionality reduction, comprehensive evaluation using Ridit analysis
	Grey system theory	Grey statistics, grey clustering, grey comprehensive evaluation, grey correlation analysis, grey relational degree analysis, grey weight distribution
	Matter-element analysis	Physical cognitive models, multivariate analysis, grey modeling, FHW (factor-weight-hierarchy) decision-making system
Computer-based methods	Artificial neural networks, expert systems, computer simulation, systems dynamics, decision support systems	

Source: Qiu J P, Wen T X. Evaluation science: Theory, Method and Practice [M]. Beijing: Science Press, 2010:114

Algorithms and Models

In Notes on the Synthesis of Form, Christopher Alexander pointed out that when dealing with problems involving conceptual procedures and patterns, mathematics as a research tool demonstrates great power.[24] With the advancement of computer technology and the development of mathematical methods, the application of mathematics in landscape architecture has become increasingly widespread. When combined with traditional qualitative research, the exploration and practice of quantitative analysis have also grown richer. The use of mathematical methods not only renders the research and design processes of landscape architecture more rational and objective but also provides new and reliable solutions to previously intractable complex problems. In addition to evaluation and analysis methods such as AHP and fuzzy mathematics, an even more important application of mathematics in landscape architecture is the design implementation of algorithms based on mathematical models supported by computer platforms.

An algorithm is an accurate and complete description of a solution strategy, represented by a series of clear instructions for solving a problem. It embodies the strategic mechanism of problem-solving in a systematic manner. The British mathematician Alan Mathison Turing proposed the Turing machine model in the twentieth century—an abstract model of a hypothetical computer—which addressed the definition of algorithms and laid the foundation for the logical operation of modern computers. The implementation of parametric landscape planning and design must rely on computer platforms, making algorithms one of the most essential tools in parametric design. For specific problems, based on logical model construction, algorithm combinations generate algorithmic models into which parameter values are input, and results are produced through computer computation. This constitutes the operational process of the parametric model for landscape planning and design (Fig. 3.17). In the parametric landscape planning and design process, the model primarily serves as a vehicle for logical expression, where logic is derived from

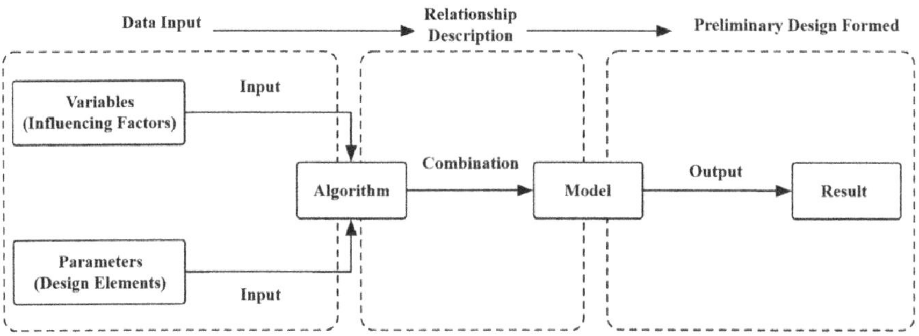

Fig. 3.17 Operation process of the parametric model for landscape planning and design

[24] Alexander C. Notes on the Synthesis of Form[M]. Harvard University Press, 1964.

specific design problems; algorithms, as implementation tools based on models, constitute the computational logic of mathematical models and must be supported by software and computer platforms to generate results through computation.

3.5.2.2 Software Platform

The parametric software platform for landscape architecture planning and design should be easy to use, versatile, highly compatible with specialized software, and capable of providing interfaces for programming tools. Based on these requirements, the author selects ArcGIS and Civil 3D, two relatively mature software platforms, as the foundational software for parametric landscape architecture planning and design. Both ArcGIS and Civil 3D integrate well with general-purpose drawing software like AutoCAD, facilitating the implementation and promotion of parametric landscape architecture planning and design.

ArcGIS

The emergence and development of GIS is relatively recent. In the early twentieth century, Charles Eliot applied geographical working methods in large-scale park and boulevard design projects in the United States, laying the initial conceptual groundwork for GIS, albeit without digital realization at the time. Notably, the work of Ian McHarg in the mid-1960s formed the basis of today's automated land planning modules. Landscape architects such as Philip Lewis, Angus Hills, and Carl Steinitz also made significant contributions to the development of GIS. Jack Dangermond, a student of Steinitz and a landscape architect himself, founded ESRI, whose product ArcGIS has become one of the most widely used GIS software systems today. Since most human activities occur on the Earth's surface, GIS—as a system for processing, analyzing, and managing geographic information—has wide applications across industries. The scope of research in the field of landscape architecture typically lies on the surface of the Earth and involves environments with strong natural characteristics, making GIS an essential research tool in the field. In the 1990s, Karen C. Hanna published GIS in Site Design and GIS for Landscape Architects, which elaborated on the application of GIS in landscape architecture. With the rise of the concept of geodesign, GIS, as its core tool, has become the subject of extensive research and discussion. Currently, GIS is widely used in traditional areas of landscape architecture, such as accessibility analysis and land suitability evaluation, with most applications focused on pre-design analysis and assessment.

The author's choice of ArcGIS as the parametric platform for landscape architecture planning and design is based on the following points:

1. Suitable for medium and large-scale landscape planning and design. From small roadside green spaces to large national parks, the field of landscape architecture

research covers a broad scale. Most planning and design research focuses on scenic areas, forest parks, and urban parks, which fall under medium to large scales. As a geographic information database, ArcGIS excels at analysis and planning at these scales.

2. Powerful site analysis capabilities. Landscape architecture planning and design based on the coupling principle emphasizes site-oriented, intensive, and reduced design, which requires a thorough understanding of the site. ArcGIS not only excels in site analysis but also features a geographic information database that accurately expresses the environment and effectively aids designers in interpreting the site. This feature aligns precisely with the needs of design under the "coupling" concept.
3. Numerous specialized modules allow for the management and integration of spatial information data (Fig. 3.18). ArcGIS integrates relevant algorithms and provides users with numerous application models to address specialized problems, while presenting the interface in a visual, user-friendly manner. ArcGIS also offers the ArcPy algorithm tool, which enables efficient command execution using Python, providing an effective means for solving design problems with programming algorithms (Fig. 3.19). Designers can therefore write suitable algorithms in computer languages according to design needs, describe relationships among design elements, and generate targeted algorithm models—i.e., "parametric modeling"—to establish logical relationships between input and output, realizing a computer-driven process and method of "generating different output results by adjusting inputs." For example, the Arc Hydro model, jointly developed by ESRI and CRWR (Center for Research in Water Resources at the University of Texas at Austin), is a collection of data models and toolsets that support geographic analysis and time-series data analysis for hydrological simulations in the ArcGIS environment. Hence, ArcGIS offers a potential platform for the realization of parametric landscape architecture planning and design.

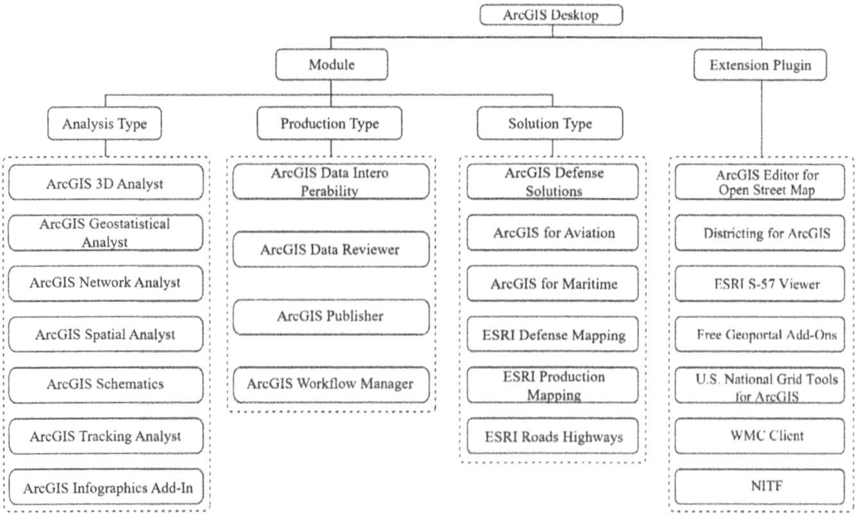

Fig. 3.18 ArcGIS desktop extension modules and plugins

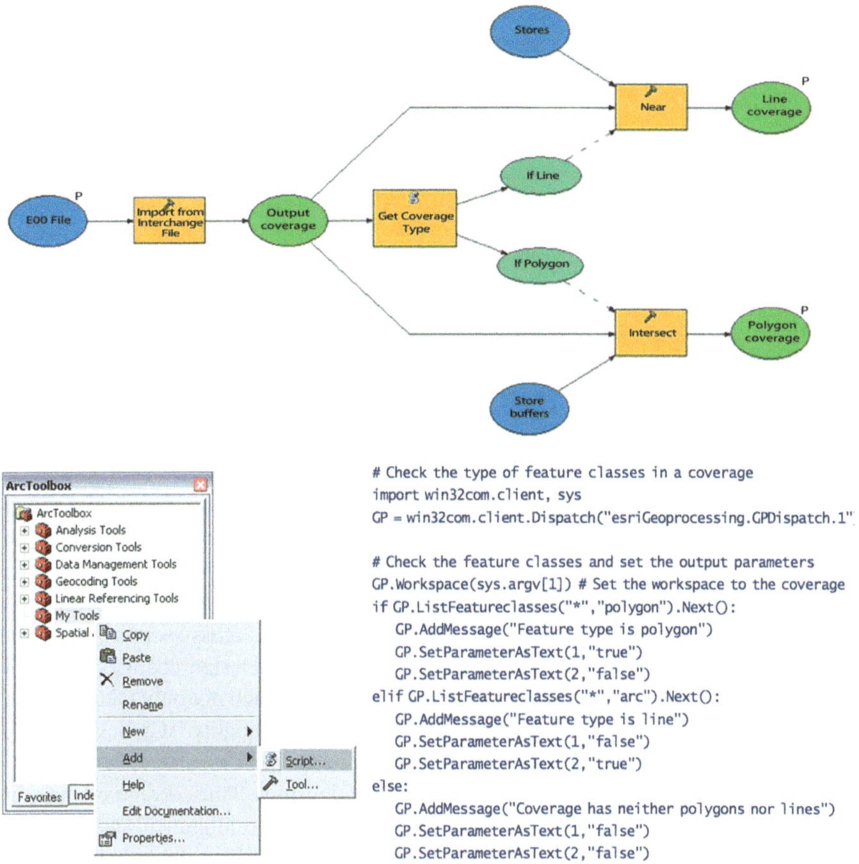

Fig. 3.19 Diagrammatic models and python scripting tools provided by ArcGIS. Source: Corey Tucker. *Writing Geoprocessing Scripts With ArcGIS* [M]. ESRI, 2004: 48

4. Linking databases with graphics facilitates design and analysis. In addition to graphic operations, one of ArcGIS's greatest advantages is its spatial data management capability, allowing linkage between data and graphics. This enables algorithm-generated results to be intuitively presented, aiding designers in judgment and analysis.
5. A relatively independent platform that can interface with various software. ArcGIS operates as a relatively independent platform. ArcGIS Desktop not only seamlessly integrates with 3D modeling and visualization software from the same company—such as CityEngine and ArcScene—but also interfaces with remote sensing software like ENVI and Erdas, as well as engineering software like AutoCAD and Civil 3D. This makes ArcGIS an efficient parametric software collaboration platform, enabling smooth data transmission from analysis to design and construction, and facilitating the implementation of parametric processes throughout the landscape architecture planning and design workflow.

In Landscape Planning and Environmental Impact Design, Tom Turner states: "When new technology first appears, it is often used to do what old technology already does. Printing was first used to reproduce the Bible; the earliest cars were seen as carriages without horses; the first tanks were thought of as moving guns. Of course, new technologies may also be applied in new ways—that's the challenge. GIS will accomplish arithmetic calculations previously done by traditional tools: it will make complex operations easier. But GIS cannot turn planning from a judgment-based art into a deductive science. As with all computer programs, there is a direct relationship between input and output."[25] ArcGIS provides scientific and precise judgment bases for design, but it cannot replace the entirety of the design process. Therefore, parametric design methods should not be mechanically interpreted as a method of "automatically" generating design schemes.

Civil 3D

Civil 3D's most important and prominent advantage lies in its three-dimensional dynamic design, which includes two aspects: first, its working mode is based on 3D model design; second, it achieves design process automation through the interaction of intelligent objects. For instance, a traditional road project design team needs significant time to ensure that changes to terrain, alignment, profiles, cross-sections, and other design data are correctly transmitted across all elements. Design changes inevitably lead to repeated drafting and recalculating of quantities, which not only takes time and lowers efficiency but also makes it difficult to ensure design quality.[26] Civil 3D provides rich design objects—such as surfaces, slopes, alignments, profiles, and cross-sections—all placed within a single model with dynamic relationships. Thus, design changes are automatically propagated. For example, modifying the road alignment automatically updates related profiles, cross-sections, and earthwork quantities. Table 3.4 shows the objects that can be updated when each object type is edited. Figure 3.20 illustrates the interconnections between AutoCAD Civil 3D intelligent objects.

Table 3.4 Update relationships of major civil 3D objects

When editing this object type...	Can update these objects...	When editing this object type...	Can update these objects...
Point	Surface	Grading	Surface, road
Surface	Grading, section, pipe network, road	Component	Assembly, road
Parcel	Grading, road	Assembly	Road
Alignment	Grading, parcel, road, section, pipe network	Pipe network	Surface, alignment
		Feature line	Grading
Section	Intersection	Sample line	Section, cut/fill map

Source: Ren Y. Civil 3D 2013, Application Bible [M]. Shanghai: Tongji University Press, 2013:2

[25] Turner T. Landscape planning and environmental impact design[M]. Routledge, 2004.

[26] Ren Y. Civil 3D 2013, Application Bible [M]. Shanghai: Tongji University Press, 2013:2

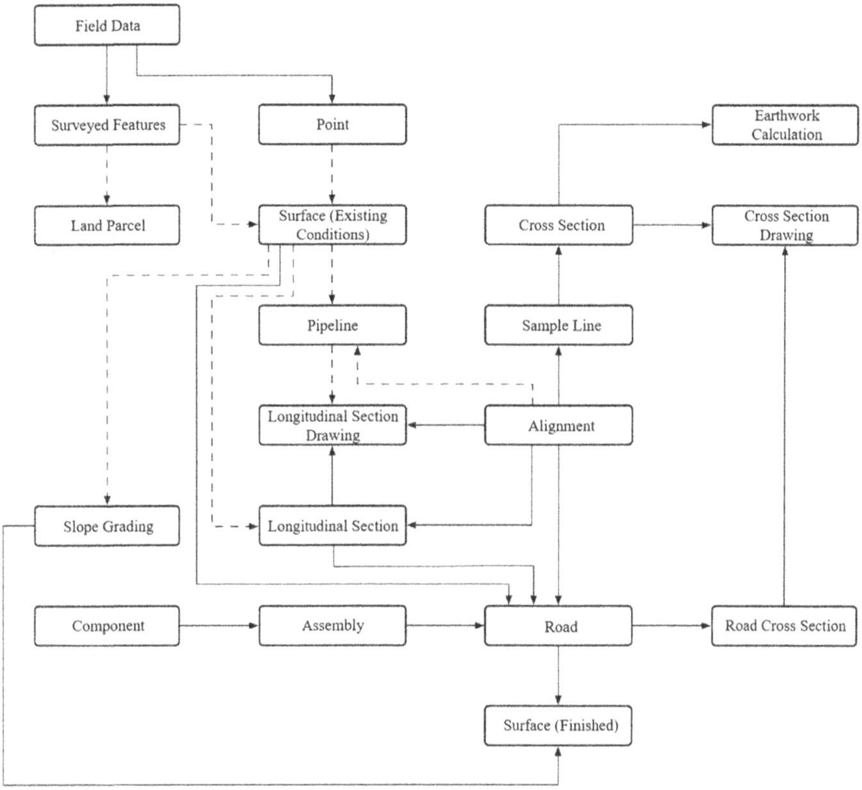

Fig. 3.20 Relationships between civil 3D objects. Redrawn from: Ren Y. Civil 3D 2013, Application Bible [M]. Shanghai: Tongji University Press, 2013:2

Based on the Civil 3D software platform, a three-dimensional digital model containing all design elements within the design scope can be rapidly created, forming data relationships among the design elements. At each stage of design, the 3D digital model can dynamically reflect the overall design results after any changes and quickly calculate data such as earthwork volumes, generating corresponding charts to provide accurate references for designer decision-making. In landscape architecture planning and design, Civil 3D effectively facilitates parametric terrain design, including the construction of 3D terrain models, slope and aspect analysis, surface runoff analysis, vertical design of roads, water systems, and sites, earthwork volume calculation, and soil balancing.

ArcGIS excels in medium to large-scale analysis and design, while Civil 3D offers strong capabilities from small to medium-scale design through to construction, making up for ArcGIS's shortcomings. The collaboration between the two facilitates the development of a parametric system for the entire landscape architecture planning and design process and represents a meaningful and positive exploration of LIM (Landscape Information Modeling). AutoCAD serves as a medium to achieve seamless integration between ArcGIS and Civil 3D. Comparing the three: first, in terms of graphic expression, AutoCAD has strong drafting capabilities, ArcGIS is capable of map creation and thematic analysis but is relatively weak in design drawing, while Civil 3D is similar to AutoCAD but superior in dynamic feedback of related drawings; second, in terms of analytical capability, ArcGIS specializes in data analysis and management, Civil 3D focuses on engineering-based analysis with strong data interactivity, while AutoCAD is weakest in this area; finally, for parametric landscape architecture planning and design, the strengths of ArcGIS and Civil 3D in analysis, data management, and dynamic feedback support the realization of parametric methods. Given that each software has its focus and strengths, they must be collaboratively employed at different stages of parametric landscape architecture planning and design.

3.5.3 Parametric Landscape Planning and Design Process Based on the Coupling Principle

The key to establishing a parametric design process lies in the construction of logical procedures. The parametric landscape planning and design process is fundamentally consistent with the landscape planning and design process based on the coupling principle, consisting of three parts: site cognition, scheme generation, and scheme optimization, while also reflecting the dynamic characteristics of iteration and feedback inherent in parametric design. "Coupling" emphasizes design originating from the site, and the author's research on the parametric landscape planning and design process focuses on scheme generation, encompassing part of the site cognition process. The progression from site cognition to scheme generation is a gradual and coherent logical process.

The analysis and design models correspond to the design stages, with close interrelations among the models, unified within the logical procedure. Parametric evaluation and planning design models are constructed based on the processes of site cognition and scheme generation. Among them, the parametric evaluation model is composed of the ecological sensitivity evaluation model and the land-use suitability evaluation model. The generation of buildable areas requires the assessment of land-use suitability based on ecological sensitivity evaluation, thus the two models are sequentially connected. The parametric planning design model consists of a project positioning model, road alignment model, quasi-natural waterscape construction model, and vertical design model, covering the main specialized content of landscape planning and design. The project positioning model includes the project suitability evaluation model and project siting model, aiming to determine appropriate project content through the mutual adaptation between the project and the site, and to define the siting scope within the site. The vertical design model is implemented following the road alignment model and quasi-natural waterscape

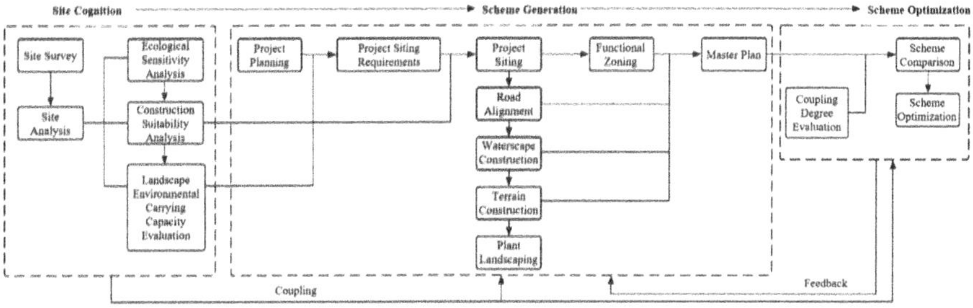

Fig. 3.21 Parametric landscape planning and design process based on the principle of coupling

construction model, focusing on the deepening of vertical design for roads, water systems, and site, as well as the regulation of total earthwork balance within the site, forming a feedback mechanism to the early design stages (Fig. 3.21).

The six specialized models correspond to the major specialized design content in landscape planning and design, all belonging to the parametric landscape planning and design process, and are tightly interrelated in logic. The purpose of specialized research is to analyze and interpret the parametric design process more clearly. The construction of each specialized model is guided by the coupling principle, where "coupling" is implemented within each specialized domain, is embodied in the correlation and coupling effect between specialized design and site, and is also reflected in the overall coordination among different specialized domains. This constitutes a systematic task, unified within the parametric landscape planning and design system.

Chapter 4
Ecological Sensitivity Evaluation Model for Land

Ecological sensitivity evaluation originated in the 1930s and, after more than half a century of development, has been extensively studied in various fields such as urban planning, landscape architecture, and ecology. With the advancement of digital technologies represented by computer technology, ecological sensitivity evaluation methods, based on the blueprint of map overlay methods, have made significant progress in terms of technical means, tools, and methodologies. Ecological sensitivity evaluation is the primary step in landscape planning and design, embodying the principle of ecological priority. The parametric ecological sensitivity evaluation model is based on the ArcGIS platform and introduces a systematic evaluation concept. Through the selection of appropriate parameters and the adjustment of their values, the model can be applied to various landscape environments. This chapter begins with the significance and value of ecological sensitivity in landscape planning and design, constructs an ecological sensitivity evaluation model, and illustrates the application of the model through a practical case study of the northern area of Niushou Mountain Scenic Area in Nanjing.

4.1 Ecological Sensitivity and Landscape Planning and Design

In a natural state, the various components of an ecosystem are coupled and coordinated, maintaining the relative balance of the entire ecological system. When disturbances imposed on the ecosystem exceed a certain threshold, the balance is disrupted, resulting in ecological and environmental issues. The "Interim Technical Regulations for Ecological Function Regionalization" issued by the former State Environmental Protection Administration in 2002 defined ecological sensitivity as: the likelihood and degree of ecological problems occurring in a given area, used to reflect the potential ecological consequences of human activities. From the literal

Y. Yuan, *Parametric Landscape Architecture Planning and Design*,
https://doi.org/10.1007/978-981-95-8011-8_4

meaning, high sensitivity indicates low resistance of the ecosystem to disturbances and weak self-regulation capability, meaning even minor disturbances may lead to serious ecological consequences; low sensitivity implies that the ecosystem has a certain resistance to disturbances and can better adapt to changes in external conditions. Research and evaluation of ecological sensitivity provide a basis for the division and planning of ecological function zones, ensure graded land use, help protect ecologically fragile and sensitive areas, and avoid ecological disasters.

4.1.1 Ecological Sensitivity Evaluation Methods

In the 1930s, with the development of ecological planning and scientific and technological advances, the United States developed a land-use suitability analysis method system, assessing the site from the two aspects of land and site to help determine the impact of land development on the regional ecological environment. Most notably, Ian McHarg and his colleagues and students, based on research and practical experience, summarized the map overlay method, which is thoroughly and systematically discussed in McHarg's book Design with Nature. The map overlay method originated from the land-use suitability analysis technique used by Warren H. Manning, who employed factor analysis maps as the basic operation method. This approach is intuitive and visual. In the 1870s, Manning participated in the planning of the Boston Metropolitan Park System and was influenced by Charles William Eliot, a pioneer in overlay analysis, particularly his concepts on landscape classification and overlay methods. Manning drew a vegetation distribution sketch of the region overlaid with road layouts, topography, and hydrological features, thereby developing a resource-based planning model.[1] In 1912, during the planning of Billerica, a town near Boston, Manning produced a series of maps including roads, cultural resources, topography, land ownership divisions, soils, forest cover, and existing and future reserved lands, and used transparent plates for overlay analysis (Fig. 4.1). Carl Steinitz considered this one of the earliest comprehensive examples of using overlay analysis techniques.[2]

Following Eliot and Manning, overlay analysis technology was applied in many research works but failed to form a systematic planning method and lacked theoretical explanation of fundamental principles. In 1943, L. B. Escritt, in Regional Planning: An Outline of the Scientific Data Related to Planning in Great Britain, introduced how to generate geographic vertical models using the overlay method and how to use such models to analyze sites based on specific requirements

[1] Charles A B, Robin K. Pioneers of American landscape design[M]. McGraw-Hill Inc, 2000:237.

[2] Guo W, Hou X L. The Philosopher of Land-American Landscape Architect Warren H. Manning[J]. Chinese Landscape Architecture, 2010(3): 58–61.

Fig. 4.1 Manning's 1912 planning analysis of Billerica town. Source: Guo W, Hou X L. The Philosopher of Land-American Landscape Architect Warren H. Manning[J]. Chinese Landscape Architecture, 2010(3): 58–61

(Fig. 4.2).[3] Later, a 1950 publication, Town and County Planning Textbook, included an article by Jacqueline Tyrwhitt that clearly discussed overlay analysis techniques.[4] In the 1960s, McHarg effectively improved the overlay analysis technique by integrating suitability methods with theory. His work holds significant historical importance in the development of land-use suitability analysis and is one of the earliest systematic methods of overlay analysis, serving as the blueprint for most subsequent methods. The map overlay method first uses a matrix table to analyze various ecological factors in relation to the intended human use (Table 4.1), dividing evaluation maps into five levels based on color intensity, drawn on transparent sheets. These categorized evaluation maps are then overlaid and photographed to produce a final composite image reflecting the overall evaluation result (Fig. 4.3). Due to technical limitations at the time, this method had reached its practical limit: in McHarg's map overlay method, converting grayscale tones into equally valued colored maps posed a cartographic challenge, and combining various shades was also difficult.[5] Additionally, drawing numerous maps was a major challenge in manual overlay analysis. The map overlay method allows comprehensive analysis of factors from different dimensions, such as social and natural ones, but its drawbacks include equal-weight overlaying that ignores the importance of different factors and the vagueness in factor selection. Since the roles of each factor vary in planning, equal weighting reduces the scientific validity of the method. Moreover, as the number of

[3] Carl S A. Framework for geodesign:changing geography by design [M]. Environmental Systems Research Institute Inc., 2013:66–67.

[4] Frederick S. The living landscape:an ecological approach to landscape planning[M]. McGraw-Hill Inc., 1991:202.

[5] McHarg I L. Design with nature[M]. New York: American Museum of Natural History, 1969.

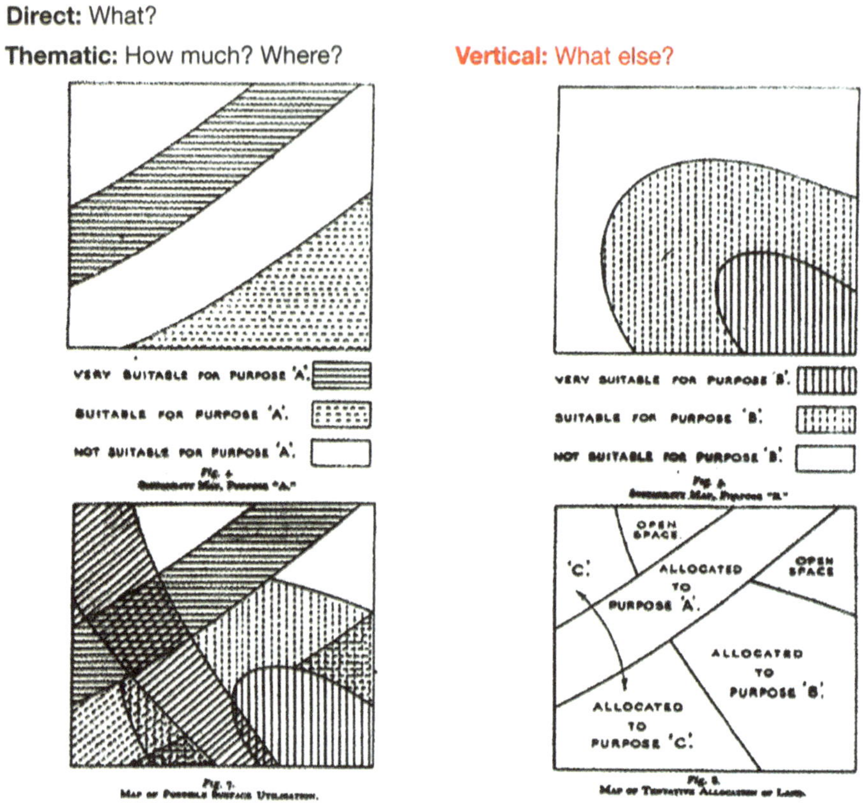

Fig. 4.2 Eskritt's layer overlay method. Source: Carl Steinitz A. *Framework for Geodesign: Changing Geography by Design* [M]. Environmental Systems Research Institute Inc., 2013: 66

factors increases, the method of using different color shades to represent suitability levels and perform overlay becomes cumbersome and makes it difficult to distinguish subtle differences in color on the composite map.

Also belonging to the category of direct overlay methods is the equal-weight summation method, which calculates a comprehensive evaluation value by quantitatively grading each factor and summing them. Unlike the traditional map overlay method, the equal-weight summation method evolved with the advancement of computer technology and scientific progress (Fig. 4.4). The calculation formula is as follows:[6]

$$V_{ij} = \sum_{k=1}^{n} B_{k_{ij}}$$

[6]Huang G Y, Chen Y. Theory, Planning and Design Methods of Ecological City [M]. Beijing: Science Press, 2003:103–104

Table 4.1 McHarg uses a matrix table to analyze the value of various ecological factors and human expected utilization

Ecological factor	Grading criteria	Sequence I	II	III	IV	V	Land use suitability				
							C	P	A	R	I
Climate											
Air pollution	Occurrence rate: High → low	High	Middle	Low		Lowest		●	●	●	
Tidal Marsh intrusion	Occurrence rate: High → low	High (recorded)	Highest (predicted)			Above flood line			●	●	●
Geology											
Geologically unique landforms with scientific and educational significance	Degree of rarity: High → low	1. Ancient lake bed 2. Drainage outlet	1. Terminal moraine 2. Ice front boundary 3. Glacial erratic traces	Serpentine hill	Exposed rock face	1. Beach 2. Buried valley 3. Clay pit 4. Gravel pit	●	●		●	
Basic conditions	Compressive strength: High → low	1. Serpentinite 2. Diabase	Shale	Cretaceous deposits	Silted-up swamp	Herbaceous marsh and forested swamp				●	●
Landform											
Geologically unique landforms with scientific and educational significance	Degree of rarity: High → low	Kames and kettles in terminal moraine	Exposed rock wall	Bayfront morainic cliffs and glacial lakes	Fracture between serpentine ridges		●	●			
Landforms with scenic value	Uniqueness: Most prominent → average	Serpentine ridges and headlands	Beach	1. Cliff 2. Enclosed valley	1. Cliff trail 2. Headland 3. Knob-and-kettle wetlands	In an evaluative or comparative context	●	●	●		
Scenic water features	Uniqueness: Most prominent → average	Bay	Lake	1. Pond 2. River		1. The narrows 2. Kill Van Kull 3. Arthur kill	●	●			

(continued)

Table 4.1 (continued)

Ecological factor	Grading criteria	Sequence I	II	III	IV	V	Land use suitability C	P	A	R	I
Riverside land with scenic waterscape	Degree of vulnerability: Moderate → average	Swamp	1. River 2. Pond	Lake	Bay(River Bay)	1. The narrows 2. Kill Van Kull 3. Arthur kill	●		●	●	●
Coastal beach along the bay	Degree of vulnerability: Moderate → average	Morainic cliffs	Small bay	Beach			●		●	●	
Surface water presence	Ratio of water to land: High → low	Marsh and wetland	Limited drainage area	Dense river and wetland network	Moderate density river and wetland network	Sparse river and wetland network			●	●	●
Slope	Gradient: Steep → gentle	> 25%	25%–10%	10%–5%	5%–2.5%	2.5%–0%			●	●	●
Hydrology											
Water activities	Navigable waterway: Deep → shallow	The narrows	Kill Van Kull	Arthur kill	Fresh kill	Raritan Bay					●
Commercial vessels leisure vessels	Range of navigable waters: Maximum → minimum	Raritan Bay	Fresh kill	The narrows	Arthur kill	Kill Van Kull		●	●		
Freshwater (brackish water) positive recreational activities (swimming, boating, sailing, etc.)	Range of navigable waters: Maximum → minimum	Silver Lake	1. Clove Lake 2. Grasmere Lake 3. Ohrbach Lake 4. Arbutus Lake 5. Wolfes pond	Other ponds	River			●	●		

Source: McHarg I L. Design with nature[M]. New York: American Museum of Natural History, 1969

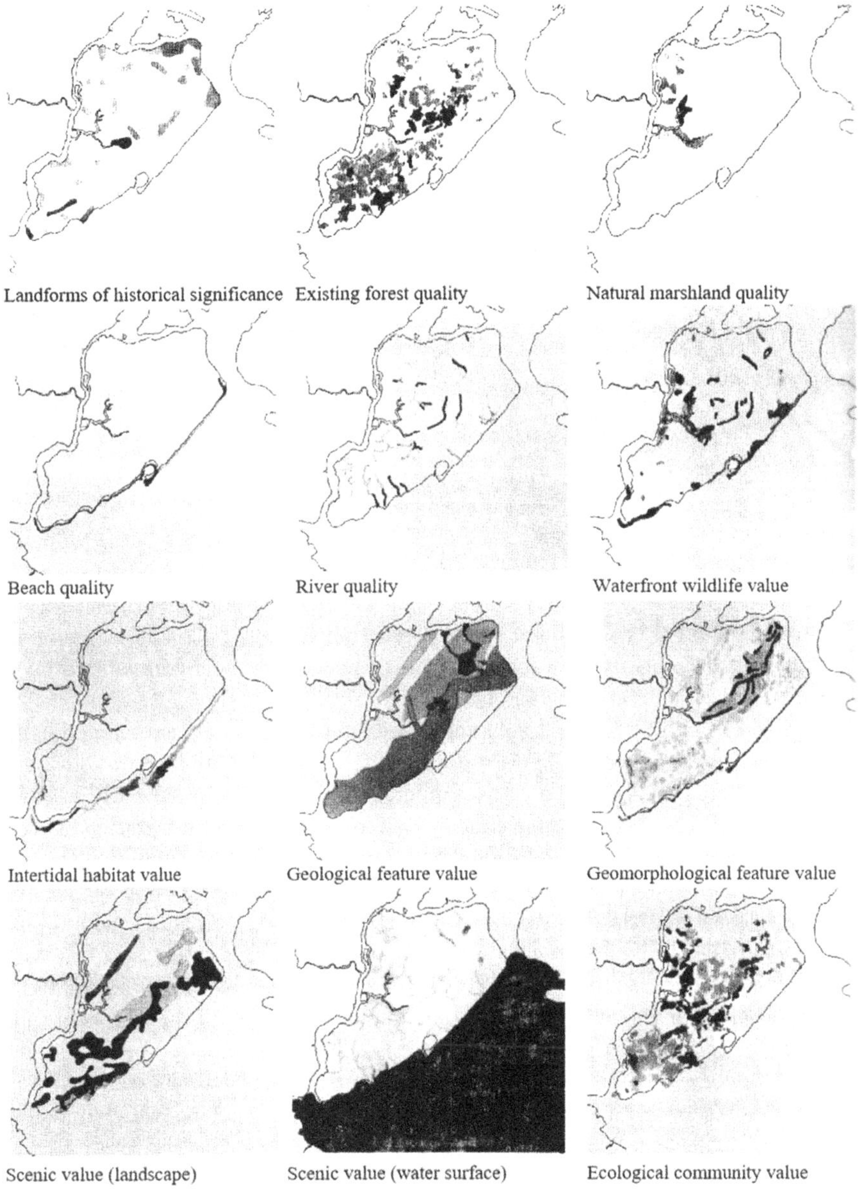

Fig. 4.3 Comprehensive evaluation result obtained from overlay of evaluation maps. Source: McHarg I L. Design with nature[M]. New York: American Museum of Natural History, 1969

In the formula:

i—Parcel number (or grid number);

j—Land use type number;

k—The number of ecological factors affecting the land use type *j*;

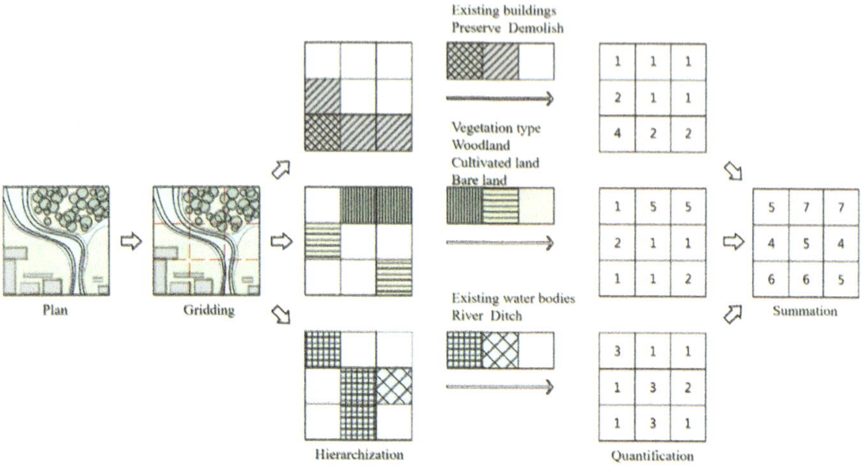

Fig. 4.4 Diagram of factor equal weight summation method

n—The total number of ecological factors affecting the land use type j;

$B_{k_{ij}}$—The suitability evaluation value of the $k-th$ ecological factor for the $i-th$ parcel with land use type j (single factor evaluation value);

V_{ij}—The comprehensive evaluation value of the $i-th$ parcel with land use type j (the ecological suitability of the j land use type).

Computer technology effectively addressed many difficulties encountered in the map overlay method. In 1967, Carl Steinitz and colleagues at Harvard University first applied computer technology to ecological planning.[7] Steinitz's early research made a significant and invaluable contribution to the successful integration of computer technology with land-use suitability analysis and laid the foundation for subsequent studies and practices.[8] The weighted scoring method is one of the new methods developed based on the map overlay method. This method incorporates weights into the evaluation process, assigning different weights to factors of varying importance for different types of land use, thereby overcoming the disadvantage of equal weighting in direct overlays. Another major advantage of the weighted summation method is that it grids, categorizes, and quantifies the graphics, making it suitable for computer applications. In the above-mentioned methods, the factors are relatively independent and fail to reflect inter-factor relationships, which may lead to bias in actual evaluation results. The widely used method today involves using the Analytic Hierarchy Process (AHP) to evaluate site factors and determine weights, followed by overlay and analysis of factor layers through GIS software platforms. This method accommodates the interrelationships among various factors in the system, establishing hierarchical correlations for evaluation, and is therefore more reasonable and scientific.

[7] Ndubisi F. Ecological planning: a historical and comparative synthesis[M]. JHU Press, 2002.

[8] Frederick S. The living landscape:an ecological approach to landscape planning[M]. McGraw-Hill Inc., 1991:213.

4.1.2 The Significance of Ecological Sensitivity Evaluation in Landscape Architecture Planning and Design

The selection of factors for ecological sensitivity evaluation varies according to region, scale, subject, and function. The *Provisional Technical Code for Ecological Function Zoning* issued by the State Environmental Protection Administration (hereinafter referred to as the "Code") regards ecological sensitivity as one of the bases for ecological function zoning. Its evaluation scope covers the entire national territory, and the content includes various aspects such as soil erosion sensitivity, desertification sensitivity, habitat sensitivity, and acid rain sensitivity, encompassing a vast scale and a multitude of factors. In terms of scale, the research scope of landscape architecture planning and design is mostly focused on outdoor spaces intended for human recreation. Therefore, ecological sensitivity evaluation in landscape architecture planning and design is relatively smaller in scale compared to that at the national land domain level.

At the national land domain level, ecological sensitivity research also involves many special geographic environments such as karst landforms and the Loess Plateau. These unique and distinctive landforms differ greatly from the objects typically involved in landscape architecture planning and design. The objects in landscape architecture planning and design are mostly located in urban or suburban areas, with scenic environments as the main component and possibly including parts of the built environment, thus belonging to the category of scenic garden environments. The types of ecological sensitivity evaluation factors in landscape architecture planning and design differ from those in evaluations at the national land domain level. Functionally, ecological sensitivity research at the national level involves multiple functions such as water conservation, windbreak and sand fixation, and biodiversity. In contrast, landscape architecture planning and design is mostly aimed at recreation and leisure, with relatively single functions, resulting in significant differences in the selection of ecological sensitivity factors. The ecological sensitivity factors in landscape architecture planning and design are specific to the planning and design level and are generally more concrete.

Ecological sensitivity evaluation has also been widely applied in ecological city planning. Compared with the method in the *Code*, which emphasizes systematic natural characteristics, ecological sensitivity evaluation in ecological city planning tends to focus more on socio-economic characteristics (see Table 4.2). Most of the sites in landscape architecture planning and design are dominated by scenic environments; hence, compared with ecological sensitivity factors in ecological city planning, they lean more towards scenic environmental factors and less towards human activity factors such as exhaust emissions, wastewater, and population density.

Placing ecological sensitivity analysis and evaluation at the forefront of landscape architecture planning and design emphasizes planning and design based on ecological protection of the site. Ecological sensitivity evaluation based on thorough site investigation and analysis reflects the coupling between design and site and also responds to

Table 4.2 Ecological sensitivity evaluation factors for urban areas

Primary indicator	Weight	Secondary indicator	Weight	Tertiary indicator	Grading Criteria: extremely sensitive	Sensitive	Moderately sensitive	Slightly sensitive
Current state of the regional natural environment	0.1	Flooding	0.6	Elevation (m)	<10	10 ~ 50	50 ~ 100	> 100
			0.4	Vegetation type	Without Vegetation, Bare land	Shrubland, grassland	Cultivated land	Coniferous Forest, broadleaf Forest
	0.2	Acid deposition	0.2	Soil type	Red Ferralsol	Red soil, Fluvo-aquic soil	Paddy soil	Vegetable garden soil
			0.3	Vegetation and soil use	Coniferous Forest	Shrubland	Broadleaf Forest	Farmland, orchard
			0.3	Precipitation(mm)	> 1900	1800 ~ 1900	1700 ~ 1800	< 1700
			0.3	Rock type	Granite, gneiss, sandstone	Sandstone, carbonate rock	Mudstone, marlstone	Limestone, dolomite
	0.3	Nature reserve		Main wetland type	√			
Intensity of human activities	0.3	Level of environmental pollution	0.3	Degree of air pollution	> 3	2 ` 3	1 ` 2	< 1
			0.3	Water quality grade	V	IV	III	II,I
			0.2	Wastewater discharge intensity (t/(m^2·a))	> 0.30	0.20 ` 0.30	0.10 ` 0.20	< 0.10
			0.1	Emissions intensity of exhaust gases (m^3/(m^2·a))	> 0.012	0.008 ` 0.012	0.004 ` 0.008	< 0.004
			0.1	Dust emission intensity (kg/(ma))	> 0.30	0.20 ` 0.30	0.10 ` 0.20	< 0.10
	0.1	Population		Population density	> 0.0030	0.0021 ` 0.0030	0.0012 ` 0.0021	< 0.0012

Adapted from: Yang Z F, Xu L Y. Urban Ecological Planning [M]. Beijing: Beijing Normal University Press, 2008:179

the "spirit of place." Through ecological sensitivity evaluation, scientific support is provided for delineating ecologically sensitive areas within the site. The gradient classification of ecologically sensitive areas clarifies for designers the areas that must be strictly protected, those requiring protection and moderate restoration, and those that are suitable for construction. This not only actively and proactively protects the ecological environment but also makes the design more reasonable and well-founded. Only by conducting land-use suitability evaluation based on ecological sensitivity evaluation can ecological protection be effectively implemented in practice.

4.2 Construction of the Ecological Sensitivity Evaluation Model

Different sites, i.e., different evaluation objects, require different evaluation factors; changes in functional requirements likewise lead to changes in the selection of evaluation factors, ultimately resulting in different evaluation outcomes. This reflects the parametric nature of landscape architecture planning and design. Selecting suitable evaluation factors based on specific sites and functional requirements ultimately leads to the classification of ecological sensitivity zones, a process that also reflects the "coupling" of site and design. Evaluating the ecological sensitivity of a site is the first step in parameterized landscape architecture planning and design based on the coupling principle.

The ecological sensitivity evaluation in landscape architecture planning and design is based on the Analytic Hierarchy Process (AHP). AHP is a multi-objective decision-making method combining qualitative and quantitative analysis, proposed by American operations researcher Thomas L. Saaty in the early 1970s. It decomposes a complex problem into several components and forms a hierarchical structure according to dominance relationships. Then, through pairwise comparisons, the relative importance of each factor (including indicators and schemes) is determined, the weights of each factor are calculated, and finally, different decision schemes are ranked based on these weights.[9] When applying AHP for evaluation, it is first necessary to establish a systematic hierarchical structure. On this basis, pairwise comparison judgment matrices (positive reciprocal matrices) are constructed, and for a given criterion, the weights of each alternative are calculated. The ranking weights of the current level elements concerning the overall objective are then calculated, followed by a consistency check of the matrix. The ecological sensitivity parametric evaluation model adopts scientific evaluation methods such as AHP and the Delphi method. The operational platform is ArcGIS, and the model comprises five stages: screening of evaluation factors, construction of the hierarchical evaluation model, determination of evaluation factor weights, overlay analysis, and generation of ecological sensitivity classification (Fig. 4.5).

[9]Qiu J P, Wen T X. Evaluation science: Theory, Method and Practice [M]. Beijing: Science Press, 2010:114.

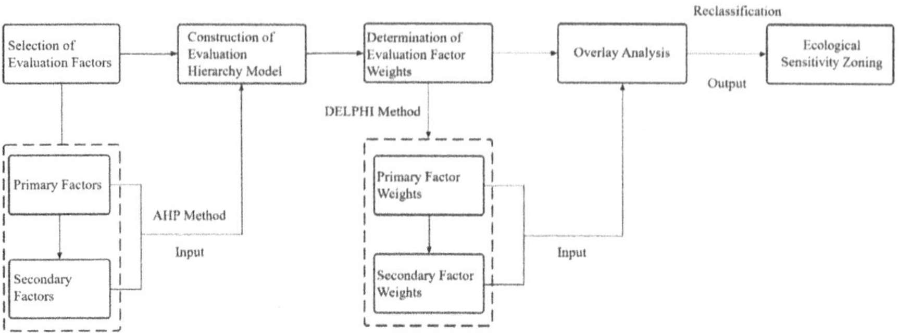

Fig. 4.5 Parametric evaluation model of ecological sensitivity

4.2.1 Screening of Ecological Sensitivity Evaluation Factors

Evaluating and judging a matter based solely on a single indicator is not reasonable. Therefore, evaluation is a comprehensive judgment under the interaction of multiple factors, often involving multiple aspects, requiring the aggregation and synthesis of various indicator information to reflect the evaluation objective holistically. Evaluation factors should cover all aspects of the subject to fully reflect the information. However, the selection of evaluation factors should also follow the principle of "appropriateness"—fewer and simpler factors are preferable. When using AHP for evaluation, the constructed index system should have clear hierarchical levels and avoid overlap. Each index at each level should be representative and have a real impact on ecological sensitivity to reduce redundancy and complexity and facilitate the evaluation process. In addition, the selection of evaluation indicators should fully consider the differences and comparability among various factors, which helps make the weight determination process more objective.

Many factors influence ecological sensitivity, such as elevation, slope, vegetation, soil, and geology, and these factors vary across regions. From the site perspective, primary indicators for ecological sensitivity evaluation in landscape architecture planning and design are mainly composed of natural environmental factors, with a small number of human activity factors. The natural factors correspond to five primary indicators: topography and landform, soil, hydrology, vegetation, and animals. In the category of "human activities," the factor that affects ecological sensitivity in scenic garden environments is mainly the current construction situation. Thus, the corresponding primary indicator is "current construction." Table 4.3 lists the common factors, classifications, and grading criteria involved in ecological sensitivity evaluation in landscape architecture planning and design. The secondary factors listed are the most commonly used evaluation factors in landscape architecture planning and design. Depending on the specific site, the participating factors may differ slightly from those listed in the table. Similarly, the specific grading quantity and types in the "grading criteria" column should be adjusted according to practical operational conditions. For example, in the case of water bodies, the required buffer

Table 4.3 Ecological sensitivity evaluation factors in landscape and forest planning and design

Evaluation objective	Indicator type	Primary indicator	Weight	Secondary indicator	Weight	Classification criteria		
						Highly sensitive	Moderately sensitive	Low sensitivity
Ecological sensitivity evaluation in landscape architecture planning and design	Spatial (S)	Topography and landform	W1	Relative elevation	W_{11}	>100 m	50–100 m	<50 m
				Slope	W_{12}	>30°	15°–30°	<15°
				Aspect	W_{13}		Northwest, northeast	South, southwest, southeast
	Ecological (E)	Soil	W2	Soil type	W_{21}	Sandy loam, silt loam, loamy sand, silty clay loam, clay loam	Loamy sand, clay	Gravel, sand, coarse sand, fine sand, clay
				pH value	W_{22}	<4.5, >9.5	4.5–6.5, 7.5–9.5	6.5–7.5
				Organic matter content	W_{23}	>20%	10%–20%	<10%
		Hydrology	W3	Water quality	W_{31}	Good	Fair	Poor
				Water body buffer zone	W_{32}	<10 m	10–50 m	>50 m
				Watershed buffer zone	W_{33}	<50 m	50–150 m	>150 m
		Vegetation	W4	Vegetation coverage	W_{41}	Dense	Sparse	Bare
				Vegetation type	W_{42}	Coniferous forest, mixed coniferous forest	Broadleaf forest, shrubland, grassland	Farmland
				Growth condition	W_{43}	Good	Fair	Poor
		Animals	W5	Rare species habitat	W_5	National-level	Provincial or other regionally protected species	No protected species
	Behavioral (B)	Existing development	W6	Land use type	W_6	Designated protection zone	Undeveloped land outside protected zones	Developed land

zones differ across spatial scales. When defining buffer zones, it is inappropriate to apply standards used for large lakes such as Taihu or Poyang Lake directly to small-scale water surfaces like ponds; the evaluation must consider the actual situation.

4.2.1.1 Topography and Landform

Topography and landform describe the undulations of the terrain, i.e., the form of the earth's surface, which is generally divided into six basic types: plateau, mountain, plain, hill, rift valley system, and basin. Ecological sensitivity evaluation in scenic garden environments generally includes three aspects: relative elevation, slope, and aspect. In general, large-scale scenic garden environments contain elevation ranges with significant differences, and ecological sensitivity tends to decrease from higher to lower elevations. Areas with steeper slopes have poorer surface stability and are more difficult to restore after disturbances; hence, steeper slopes imply higher ecological sensitivity. South-facing areas in the Northern Hemisphere receive better sunlight, which promotes plant growth, resulting in stronger ecological recovery and thus lower sensitivity.

4.2.1.2 Soil

Soil is a critical component of the habitat and significantly affects its restoration. Soil types vary across regions. For example, the zonal soil types along the latitude from north to south in the Eurasian continent are: Gelisol – Podzol – Grey Forest Soil – Chernozem – Chestnut Soil – Brown Soil – Desert Soil – Alpine Soil – Red Soil – Laterite. On the west coast of continents, from north to south: Gelisol – Podzol – Brown Soil – Kastanozem – Desert Soil. On the east coast of continents, from north to south: Gelisol – Podzol – Brown Soil – Red/Yellow Soil – Laterite. Various physical and chemical properties are used to describe soil conditions, such as aggregate structure, porosity, pH, and organic matter content. Different soil types have distinct physical and chemical properties. The soil characteristics of a site reflect certain aspects of the habitat. Ecological sensitivity evaluation in scenic garden environments primarily involves three basic soil indicators: soil type, pH, and organic matter content. These indicators are not only easily obtainable in practical work but also reflect the soil's influence on plant growth.

4.2.1.3 Hydrology

Hydrological indicators reflect the status of water bodies in the study site, including existing water bodies and catchment conditions. Natural water bodies and large artificial water bodies with stable habitats must be protected. Areas with artificial disturbances that are too close to water bodies will inevitably affect their habitats.

Therefore, ecological sensitivity evaluation must examine the relationship between human activities and natural water bodies, introducing the concept of a "water body buffer zone." The study of catchments should also use buffer zone delineation to protect these areas and avoid excessive human interference in natural hydrological processes, which can cause ecological imbalances.

4.2.1.4 Vegetation

Vegetation factors refer to the various types of plants growing on the site, described by three specific evaluation indicators: vegetation coverage, vegetation type, and growth condition. Vegetation coverage reflects plant growth in the area. Well-covered areas are labeled "dense," while poorly covered areas are described as "bare ground." The better the coverage, the higher the vegetation density and diversity, and the better the growth, indicating high ecological sensitivity and unsuitability for major artificial disturbances. Different vegetation types also have different ecological sensitivities. Primary vegetation is more sensitive than secondary or artificial vegetation due to its lower stability and greater susceptibility to disturbance. According to vegetation succession principles, higher-order plant communities are more stable but harder to restore once damaged, resulting in higher ecological sensitivity. Based on this, sensitivity levels can be assigned to different vegetation types. Likewise, areas with good vegetation growth should be given corresponding protection.

4.2.1.5 Animals

Animal factors in ecological sensitivity evaluation primarily refer to designated animal protection areas and their levels. The rarer the species in a region, the higher the protection level, and therefore the higher the ecological sensitivity. For example, parts of the Niushou Mountain area in Nanjing are breeding and habitat grounds for the nationally protected species *Bhutanitis thaidina* (Chinese Windmill butterfly). Thus, these areas are designated as extremely sensitive ecological zones in the evaluation and are subject to strict protection.

4.2.1.6 Current Construction

Current construction factors reflect existing human activities within the evaluation site. Areas with more frequent human activity generally have lower ecological sensitivity. In ecological sensitivity evaluation for scenic garden environments, land use types within the site are roughly categorized, and specific grading standards should be flexibly set according to the site's actual conditions and design requirements.

4.2.2 Construction of the Hierarchical Model for Ecological Sensitivity Evaluation

Based on the ecological sensitivity evaluation factor table for landscape planning and design listed in the previous section, a hierarchical model for ecological sensitivity evaluation in landscape planning and design can be constructed using the Analytic Hierarchy Process (Fig. 4.6). This hierarchical model is composed of three main parts: the goal level, the element level, and the indicator level. The goal level refers to the objective to be achieved through the evaluation. The element level reflects the factors affecting ecological sensitivity and consists of three types and six specific factors. The final level is the indicator level, which mainly describes the indicators of the six specific factors. For example, the topography and landform factor can be reflected by three indicators: relative elevation, slope, and aspect.

4.2.3 Determination of Weights of Ecological Sensitivity Evaluation Factors

Based on the model constructed in the previous step, a judgment matrix is constructed to determine the importance of each evaluation factor at the element level and the indicator level within the evaluation system. In the constructed hierarchical model, let a certain factor at the element level be *C*, and the subordinate factors of the next level be $u_1, u_2, ..., u_n$. The relative importance of these factors to *C* is the weight W. In the context of landscape planning and design, the importance of $u_1, u_2, ..., u_n$ to *C* cannot be directly quantified, so this step requires professionals or experts

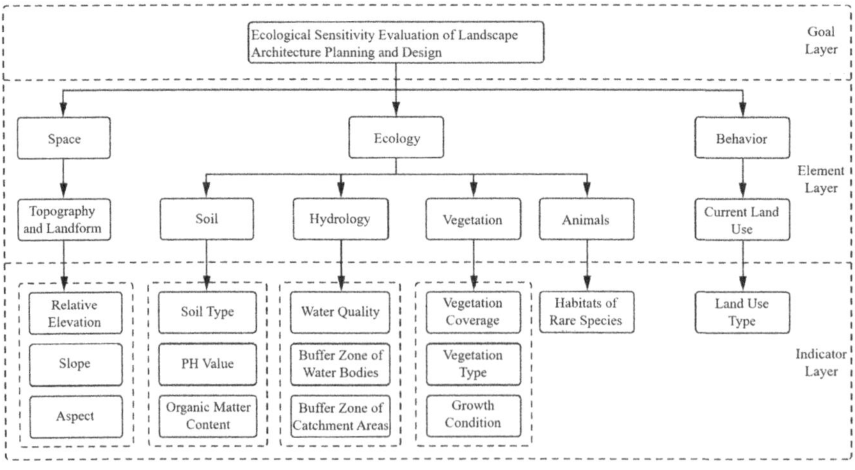

Fig. 4.6 Hierarchical model for ecological sensitivity evaluation in landscape planning and design

Table 4.4 Quantitative scale value table for ranking

Scale description	Assigned value
Two indicators are equally important	1
One indicator is slightly more important	3
One indicator is more important	5
One indicator is significantly more important	7
One indicator is extremely more important	9
Intermediate values between the above scales	2, 4, 6, 8
One indicator is less important than another	Reciprocals of the above values

to compare the factors at each level, and construct corresponding judgment matrices. This study adopts the commonly used 1–9 scale for AHP, which respectively represents the comparisons of "equally important", "slightly more important", "important", "more important", and "extremely important". If further subdivision is needed, the values 2, 4, 6, and 8 can be used to interpolate, forming a more precise nine-point scale (Table 4.4).

Accordingly, for element *C*, *a* comparison of the relative importance among *n* elements can yield a judgment matrix:

$$A = \left(a_{ij}\right)_{n \times n}$$

Among them, a_{kj} represents the ratio scale of the importance between elements u_i and u_j with respect to *C*. The judgment matrix *A* has the following properties:$a_{ij} > 0$, $a_{ji} = 1/a_{ij}$, and $a_{ii} = 1$.If all elements in the judgment matrix *A* satisfy $a_{ij} \cdot a_{jk} = a_{ik}$, then *A* is called a *consistency matrix*. Taking the topographic and geomorphic factors from the element layer as an example, a judgment matrix is established using three factors: *relative elevation*, *slope*, and *aspect*. Based on the pairwise comparisons of the importance among these three factors, the corresponding matrix is determined as shown in Table 4.5. Here, $W_{11} + W_{12} + W_{13} = 1$, and X_1, X_2, and X_3 are three values selected from $\{1, 3, 5, 7, 9\}$.

After establishing the judgment matrix, it is necessary to determine the relative weights $w_1, w_2, w_3, \ldots, w_n$ of $u_1, u_2, \ldots, u_n$ with respect to element *C*. In vector form, this is written as $w = (w_1, w_2, \ldots, w_n)^T$. The author adopts the root method to calculate the weights. In addition to using mathematical models for computation, software tools such as Yaahp (Yet another AHP software) based on the Analytic Hierarchy Process can also assist in reducing the computation time and improving accuracy. Using the root method, the geometric mean of each row vector in the judgment matrix is first calculated, and then normalization is performed to obtain the weight vector. The formula is as follows:

Table 4.5 Ecological sensitivity evaluation—judgment matrix for topography and geomorphology evaluation

Topography & geomorphology	Relative Elevation	Slope	Aspect	Weight
Relative elevation	1	X_1	X_2	W_{11}
Slope	$1/X_1$	1	X_3	W_{12}
Aspect	$1/X_2$	$1/X_3$	1	W_{13}

$$W_1 = \frac{\left(\prod_{j=1}^{n} a_{ij}\right)^{\frac{1}{n}}}{\sum_{k=1}^{n}\left(\prod_{j=1}^{n} a_{jk}\right)^{\frac{1}{n}}}$$

1. Multiply the elements of matrix *A* by columns to obtain a new vector;
2. Take the n-th root of each component of the new vector;
3. Normalize the resulting vector to obtain the weight vector.

After the weights of the indicator level are calculated, the importance of each element level factor to the goal level can also be determined using the above method, ultimately resulting in the weights of each hierarchical factor in the ecological sensitivity evaluation model.

Due to the complexity of landscape planning and design, it is not required that $a_{ij} \cdot a_{jk} = a_{ik}$ must strictly hold. In judging the relative importance of factors, inaccuracies may arise, leading to weights that do not truly reflect the importance of each factor to the upper criterion. In AHP, a quantitative measure of inconsistency is often used, namely the Consistency Index (*C.I.*) and the Random Index (*R.I.*), to conduct a consistency test. The first step is to determine the consistency index *C.I.*, which is calculated as follows:

$$C.I. = \frac{\lambda_{\max} - n}{n-1}$$

The maximum eigenvalue λmax of the matrix can be calculated using the following formula:

$$\lambda_{\max} = \sum_{i=1}^{n} \frac{(AW)_i}{nw_i} = \frac{1}{n}\sum_{i=1}^{n} \frac{\sum_{i=1}^{n} a_{ij} w_j}{w_i}$$

Next, it is necessary to refer to the average random consistency index (R.I.). The values of the average *R.I.*, obtained by calculating 1000 randomly generated positive reciprocal matrices of order 1 to 10, are shown in Table 4.6.

When the ratio of C.I. to R.I. (Consistency Ratio, *C.R.* = *C.I./R.I.*) is less than 0.1, the judgment matrix can be considered to have acceptable consistency. Otherwise, the matrix needs to be adjusted to meet the consistency check requirements. This process can also be completed using AHP software such as Yaahp.

Table 4.6 Average random consistency index (R.I.) for 1–10 order positive reciprocal matrices

Matrix order n	1	2	3	4	5	6	7	8	9	10
R.I.	0	0	0.58	0.9	1.12	1.24	1.32	1.41	1.45	1.49

Source: Qiu J P, Wen T X. Evaluation science: Theory, Method and Practice [M]. Beijing: Science Press, 2010:114

Table 4.7 Five categories of ecological sensitivity

No.	Name	Grade	Description
1	Category I—highly sensitive area	Highly sensitive	Areas highly sensitive to human activities; even minor disturbances may cause significant damage, leading to ecosystem degradation. These areas require strict conservation measures and protection
2	Category II—sensitive area	Sensitive	Areas with moderate sensitivity to human activities; ecosystem recovery is difficult, and development should be carefully planned to minimize damage
3	Category III—low sensitivity area	Low sensitivity	Areas with relatively stable ecosystems, capable of withstanding some human disturbance, but the ecosystems still require restoration if disrupted
4	Category IV—weak sensitivity area	Weak sensitivity	Areas with flexible ecological conditions, capable of withstanding a certain level of human activities
5	Category V—insensitive area	Insensitive	Areas with low ecological sensitivity, suitable for multiple land uses and human activities

4.2.4 Classification of Ecological Sensitivity

After determining the relative weights of the indicator-level and element-level factors, the corresponding analysis maps derived using ArcGIS software can be weighted and overlaid. For landscape planning and design purposes, dividing the ecological sensitivity zones into three categories—low sensitivity, moderate sensitivity, and high sensitivity—is sufficient to meet the design requirements. In special cases, five or more categories can be used (Table 4.7). The more detailed the classification, the more precisely the ecological sensitivity of the site can be described, which is beneficial for design-related site assessment. However, it should be noted that increasing the number of classifications will lead to a multiplied workload, so it is crucial to choose an appropriate number of evaluation levels based on the needs of the design.

4.3 Application of the Parametric Model of Ecological Sensitivity Evaluation

Niushou Mountain is located in the northern part of Jiangning District, Nanjing, Jiangsu Province. It is part of the southern branch of the western section of the Ningzhen Mountain range and belongs to the low mountain and hilly areas along

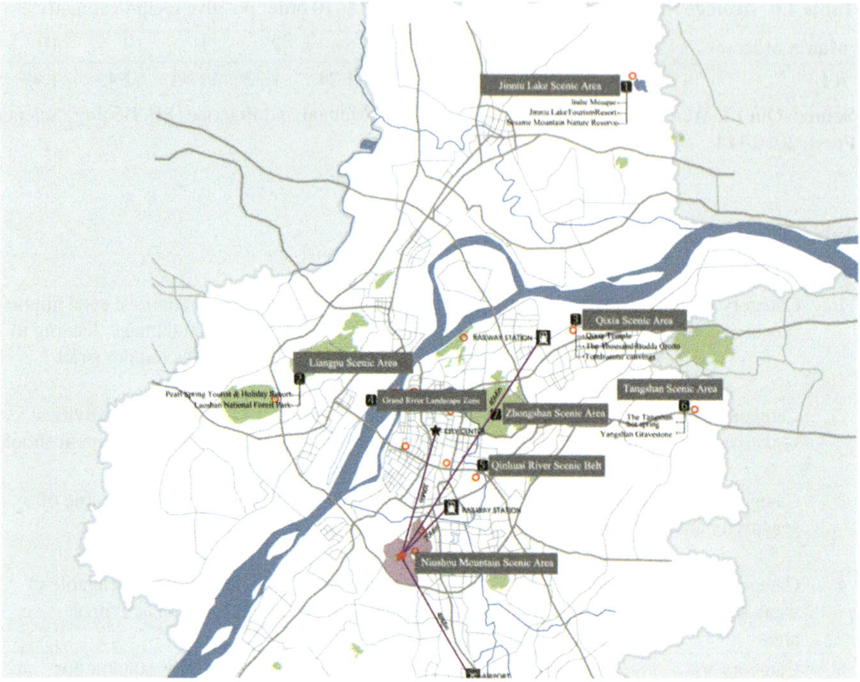

Fig. 4.7 Location map of niushou mountain scenic area

the Yangtze River. Since ancient times, Niushou Mountain has been a renowned destination for spring outings and flower viewing. "The Misty Peaks of Niushou" was one of the "Forty-eight Views of Jinling". In addition to its picturesque natural scenery, Niushou Mountain is also rich in historical and cultural resources and is the ancestral site of one branch of Zen Buddhism—the Niutou Sect. According to the *Master Plan of the Niushou-Zutang Scenic Area (2003–2020)*, the Niushou Mountain Scenic Area will be developed into a municipal-level scenic area that "relies on historical relics and famous sites, features spring scenery, forested hills, and pastoral landscapes, and integrates natural ecological sightseeing, historical and cultural education, and leisure vacation" (Fig. 4.7). This study takes the northern part of the Niushou Mountain Scenic Area as an example to discuss the application of the parametric evaluation model of ecological sensitivity in landscape planning and design.

In the entire process of landscape planning and design, ecological sensitivity evaluation based on thorough site investigation and analysis is the foundation and prerequisite of all planning work. The planning and design must begin with comprehensive site research, collecting natural factor data such as topography and landforms, soil, hydrology, flora and fauna, as well as understanding human activities on the site, with a focus on the current development status for ecological sensitivity evaluation. These collected data form the basis for establishing the hierarchical evaluation model and should closely align with the objectives of the ecological sensitivity evaluation. According to the parametric evaluation model of ecological sensitivity, the specific process includes: constructing a hierarchical model and determining weights using the Delphi Method and AHP; overlay analysis of factor layers using the layer superposition method; and ultimately producing an ecological sensitivity zoning map. The primary software used in the evaluation process includes Yaahp for AHP and ArcGIS.

4.3.1 Construction of the Hierarchical Structure Model for Ecological Sensitivity Evaluation

The first step in ecological sensitivity evaluation is to construct a hierarchical structure model. The evaluation of ecological sensitivity in the northern part of the Niushou Mountain Scenic Area involves six factors: topography and landform, soil, hydrology, fauna, flora, and current development. These six aspects form the element level of the evaluation. Specific indicators for each element must also be determined, as shown in Table 4.8.

It should be noted that the element "current development" has only a single indicator. Therefore, when using Yaahp software to construct the hierarchical evaluation model, the indicator level for this element should be omitted to avoid confusion in the hierarchical relationship, which may prevent proper model construction in the software. Yaahp allows for the convenient and rapid determination of the weights for each level in the ecological sensitivity evaluation of landscape planning and design, preparing parameters for the subsequent overlay analysis and avoiding redundant calculations. Due to the characteristics of Yaahp, at least two "alternative scheme" layers must be set. Thus, two placeholder "alternatives" are temporarily included during operation to meet Yaahp's modeling requirements (Fig. 4.8). In the next step of constructing the judgment matrix, the weights of the two "alternatives" are set equally to 0.5, indicating that they are of equal importance and will not affect the weight calculation of the element and indicator levels.

4.3.2 Determination of Weights for Each Evaluation Factor

Based on the hierarchical structure model, an expert panel is established using the Delphi Method. Experts compare the relative importance of factors at each level, and the results are entered into the "Judgment Matrix" tab of Yaahp software to construct the judgment matrices for each level, from which the weight table for each level is derived (Table 4.9).

From the weight table, it is evident that among the primary factors at the element level, the weights of topography and flora are relatively high, totaling 0.8602,

Table 4.8 Ecological sensitivity evaluation factors for the northern region of niushan scenic area

Evaluation objective	Indicator type	Primary indicator	Weight	Secondary indicator	Weight	Classification criteria		
						Highly sensitive	Moderately sensitive	Low sensitivity
Ecological sensitivity evaluation in landscape architecture planning and design	Spatial (S)	Topography and landform	W1	Relative elevation	W_{11}	>100 m	50–100 m	<50 m
				Slope	W_{12}	>30°	15°–30°	<15°
				Aspect	W_{13}		Northwest, northeast	South, southwest, southeast
	Ecological (E)	Hydrology	W3	Water quality	W_{31}	Good	Fair	Poor
				Water body buffer zone	W_{32}	<10 m	10–50 m	>50 m
				Watershed buffer zone	W_{33}	<50 m	50–150 m	>150 m
		Vegetation	W4	Vegetation coverage	W_{41}	Dense	Sparse	Bare
				Vegetation type	W_{42}	Coniferous forest, mixed coniferous forest	Broadleaf forest, shrubland, grassland	Farmland
				Growth condition	W_{43}	Good	Fair	Poor
	Behavioral (B)	Existing development	W6	Land use type	W_6	Designated protection zone	Undeveloped land outside protected zones	Developed land

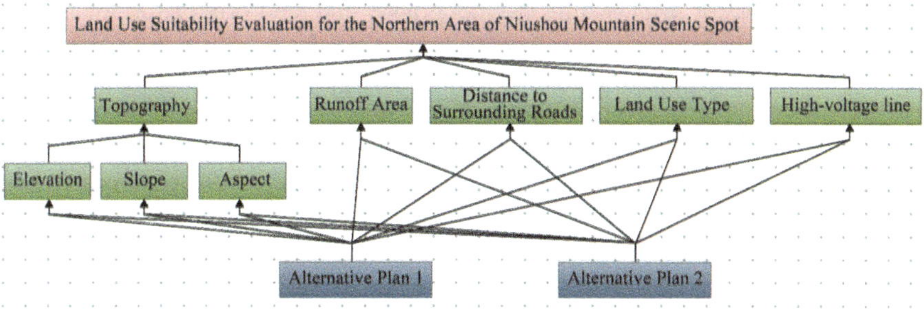

Fig. 4.8 Ecological sensitivity evaluation for the northern region of Niushan Scenic Area—hierarchical structure model (Yaahp Software)

Table 4.9 Ecological sensitivity evaluation for the northern region of Niushan Scenic area—judgment matrix and factor weights (Yaahp Software)

1. Ecological sensitivity evaluation for the northern region of niushan scenic area (consistency ratio: 0.0529;λmax: 4.4142)					
Factor	Topography	Water	Vegetation	Current Development	Wi
Topography	1	5	1	8	0.4301
Water	0.2	1	0.2	1.6	0.086
Vegetation	1	5	1	8	0.4301
Current development	0.125	0.625	0.125	1	0.0538
2. Topography (consistency ratio: 0.0517; weight: 0.4301;λmax: 3.0538)					
Topography	Relative elevation	Slope	Aspect	Wi	
Relative elevation	1	0.2	0.25	0.1018	
Slope	5	1	0.5	0.3661	
Aspect	2	4	1	0.5321	
3. Water (consistency ratio: 0.0000; weight: 0.086;λmax: 2.0000)					
Hydrology	Water body buffer zone	Watershed buffer zone			Wi
Water body buffer zone	1	3			0.75
Watershed buffer zone	0.3333	1			0.25
4. Vegetation (consistency ratio: 0.0000; weight: 0.4301;λmax: 2.0000)					
Factor	Vegetation coverage	Vegetation type			Wi
Vegetation coverage	1	0.25			0.2
Vegetation types	4	1			0.8
5. Current development (consistency ratio: 0.0000; weight: 0.0538;λmax: 2.0000)					

indicating that these two factors have a significant influence on ecological sensitivity. In contrast, the weights of hydrology and current development are relatively low at 0.086 and 0.0538, respectively, with the latter being the lowest and having the weakest influence on ecological sensitivity.

For the indicator level, within the topography and landform factor, the order of importance from high to low is: aspect, slope, and relative elevation. As the difference between the highest and lowest elevations in the northern part of Niushou Mountain Scenic Area is only 122 meters, elevation variation is not significant, and thus relative elevation has a weaker impact on ecological sensitivity. Among hydrological and plant factors, buffer zones around water bodies and vegetation types are of higher importance and carry larger weights.

4.3.3 Formation of Ecological Sensitivity Zoning

4.3.3.1 Analysis of Secondary Ecological Sensitivity Factors

Based on the collected research data, analysis maps corresponding to each indicator layer were produced using ArcGIS software. Figure 4.9 displays eight specific indicator factor maps: elevation, slope, aspect, water body buffer zones, watershed buffer zones, current development, vegetation cover, and vegetation type. The elevation map shows a general trend of higher elevation in the south and lower in the north, with two prominent ridgelines extending northward from the highest point. The slope map, in combination with the elevation map, reveals the internal slope conditions of the site—steeper slopes are mostly distributed on both sides of the ridges, while lower areas are relatively flat. The water body buffer zone raster map shows that most water bodies are located in the western part of the site, with fewer existing water bodies in the higher southeastern area. There are a few buildings scattered in the valleys between the ridges, where human activity is more frequent and ecological disturbance is higher. Regarding vegetation factors, the site's overall vegetation condition is good, with denser vegetation on both sides of the ridges.

4.3.3.2 Normalization of Ecological Sensitivity Factors

The purpose of normalizing factors is to integrate factors with different classifications and standards into a unified evaluation system to facilitate overlay analysis. In this study, each indicator factor is divided into three levels according to the grading criteria in the ecological sensitivity evaluation factor table. The reclassification function of ArcGIS software is used for normalization. After this process, each of the eight maps shown in Fig. 4.10 is divided into three levels with values of 1, 2, and 3.

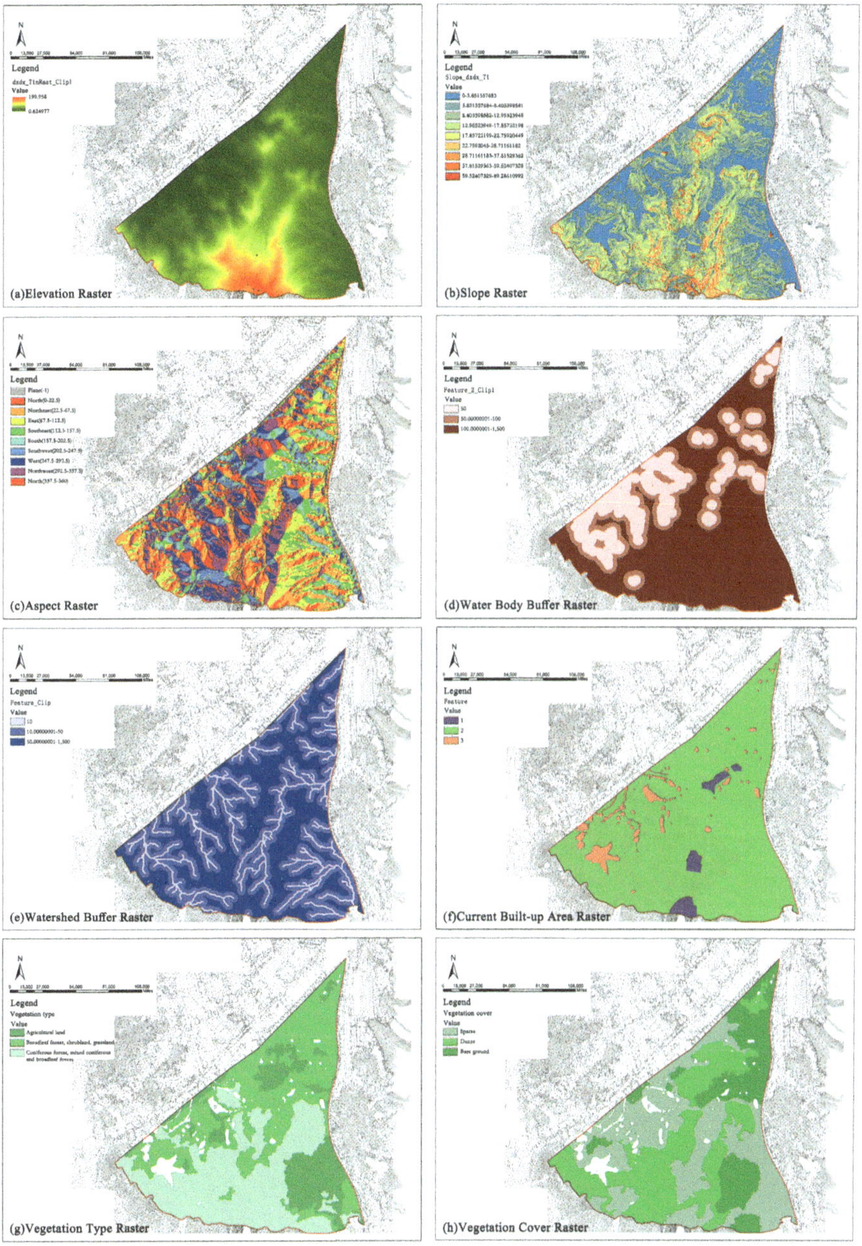

Fig. 4.9 Secondary Factor Analysis Map of Ecological Sensitivity Evaluation in the Northern Area of Niushou Mountain Scenic Area

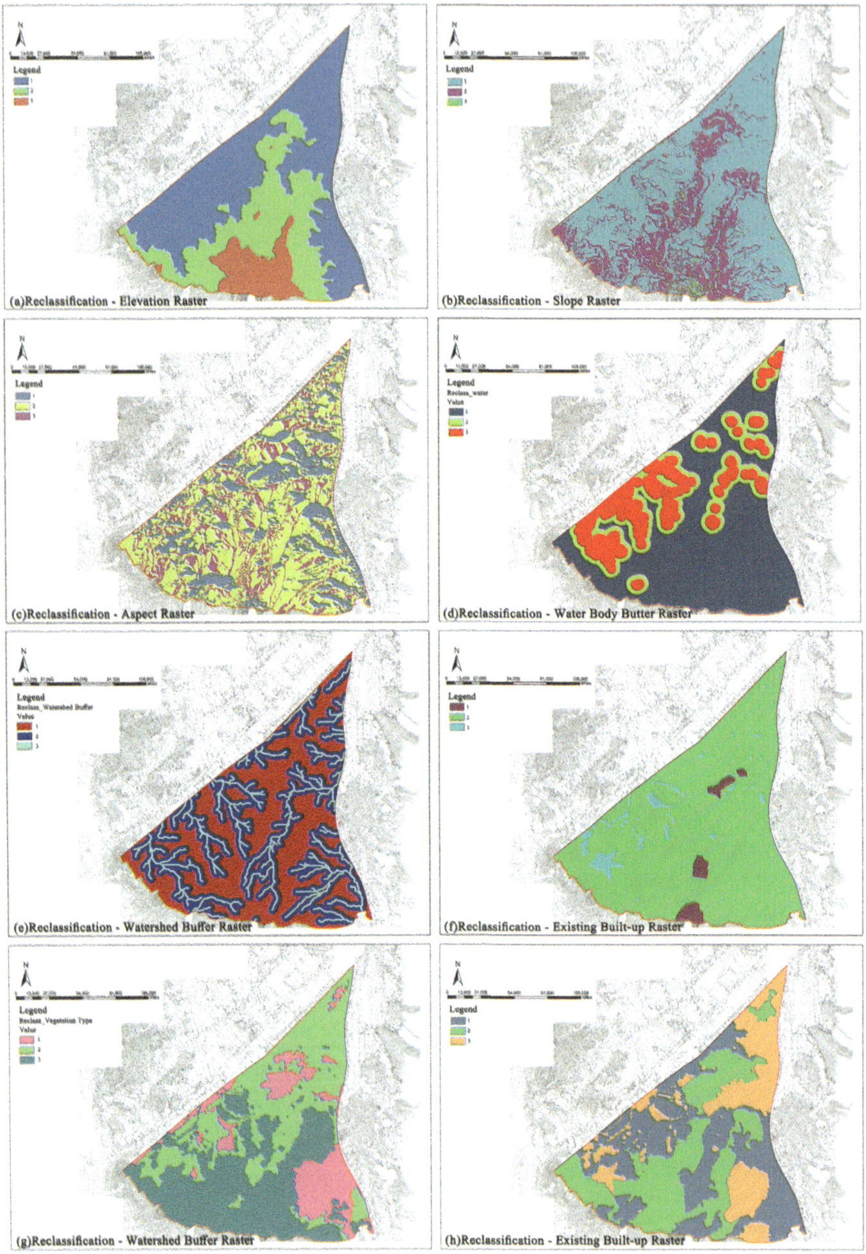

Fig. 4.10 Reclassified map of secondary factors for ecological sensitivity evaluation in the northern area of Niushou mountain scenic area

4.3.3.3 Analysis of Primary Ecological Sensitivity Factors

Primary factors are the element-level factors, and their analysis is based on the analysis of secondary factors. The reclassified and normalized secondary indicator maps are weighted and overlaid according to items 2–4 in Table 4.9 to obtain the analysis results for primary factors. These reflect the influence of topography, water bodies, and vegetation on ecological sensitivity across the site. Figure 4.11 shows that for the topography factor, the second sensitivity level dominates, indicating that most areas fall into the moderately sensitive category. For water bodies, non-sensitive areas are dominant. The vegetation factor map reveals that there are large sensitive areas within the site.

4.3.3.4 Formation of Ecological Sensitivity Zoning

By overlaying the primary factor analysis maps with their respective weights, the ecological sensitivity zoning map for the entire study area is obtained (Fig. 4.12). In this study, ecological sensitivity is divided into three levels: low sensitivity, moderate sensitivity, and high sensitivity. Highly sensitive areas have good or fragile ecological environments and are subject to strict protection, prohibiting any development or construction activities. Such areas in the northern part of the Niushou Mountain Scenic Area are relatively few and dispersed, mostly concentrated in higher elevations and water-rich areas. The second level is moderately sensitive areas, which can tolerate a certain degree of human disturbance and have some capacity for ecological recovery. These areas account for the majority of the site. Low sensitivity areas are less affected by human activity and can accommodate certain development activities. They also occupy a significant portion of the site.

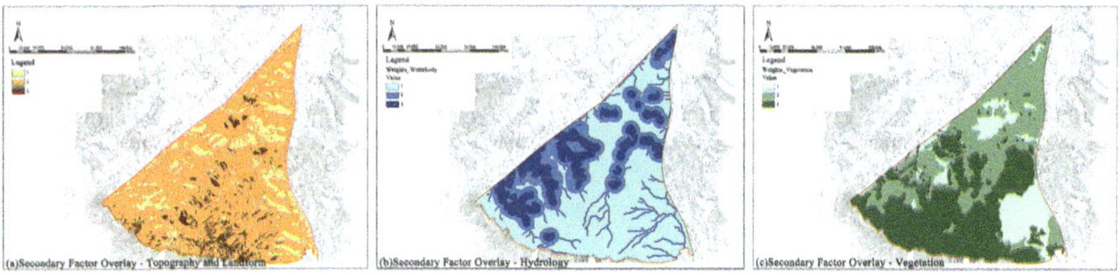

Fig. 4.11 primary factor analysis map of ecological sensitivity evaluation in the northern area of Niushou mountain scenic area

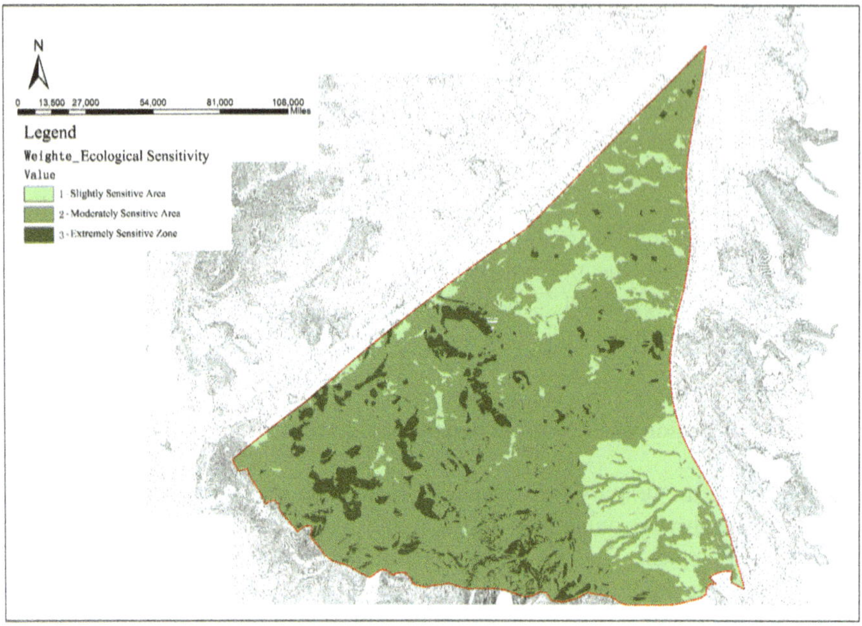

Fig. 4.12 Ecological sensitivity zoning of the northern area of Niushou mountain scenic area

Chapter 5
Land Use Suitability Assessment Model

Since landscape environments serve human needs, they require functional development. Land use suitability assessment is the subsequent stage following ecological sensitivity analysis. Unlike ecological sensitivity analysis, which prioritizes ecological considerations, land use suitability assessment evaluates the site from a construction-oriented perspective and classifies land based on its potential for development. While the technical approaches and methods of both assessments are broadly similar, the selection of evaluation factors differs due to their respective emphases. This chapter begins by defining the concept of land use suitability assessment and its relevance to landscape planning and design. It then proposes a parametric evaluation model for assessing land suitability and demonstrates its application through case studies.

5.1 Land Use Suitability and Landscape Architecture Planning and Design

Construction land generally refers to land designated for the development of buildings, structures, and their associated functional spaces. In scenic areas and other landscape environments, construction land typically encompasses several categories, including land for tourism facilities, social and residential services, transportation, and infrastructure. Unlike cities, which are dominated by artificial environments, construction land in landscape environments is embedded within natural settings. It plays a vital role in supporting transportation, services, and other operational functions within scenic areas, making it an essential component of the overall landscape system. This study focuses on land allocated for functional facilities within landscape environments—specifically, areas used for management, dining, accommodation, exhibitions, and related services. It excludes land designated for independent residential communities, production and administrative facilities, and large-scale

Y. Yuan, *Parametric Landscape Architecture Planning and Design*,
https://doi.org/10.1007/978-981-95-8011-8_5

infrastructure such as transportation and communication systems that, although serving tourists, are managed separately. As the term suggests, land use suitability evaluation refers to assessing whether a given parcel of land is appropriate for development. In landscape planning and design, such evaluation aims to identify areas within predominantly natural environments that are suitable for functional construction. It also includes areas that may require landscape enhancement or optimization. Importantly, land use suitability evaluation in landscape architecture must be grounded in prior ecological sensitivity assessment. Planning and development should be conducted with ecological protection as a fundamental principle, ensuring that construction activities align with the preservation of environmental integrity.

5.2 The Significance of Land Use Suitability Evaluation

Land designated for construction within natural environments arises from human production and living needs, serving various human activities. Due to the limited reversibility of land, construction activities inevitably alter and degrade the physical and biochemical properties of land with natural attributes. Once converted to built land, it becomes extremely difficult to restore the land's original ecological condition. For example, when agricultural land is developed, significant changes occur in its physicochemical properties—such as soil porosity, structure, compaction, pH, and organic matter—and its biological environment may also be disrupted. Restoring these conditions often requires a prolonged process and high costs. Therefore, from the perspective of landscape protection, designating construction land within scenic and landscape environments must be approached with caution.

From the perspective of human use, evaluating land use suitability based on a thorough investigation of the natural environment can help maximize construction benefits and promote the efficient use of land resources. Landscape planning often involves large, complex sites characterized by a mix of natural and cultural elements. Relying solely on subjective experience in such cases can lead to inconsistencies or oversights. Land use suitability analysis conducted on platforms such as ArcGIS enables quantitative comparisons based on site surveys and analyses, offering greater accuracy than traditional planning approaches. When integrated with prior ecological sensitivity assessments, such evaluations provide clear guidance for planners, helping delineate construction boundaries more precisely and identify areas suitable for development. This process also lays a foundation for subsequent detailed design stages. Although advances in construction technology have made most land technically buildable, the efficiency and sustainability of such development remain key concerns. Scientific evaluation allows multiple factors to be comprehensively considered, thereby helping to avoid ecologically sensitive areas, geologically unstable zones, and topographically complex terrain. Ultimately, this approach minimizes environmental damage, meets human needs, and supports compact, sustainable landscape design.

5.3 Construction of the Land Use Suitability Evaluation Model

The process of land use suitability evaluation closely parallels that of ecological sensitivity assessment, as both employ the Delphi method and the Analytic Hierarchy Process for analysis. Consequently, the structure of the two evaluation models is highly similar. The evaluation consists of four key stages: identification of evaluation factors, construction of the hierarchical evaluation model, determination of factor weights, and overlay analysis. This process ultimately yields a classification of land use suitability for the site, as illustrated in Fig. 5.1.

5.3.1 Identification of Evaluation Factors for Land Use Suitability

The selection of evaluation factors for land use suitability in landscape architecture is fundamentally based on construction feasibility. These factors are primarily categorized into three main types: spatial, ecological, and behavioral—maintaining consistency with the classification used in ecological sensitivity assessment. Each category is further divided into primary and secondary factors. For example, the primary spatial factor is topography, which includes the secondary factors of elevation, slope, and aspect.

In addition to listing all the relevant primary and secondary factors used in land use suitability evaluation for landscape architecture, Table 5.1 also presents the classification criteria for secondary indicators based on suitability levels. Depending on the specific site context and design objectives, the selection of both primary and secondary factors, as well as their grading standards, must remain adaptable and responsive to situational needs.

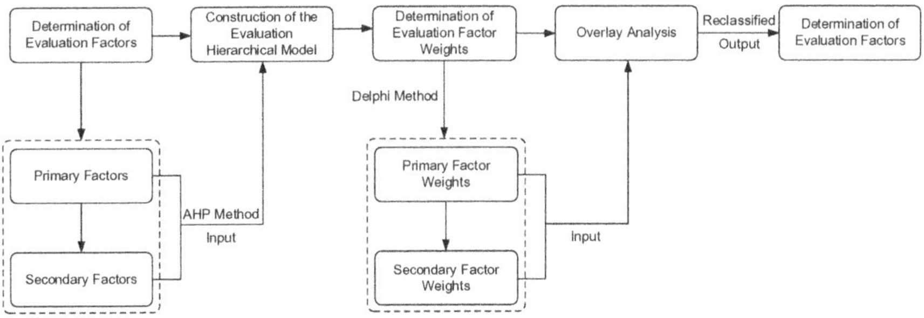

Fig. 5.1 Parametric evaluation model for land use suitability

Table 5.1 Classification of evaluation factors for land use suitability in landscape planning and design

Evaluation objectives	Indicator type	Primary indicator	Weight	Secondary indicator	Weight	Classification criteria		
						Highly suitable for construction	Moderately suitable for construction	Unsuitable for construction
Land use suitability evaluation for landscape architecture planning and design	Spatial	Topography	W_1	Elevation	w_{11}	< 50 m	50 ~ 100 m	> 100 m
				Slope	w_{12}	< 5%	5% ~ 15%	> 15%
				Aspect	w_{13}	South	East, west	North
	Ecological	Geology	W_2	Geological hazards	w_{21}	Geologically Hazard-free area	Low geological Hazard area	High geological Hazard area
				Soil bearing capacity	w_{22}	> 250 kPa	180 ~ 250 kPa	< 180 kPa
		Hydrology	W_3	Runoff area	w_{31}	No waterway area	Gully area	Flood discharge channel
	Behavioral	Transportation	W_4	Accessibility	w_{41}	Good	Fair	Poor
				Distance to surrounding roads	w_{42}	< 100 m to main roads, or < 200 m to secondary roads	100–200 m to main roads, or < 300 m to secondary roads	> 200 m to main roads, or > 300 m to secondary roads
		Current land use	W_5	Land use type	$W51$	Building land, paved areas	Roads, unused land	Farmland
		Urban planning setback	W_6	Planning setback	w_{61}	–	–	–

5.3.1.1 Spatial Factors

(1) Elevation

Landscape planning and design areas often encompass natural terrain such as mountains and water systems, resulting in significant topographic variation. Although modern construction technologies allow for development at higher elevations with minimal technical barriers, construction costs tend to increase with elevation. Therefore, areas with relatively high elevation within the planning region are generally less suitable for large-scale development. In land use suitability evaluations, relative elevation is classified and assigned values based on the specific context of the planning area. Higher elevation zones are given higher values, indicating lower suitability for construction, while lower elevation areas receive lower values, indicating higher construction suitability.

(2) Slope

From the perspective of construction economics, building in areas with steep slopes often requires extensive excavation and backfilling, thereby increasing construction complexity and costs. Furthermore, steep slopes are more prone to natural hazards such as landslides and erosion. From an environmental protection standpoint, development on steep slopes can severely impact native vegetation, soil structure, and bedrock stability, leading to increased soil erosion and ecological degradation. Therefore, in landscape environments, buildings should be sited with careful consideration of slope gradients. In general, slopes less than 3% are classified as flat, 3–10% as gentle, 10–25% as moderate, 25–50% as steep, and above 50% as very steep[1]. In the context of land use suitability evaluation, 10%, 25%, and 50% are often used as classification thresholds. The steeper the slope, the higher the assigned value, indicating lower suitability for construction (Table 5.2).

Table 5.2 Six-level slope classification by the International Geographical Union

Slope classification	Grade intervals	Detailed classification	Slope type name	Slope gradient	Vertical to horizontal ratio
0 ~ 2°	2°	0 ~ 2°	Gentle slope	0 ~ 3.5	> 28.6
2° ~ 5°	3°	2° ~ 5°	Slight slope	3.5 ~ 8.7	28.6 ~ 11.4
5° ~ 15°	10°	5° ~ 10°	Moderate slope	8.7 ~ 26.8	11.4 ~ 3.7
		10° ~ 15°	Regular slope		
15° ~ 35°	20°	15° ~ 25°	Steep slope	26.8 ~ 70	3.7 ~ 1.4
		25° ~ 35°	Very steep slope		
35° ~ 55°	20°	35° ~ 45°	Extremely steep slope	70 ~ 140	1.4 ~ 0.7
		45° ~ 55°	Precipitous slope		
> 55	–	–	Cliff	> 140	0.7 ~ 0

(3) Aspect

Most regions in China are located in the Northern Temperate Zone, where building orientation has historically been a crucial consideration. The traditional architectural principle of "facing south with the back to the north" is widely regarded as the optimal orientation. However, due to the vast geographical span of China, the optimal orientation varies slightly across different regions. In northern China, maximizing solar radiation is desirable; thus, an orientation between 5° to 15° west of due south is considered ideal. In contrast, in the non-heating areas of southern China, minimizing solar radiation is prioritized, making an orientation between 5° to 15° east of due south preferable. Generally, the best building orientations across most parts of China are due south, southeast, and southwest. During winter, the predominant wind directions are from the north, northeast, and northwest. Consequently, construction on slopes facing north, northeast, and northwest is generally less suitable. In the context of land use suitability evaluation for landscape planning and design, appropriate slope aspects should be determined based on the specific site conditions. The slope aspect of the site is then classified and assigned suitability values accordingly.

5.3.1.2 Ecological Factors

(1) Geology Geological Hazards

Geological hazards refer to destructive geological processes caused by natural or anthropogenic factors that pose risks to human life, property, and the environment. These include landslides, collapses, debris flows, ground fissures, soil erosion, earthquakes, volcanic activity, and geothermal anomalies. From the perspective of construction safety, buildings in landscape environments should avoid areas that are prone to such hazards.

Subsoil Bearing Capacity

The bearing capacity of subsoil determines its ability to support structural loads. While buildings in landscaped areas are typically low-rise or mid-rise and therefore have relatively low load-bearing requirements, it is still crucial to avoid areas with poor geological conditions—such as highly weathered rocks, silt, or marshland—to ensure structural safety and cost-effectiveness during construction.

(2) Hydrology

Hydrological factors also play a vital role in evaluating land suitability for construction within landscape planning. Water is a valuable scenic element in such environments, and buildings with proximity to water often benefit from enhanced experiential quality. However, during the rainy season or periods of flooding, excessive surface runoff or floodwaters may inundate or even destroy nearby structures. Therefore, it is essential to conduct scientific analyses of the hydrological conditions within the planning area, clearly identifying flood zones, surface runoff pathways, and catchment areas. These zones pose significant safety risks and should be excluded from potential development. The distance between proposed construction areas and these hydrologically sensitive regions should be categorized and assigned evaluation values accordingly.

5.3.1.3 Behavioral Factors

In the context of land suitability evaluation, behavioral factors refer to the imprints of human activity on the site. These include the current state of transportation, existing land use patterns, and planning setbacks required by urban regulations.

(1) Transportation

Accessibility can be broadly defined as the ease with which one can travel from one location to another. It is closely related to the convenience of transportation infrastructure. In landscape-based land suitability assessments, areas located closer to major roads are considered to have higher accessibility and are therefore more suitable for development. Moreover, roads of higher classification typically indicate better connectivity and are more conducive to construction activities.

(2) Current Land Use

Scenic areas often comprise a variety of land use types, which can generally be classified into two categories: Areas with significant human disturbance, such as construction land, orchards, and cultivated farmland; Areas that retain predominantly natural characteristics, such as forests, grasslands, and water bodies. Given that disturbed lands have already lost part of their ecological integrity, development should prioritize these areas to minimize further disruption to naturally sensitive regions. Special attention must be paid to land types with specific protections, such as prime farmland, water conservation forests, and shelterbelts, which are not suitable for development under any circumstances.

(3) Urban Planning Setbacks

Urban planning regulations often impose development setbacks that must be strictly observed. These include six primary setback zones: Road and building red lines, Greenbelt setback lines, Water body setback lines, Historic district and cultural heritage protection zones, Public infrastructure setbacks, and Safety setbacks from utilities such as high-voltage power lines, microwave towers, and railways. Even in landscape environments, these setbacks must be respected. Areas within the setback zones are deemed unsuitable for construction, and buffer zones around these lines should also be evaluated and categorized by level of suitability. For example, in the case of high-voltage transmission corridors, construction is strictly prohibited due to the associated electrical and electromagnetic hazards. Although specific design standards may not be established within natural landscapes, urban planning guidelines can be referenced. The closer a potential development site is to a high-voltage line, the less suitable it is for human activity due to health and safety concerns (Table 5.3).

Table 5.3 Corridor widths of high-voltage transmission lines

Transmission voltage level/kV	Height of high-voltage corridor/m	Transmission voltage level/kV	Width of high-voltage corridor/m
500	60 ~ 75	66、110	15 ~ 25
330	35 ~ 45	35	12 ~ 20
220	30 ~ 40		

5.3.2 Construction of the Land Use Suitability Evaluation Hierarchical Model

Similar to the ecological sensitivity evaluation model in landscape planning and design, the land use suitability evaluation model is constructed according to three levels: the target level, the element level, and the indicator level. The element level consists of three main categories: spatial, ecological, and behavioral, which are composed of six types of factors—topography, geology, hydrology, transportation, current land use, and urban planning setbacks. Each of these is further subdivided into specific factors at the indicator level (Fig. 5.2).

5.3.3 Determination of Weight for Land Use Suitability Evaluation Factors

The process of determining the weights of land use suitability evaluation factors is largely consistent with that of ecological sensitivity evaluation. The analytic hierarchy process (AHP) similarly adopts a 1–9 scale to express the relative importance between two compared factors. The weights of factors within the same category sum up to 1. For example, in Table 5.1, $W1+ W2+ W3+ W4+ W5+ W6 = 1$, and likewise $W11 + W12 + W13 = 1$. By constructing a judgment matrix through comparisons between primary and secondary indicators, the weight of each level can be obtained, reflecting the significance of each factor to land use suitability. This provides a foundation for the subsequent weight assignment and overlay analysis using the ArcGIS platform.

5.3.4 Land Use Suitability Classification

After the weights of each evaluation factor have been determined, the dimensionless base maps of each factor can be overlaid using weighted summation to produce the land use suitability classification map. The number of classification levels in the resulting map corresponds to the number of levels defined during the dimensionless processing of factors. Therefore, if a finer classification is required, more

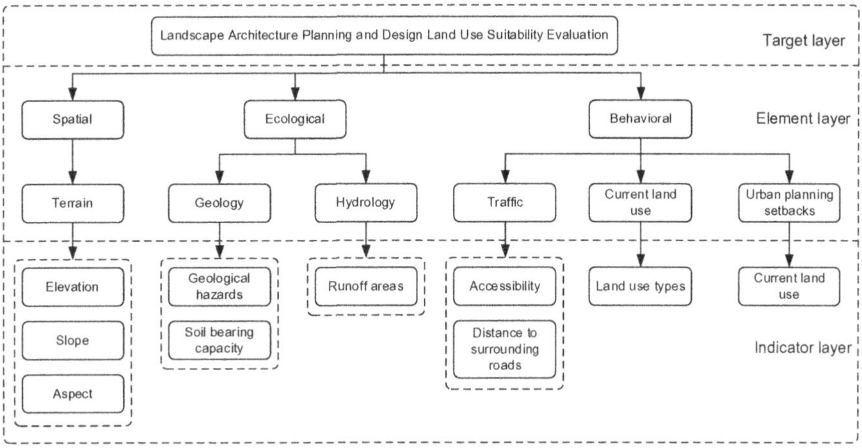

Fig. 5.2 Hierarchical model for land use suitability evaluation in landscape planning and design

levels—such as five, seven, or more—should be uniformly defined during the dimensionless step. Generally, dividing land use suitability into three levels is sufficient to distinguish between suitable construction areas, conditionally suitable areas, and areas unsuitable for development, which basically meets the needs of landscape design (Table 5.4).

5.4 Application of the Parametric Land Use Suitability Evaluation Model

As an essential component of parametric landscape architecture planning and design, land use suitability evaluation holds irreplaceable value. To maintain the coherence and integrity of the research and discussion, the author takes the northern area of the Niushou Mountain Scenic Area in Nanjing—previously mentioned—as an example to illustrate the application process of the land use suitability evaluation model.

5.4.1 Construction of the Land Use Suitability Evaluation Hierarchical Structure Model

5.4.1.1 Determination of Evaluation Factors

The land use suitability evaluation indicators for the northern area of the Niushou Mountain Scenic Area are generated based on Table 5.5, with slight adjustments made according to actual conditions. Table 5.5 presents the land use suitability

Table 5.4 Three-level classification of land use suitability

No.	Name	Level	Description
1	ClassIConstruction area	Suitable construction area	Geological conditions are relatively favorable, or the area is an existing developed zone; construction here will not cause significant ecological impact
2	Class II construction area	Suitable construction area	Construction conditions are generally acceptable, and the ecological environment is good; however, the types of construction functions are limited
3	Class III construction area	Unsuitable construction area	Construction conditions are poor, and the area is not suitable for human activities

evaluation factors and classification criteria for the planning and design of the northern area of the Niushou Mountain Scenic Area. The evaluation factors consist of three categories: spatial, ecological, and behavioral. Among them, the ecological factors include relatively few primary indicators, with only one—hydrology—corresponding to the secondary indicator of runoff area.

5.4.1.2 Construction of the Hierarchical Model

After the evaluation factors are determined, the hierarchical model is constructed in the Yaahp software, as shown in Fig. 5.3. Since all primary indicators except for topography correspond to only one secondary indicator, only the topography factor is divided into two levels of indicators. To meet the modeling requirements of the Yaahp software, two "alternative options" layers are temporarily added during the construction of the hierarchical model, with their weights both set to 0.5. This avoids calculation errors in the final results.

5.4.2 Determination of Weights for Factors at All Levels in Land Use Suitability Evaluation

Table 5.6 presents the weights of each level factor in this land use suitability evaluation, as well as the comparison of their relative importance. This evaluation still adopts the 1–9 scale to describe the degree of importance between factors. As can be seen from Table 5.6, among all the primary factors, the terrain factor and the land use type factor have a greater impact on land use suitability, with weights of 0.397 and 0.3565 respectively, together accounting for more than half of the total. In contrast, the buffer zone of high-voltage lines holds relatively low importance, followed by the runoff area factor and the distance to surrounding roads factor. Among the terrain factors, slope as a secondary factor has the greatest influence on land use

Table 5.5 Land use suitability evaluation factors for the northern area of Niushou Mountain scenic area

Evaluation objective	Primary indicators	Primary indicators	Weight	Secondary indicators	Weight	Classification criteria: Highly suitable for construction	Moderately suitable for construction	Unsuitable for construction
Land use suitability assessment for landscape architecture planning and design	Spatial	Topography	$W1$	Elevation	$^{W}11$	< 50 m	50 ~ 100 m	> 100 m
				Slope	$^{W}12$	< 5%	5% ~ 15%	> 15%
				Aspect	$^{W}13$	South	East, west	North
	Ecological	Hydrology	$W3$	Runoff area	$^{W}31$	> 50 m	10 ~ 50 m	< 10 m
	Behavioral	Transportation	$W4$	Distance to surrounding roads	$^{W}42$	< 200 m to main roads or < 100 m from secondary road	200–400 m from main roads or < 200 m from secondary road	400 m from main roads or > 200 m from secondary road
		Current land use	$W5$	Land use type	$^{W}51$	Building land, paved areas	Roads, unused land	Farmland
		Rban planning setback	$W6$	High-voltage line buffer zone	$^{W}61$	> 90 m	60 ~ 90 m	< 60 m

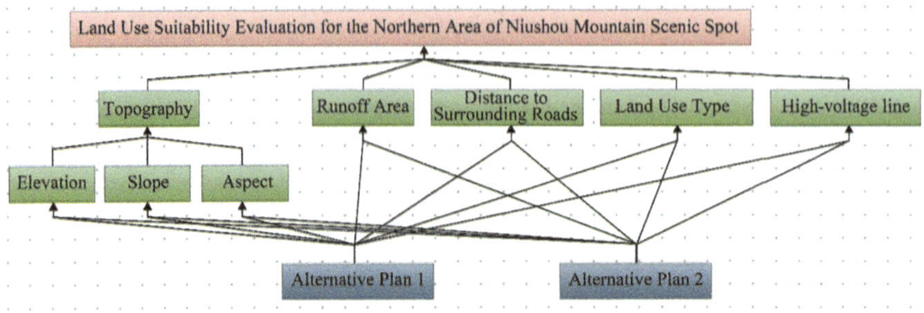

Fig. 5.3 Hierarchical structure model for land use suitability evaluation of the northern area of Niushou Mountain scenic area (Yaahp software)

Table 5.6 Land use suitability evaluation for the northern area of Niushou Mountain scenic area—Judgment Matrix and Factor Weight Table (Yaahp software)

1. Land use suitability evaluation(consistency ratio:0.0318；λmax:5.1426)						
Land use suitability evaluation	Topography	Runoff area	Distance to nearby roads	Land use type	High voltage line buffer zone	W_i
Topography	1	7	4	1	5	0.397
Runoff area	0.1429	1	1	0.3333	0.5	0.0744
Distance to nearby roads	0.25	1	1	0.2	1	0.0809
Land use type	1	3	5	1	5	0.3565
High voltage line buffer zone	0.2	2	1	0.2	1	0.0913
2. Topography(consistency ratio:0.1022；weight:0.3970；λmax:3.1063)						
Topography	Elevation	Slope	Aspect	W_i		
Elevation	1	0.1839	0.3903	0.1007		
Slope	5.438	1	5.562	0.7125		
Aspect	2.562	0.1798	1	0.1868		

suitability, accounting for 71.25% of the total weight; the slope aspect and elevation factors are relatively close, at 0.1868 and 0.1007.

5.4.3 Generation of Land Use Suitability Zoning

5.4.3.1 Analysis of Secondary Factors of Land Use Suitability

Based on the survey data and the land use suitability evaluation factor table, the basic factor raster maps were generated in ArcGIS software. In Fig. 5.4, (a) to (h) represent the raster maps of elevation, slope, aspect, runoff buffer zone, external road distance, internal road distance, current land use, and high-voltage line buffer

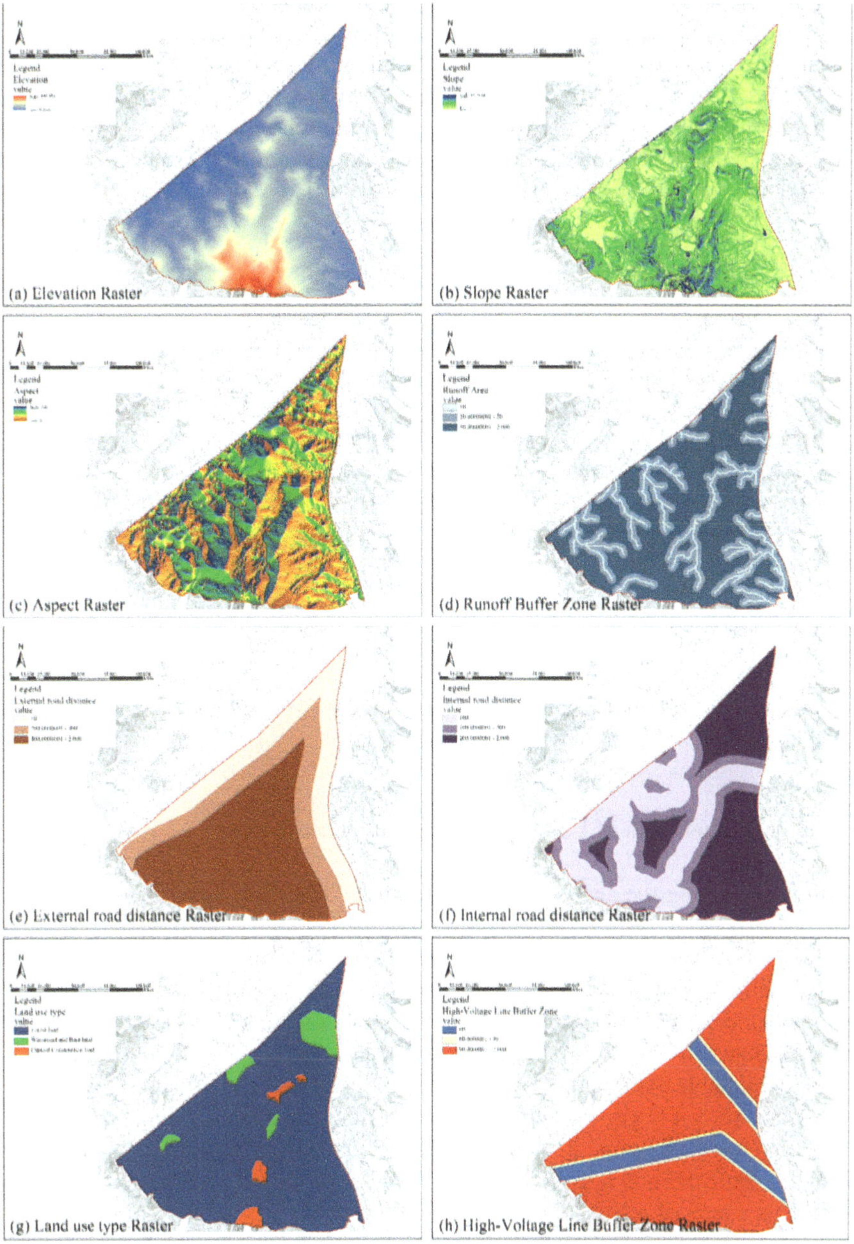

Fig. 5.4 Primary factor analysis maps for land use suitability evaluation in the northern area of Niushou Mountain scenic area

zone, respectively. When creating the runoff buffer zone, the threshold of runoff should be adjusted according to the actual evaluation needs. The land use suitability evaluation does not require a dense runoff network; identifying the buffer zones of primary runoff routes is sufficient. An overly dense runoff network may instead lead to evaluation errors. Therefore, after comparison, a runoff network with a threshold of 8000 was selected for this evaluation.

5.4.3.2 Dimensionless Processing of Land Use Suitability Factors

The dimensionless processing of factors is a key step and foundational work before factor overlaying. This land use suitability evaluation adopted a three-level scale to classify each factor and assign corresponding values, resulting in the reclassified maps of the eight factors shown in Fig. 5.5.

5.4.3.3 Analysis of Primary Factors of Land Use Suitability

Before performing the overlay analysis of primary factors, it is first necessary to generate the analysis maps of the primary factors that have secondary indicators. In this evaluation, the primary factors to be overlaid are topography and road distance, which are derived from the overlays of elevation, slope, and aspect, as well as internal and external road distances (Fig. 5.6).

5.4.3.4 Generation of Land Use Suitability Classification

After completing the analysis of all primary factors, each factor can be weighted and overlaid to ultimately generate the land use suitability classification, as shown in Fig. 5.7. In the figure, the unsuitable construction areas, buildable areas, and highly suitable construction areas can be clearly distinguished based on color differences. Among them, the highly suitable construction areas account for the smallest area, followed by the unsuitable construction areas, with the remaining areas classified as buildable. After obtaining the land use suitability classification map, project siting cannot yet begin directly. As a component of the parametric landscape planning and design process, it is closely linked to the next stage and also jointly contributes with the previous stage to the following one, reflecting the linkage and dynamic nature of parametric planning and design.

The land use suitability zoning map is generated from the perspective of construction, classifying the site based on its suitability for development. This classification is goal-specific and does not incorporate the ecological considerations from the previous stage. Therefore, the generation of buildable areas requires the exclusion of ecologically highly sensitive zones, special factors, water bodies, and other areas unsuitable for development. Two 110 kV high-voltage lines pass through the northern area of the Niushou Mountain Scenic Area. According to the Code for

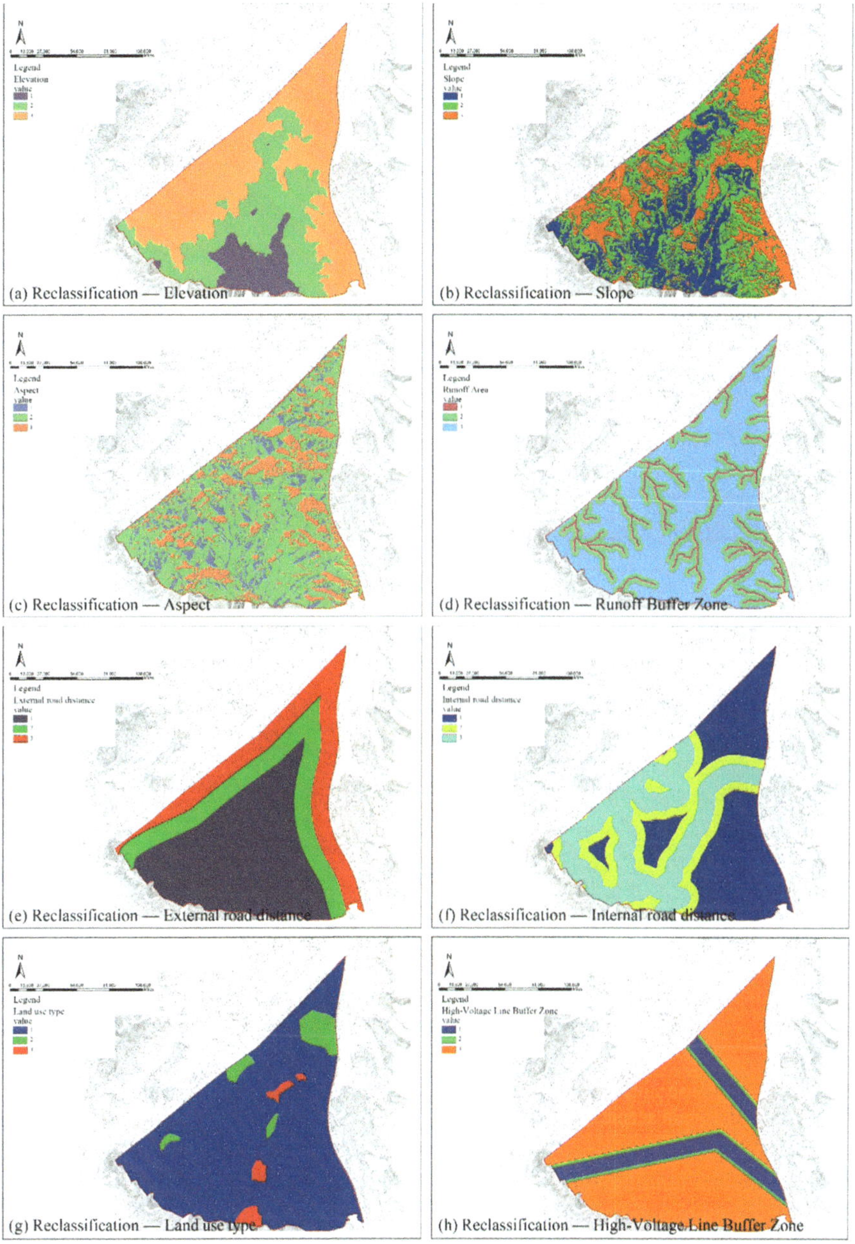

Fig. 5.5 Reclassification maps of secondary factors for land use suitability evaluation in the northern area of Niushou Mountain scenic area

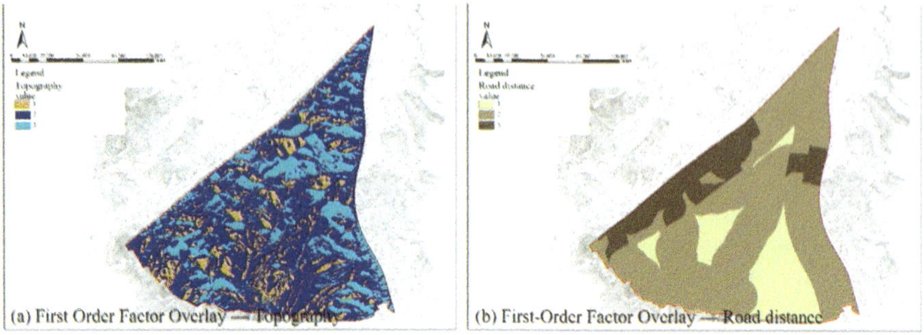

Fig. 5.6 Primary factor analysis maps for land use suitability evaluation in the northern area of Niushou Mountain scenic area

Fig. 5.7 Land use suitability zoning based on land use suitability evaluation in the northern area of Niushou Mountain scenic area

Urban Power Planning (GB/T 50293–2014), the corridor width for high-voltage lines is 25 meters. Since areas closer to high-voltage lines are less suitable for human activities, a buffer zone for the high-voltage lines needs to be established in the evaluation. Figure 5.8 presents the unsuitable construction areas to be excluded in this land use suitability evaluation, and finally, the buildable range of the northern Niushou Mountain Scenic Area is obtained (Fig. 5.9).

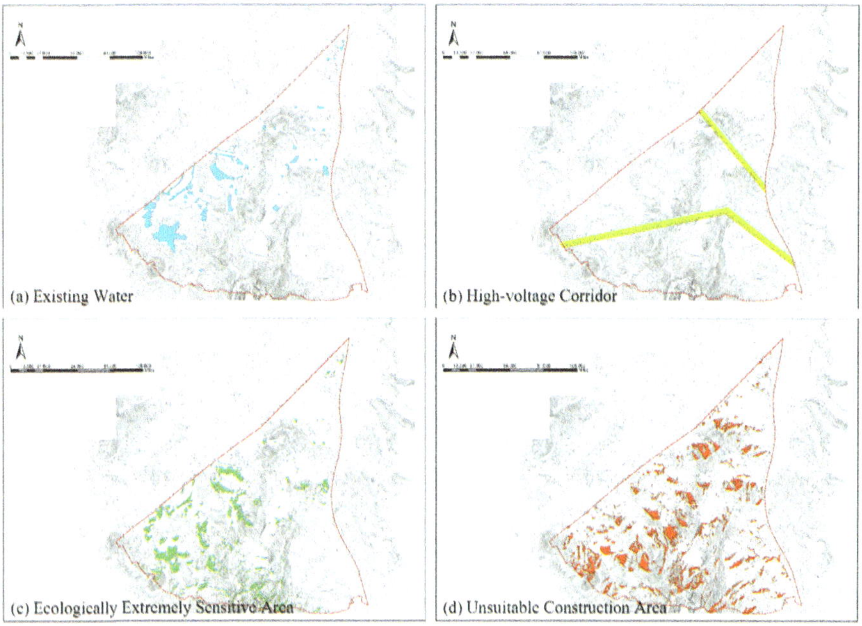

Fig. 5.8 Excluded areas

Fig. 5.9 Land use suitability evaluation in the northern area of Niushou Mountain scenic area—buildable areas

Chapter 6
Project Location Selection Model

In landscape architecture planning and design, "project" refers to "scenic spots," which are places where visitors can rest and play. The site selection of landscape architecture projects differs from general project planning and site selection in terms of content and process. This chapter first clarifies the connotation of site selection for landscape architecture projects and then points out the value and significance of project location. The positioning of the project involves establishing both the content and the location of the project. Furthermore, the study of project location includes both qualitative and quantitative aspects. Therefore, this chapter constructs a project location model based on project suitability evaluation and site selection parts and discusses the application of the model with examples.

6.1 The Value and Significance of Project Location

Based on the research and analysis of site conditions, the second step in the landscape architecture planning and design process based on the coupling principle is to couple the various design elements with the site. This process corresponds to the establishment of planning zones and is the preliminary formation of the overall plan. The coupling of elements and the site involves establishing a corresponding relationship between specific projects identified in the previous coupling process and the specific site, which is project location. Project location has two layers of meaning: one is the establishment of specific content, which corresponds to project planning, and the other is the determination of the project location, which is site selection.

The establishment of landscape architecture project content is a planning process that differs from contemporary tourism planning. It requires market consideration and site research. Planning must be forward-looking and base predictions on the potential recreational content that may occur in the site, identifying causal relationships to support and justify decision-making. Landscape architecture project

Y. Yuan, *Parametric Landscape Architecture Planning and Design*,
https://doi.org/10.1007/978-981-95-8011-8_6

planning is not a simple or "parachuted" concept; it needs to find the basis for the project's existence within the site by researching the coupling relationship between the project and the environment, ultimately establishing the project's content and positioning. The establishment of projects within the landscape environment needs to consider upper-level planning, government policies, tourism factors including market demand, visitor numbers, and target groups. Furthermore, the cultural connotation within the site is an important basis for establishing project content. Landscape architecture projects are neither without roots nor without sources; projects grounded in the site are more rational and sustainable. By classifying selected projects based on function and positioning, combined with qualitative and quantitative analysis from thorough site research on aspects such as traffic, natural resources, etc., planning zones can be determined, achieving coupling between the project and specific areas.

"Site selection" is a crucial part of landscape architecture planning and design, corresponding to the further refinement after the preliminary project planning. Landscape architecture planning and design seek the corresponding relationship with the site, constructing "project and site coupling," which means placing the appropriate project in the suitable area within the site under the guidance of the coupling principle, establishing mutual suitability between the project and the site. This coupling process must be based on the requirements for project site selection and the analysis and evaluation of the site. Traditional landscape architecture planning and design methods mainly involve on-site surveys and data collection to classify and analyze landscape resources within the design site, thus establishing site cognition. Given the varying scales in landscape architecture planning and design, traditional cognition processes struggle to systematically and comprehensively understand single-factor conditions like topography, hydrology, climate, flora, and fauna, thereby facing substantial difficulty in performing comprehensive, objective analyses on ecological sensitivity, land use suitability, and visibility. These conditions often rely on the designer's experience for judgment, leading to vague, subjective, and arbitrary site selection, which hinders achieving minimal site interference, optimal resource utilization, and design effects. The development of Geographic Information Systems (GIS) enables scientific analysis, evaluation, and management of site information, making site cognition in landscape architecture planning and design more scientific, rational, accurate, and complete.

6.2 Construction of the Project Location Model

The characteristics of project quality, site selection, and scale should be subordinate to the site. Projects that arise naturally from the site possess characteristics of harmony and symbiosis. Designs guided by "parachuted" concepts, which are not restricted by the site and detached from its context, deviate from the inherent value of landscape architecture. The term "suitability" refers to appropriateness and compatibility, representing a reciprocal process where the project and site influence each other to form a union and mutual suitability. Therefore, "mutual suitability" corresponds to "coupling," indicating not only the project's appropriateness for the site but also the site's "guidance" for the project. The project location selection model in landscape architecture planning and design includes two sub-models: the Project Suitability Evaluation Model and the Project Site Selection Model. Both sub-models are constructed based on the coupling of the project and site, representing a succession relationship. Firstly, the Project Location Selection Model establishes suitable project content for the site through the analysis and study of existing site resources, comprehensively considering market, functional, and characteristic needs, embodying the "site-project" coupling process. Secondly, based on the determined project content, the Project Site Selection Model seeks areas suitable for the project within the site based on specific project construction requirements, embodying the "project-site" coupling process (Fig. 6.1).

6.3 Sub-model One: Project Suitability Evaluation Model

The Project Suitability Evaluation Model includes three stages: Research and Analysis, Project Location, and Project Selection. Firstly, in the Research and Analysis stage, aspects affecting project development are analyzed from the

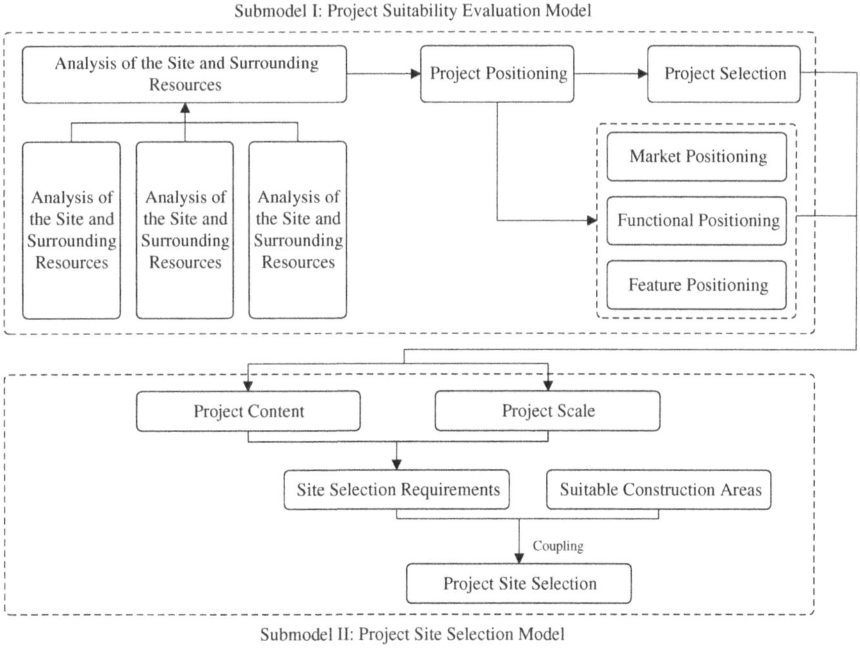

Fig. 6.1 Landscape architecture planning and design project positioning model

following layers: use by visitors, management through policies, and content via site conditions. For example, research on the visitor layer includes analyzing surrounding tourism resources and the tourist market base, projecting market demand, target audience, and expected visitor numbers, thus forming predictions for current and potential markets, i.e., market positioning in the second stage of project suitability evaluation. Based on research and analysis, it is possible to roughly estimate the market situation, functional requirements, and unique characteristics the design site will face post-construction. According to the project's market positioning, functional positioning, and characteristic positioning, various projects suitable for specific sites can be developed. Among these, projects with content directly originating from the site are considered primary projects, having the closest relation with the site; extended projects follow, having a certain degree of association with the site content; whereas derivative projects have the weakest link with the site (Fig. 6.2).

6.3.1 Project Location Research

6.3.1.1 Market Location

(1) Analysis of Surrounding Tourism Resources
Landscape architectural design sites, due to their proximity to already constructed landscape environments, share internal natural and cultural resources. The site form and climatic conditions are often very similar, resulting in a high possibility of project repetition. To prevent resource waste and market crowding caused by repetitive projects, it is necessary to analyze the tourism resources around the site. Besides, certain negative functional sites, such as cemeteries, chemical plants, etc., can affect project location negatively, hindering tourism activities. Depending on the project situation, the scope of the research should

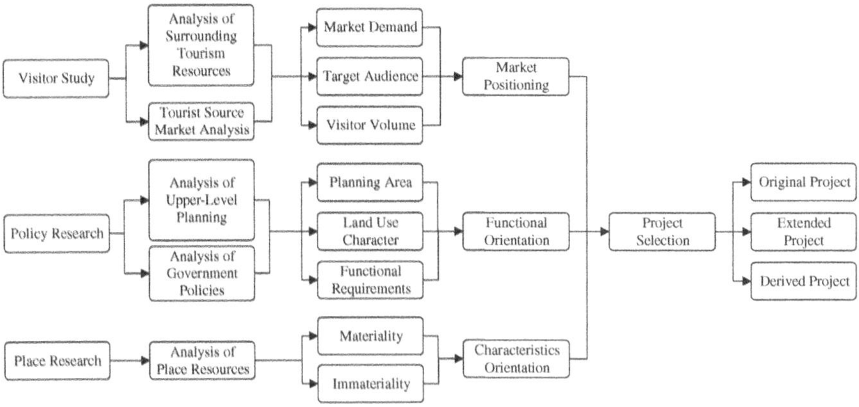

Fig. 6.2 Scenic and landscape architecture planning and design project suitability evaluation model

be adjusted accordingly, possibly expanding the focus to the surroundings of the site, the city, or even the region.

(2) Tourist Source Market Analysis

Given the service characteristics of landscape environments, tourists and visitor numbers—who are the main participants in activities—are essential factors in design. Besides economic relevance, controlling environmental capacity is crucial. The "Overall Planning Standard for Scenic Areas" (GB/T 50298–2018) specifies that tourist source analysis and visitor development scale forecasts should analyze the number and structure, temporal and spatial distribution, travel patterns, consumption conditions, as well as the development direction and goals of the tourist source market [1] (Fig. 6.3).

6.3.1.2 Functional Location

(1) Analysis of Upper-Level Planning

Upper-level planning reflects the government's development strategy and requirements, representing the regional overall and long-term interests while holding legal validity. It is a necessary reference and execution document for any lower-tier planning, including landscape architectural design. For landscape architectural planning, the analysis and research of upper-level planning help to clarify the site's locational characteristics, relationships with surround-

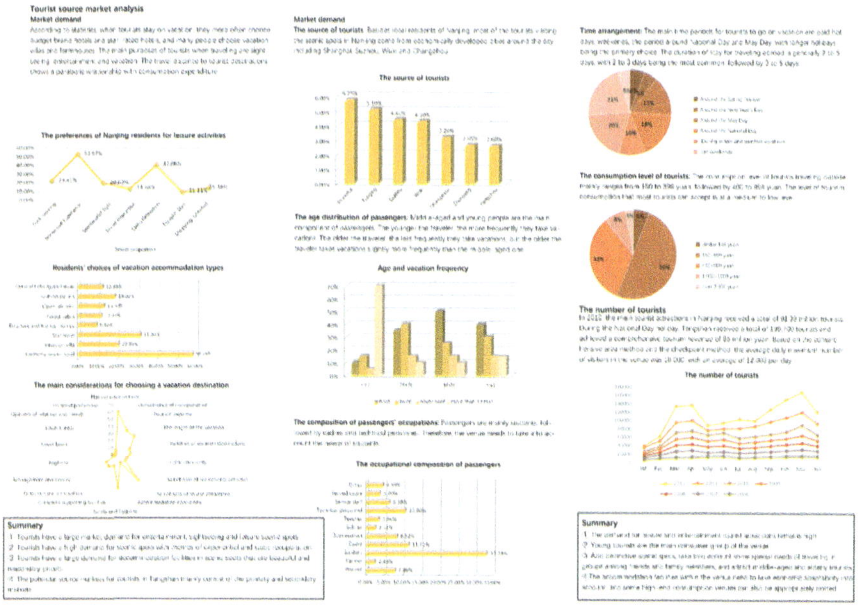

Fig. 6.3 Tourist source analysis (planning and design of Nanjing Ziqing Lake ecological tourist resort) (quoted from: Cheng Yuning's studio materials)

ing areas, traffic conditions, land use nature, and its functions within the city or region, which are prerequisites and essential conditions for landscape architectural design work.

(2) Relevant Policy Analysis
Relevant policies refer to government-guided policies related to the planning scope, such as policies and regulations promoting ecological civilization construction or guiding the tourism industry. These policies reflect the government's decision-making and development intentions, significantly guiding planning and design and influencing project location. Therefore, analyzing relevant policies is one of the key focuses in the early stage of analysis and research.

6.3.1.3 Characteristic Location

(1) Analysis of Physical Resources
Physical resources refer to distinct material entities within the site, including traces of natural movements and human activities, such as terrain and landform characteristics, plant and animal resources, and cultural relics. Relying on their material existence, physical resources can be directly converted into leisure tourism resources, forming an essential part of the site's characteristics.

(2) Analysis of Non-material Resources
Non-material resources do not have directional material carriers and rely on material carriers to be perceived by people, mainly including the accumulation of human culture, such as stories, legends, folk customs, and traditions. These resources, often unique to the site, are crystallizations of human wisdom and possess distinct characteristics. Therefore, interpreting and expressing non-material resources is an effective way to form site characteristics. For example, in the northern region of the Nanjing Niushoushan Scenic Area, the natural landscape feature "Niushou Smoke Mist" has historically been a distinctive feature of the area, integrated with Zen Buddhist cultural landscapes represented by Niutou Chan. The design tightly revolves around these non-material resources to plan leisure tourism projects, creating a scenic tourism resort combining natural and cultural landscapes.

6.3.2 Analysis of Project and Site Visitor Capacity

The "Regulations on Scenic Areas" (2006) stipulate that the overall plan for scenic areas must measure the visitor capacity of the scenic area [2]. The "Overall Planning Standard for Scenic Areas" (GB/T 50298–2018) points out that tourism service facilities should be determined based on the nature and function of the scenic area, the scale and structure of visitors, as well as land, freshwater, and environmental conditions. The standard also provides a formula for calculating ecological environment visitor capacity: Ecological environment visitor capacity = Total area × Unit scale index (Table 6.1). Using this formula, a rough estimate of the visitor capacity for the landscape environment can be made. It is important to note that the total area refers to land area available for recreation, corresponding to the constructible area derived from ecological sensitivity evaluation and land use suitability evaluation. Estimating site visitor capacity has significant implications for determining project scale. After determining the project scale, the visitor capacity of each project should be coupled inversely with the visitor capacity through adjustment and optimization to ensure basic coherence. Hence, estimating site visitor capacity plays a guiding role in the establishment and site selection of projects throughout the entire site.

6.3.3 Selection of Suitable Projects

6.3.3.1 Principles of Selection

Based on preliminary research and analysis, three categories of projects suitable for the site have been developed: original projects, extended projects, and derivative projects. However, not all projects can be implemented at the site; therefore, it is necessary to screen the initially determined projects. Since these preliminary projects are derived from qualitative judgments and analyses, this screening primarily considers the content of the projects to determine their suitability for the site. The screening principles are as follows:

1. Positivity
   Positivity refers to the positive significance of the project content, such as alignment with sustainable concepts, compliance with laws and regulations, and meeting the needs of ecological construction. The "Regulations on Scenic Areas" (2006) and the "General Planning Standards for Scenic Areas" (GB/T

Table 6.1 Ecological carrying capacity of recreational land use

Land use type	Permitted capacity and Land Use Index persons/hm^2	m^2 /persons
Coniferous Forest land	2 ~ 3	5000 ~ 3300
Broadleaf Forest land	4 ~ 8	2500 ~ 1250
Forest Park	< 15 ~ 20	> 660 ~ 500
Open woodland grassland	20 ~ 25	500 ~ 400
Meadow Park	< 70	> 140
Urban Park	30 ~ 200	330 ~ 50
Designated bathing area	< 500	> 20
Bathing water area	1000 ~ 2000	20 ~ 10
Bathing Beach	1000 ~ 2000	10 ~ 5

Source: Overall planning standard for scenic areas (GB/T 50298–2018)

50298–2018) list suitable projects, which have already been presented in Chap. 3, thus will not be repeated here. Project selection for landscape environments can refer to these two regulations.

2. Thematic Nature
   Thematic projects can revolve around the site theme, enhance the site's atmosphere, and help form site characteristics. Projects within the site should not exist in isolation but should be interconnected. The site theme is the key to closely linking the projects. Setting up projects in this manner makes the site more cohesive and distinct.
3. Originality
   Originality describes projects that directly originate from the site, generally referring to original projects. These projects' uniqueness and site-specific characteristics enable a great degree of coupling with the site. As the projects that can most effectively reflect the site's characteristics, original projects should occupy a major portion of the site.

6.3.3.2 Project Location

The positioning of projects is based on investigation and analysis and involves not only planning the content of the projects but also making a preliminary function assumption for them. Table 6.2 reflects the final outcome of the project suitability evaluation model, showcasing various aspects such as project types, project numbers, project names, project location, construction indicators, construction styles, and functional area allocation, comprising seven major items and eleven sub-items, detailing the projects within the site. The project type is a prototype for the general layout functional zoning, with similar types and functions of projects belonging to the same project type. Project location reflects the function the project is expected to undertake within the site. Based on investigation and research, the scale of the project is preliminarily judged, construction indicators are proposed, divided into four quantitative and qualitative indicators: building area, land area, number of building floors, and building form. The construction indicators are directly related to the site selection phase, proposing requirements for site selection. Functional area allocation lists the functional area distribution of the construction project, allowing for the calculation of the project's capacity based on area usage, which is a component of the overall capacity of projects within the site.

The "Project Location and Function Assumption Table" (Table 6.2) presents the preliminary research work in qualitative and quantitative forms, representing a phase outcome of the landscape architecture planning and design process based on the coupling principle. It serves as both a summary and expression of the previous phase of research and a premise for the next stage of work. As a step in the systematic process, it connects the preceding and following stages. According to the "General Planning Standards for Scenic Areas" (GB/T 50298–2018), in architectural landscape planning, all valuable existing buildings and their environments should be preserved. Various new buildings should conform to the overall needs of

Table 6.2 Project positioning and functional forecast table

Item type	ID	Name	Project positioning		Construction indicators				Architectural style	Functional zoning allocation
			Function	Target audience	Building area (m^2)	Land area (m^2)	Number of floors	Building form		
Zen retreat zone	A-1	Jingyi Mountain lodge	Health and Zen practice	People seeking high-quality experiences and inner peace in modern life	600	1500	1	Courtyard-style	Modern architecture with Zen elements	Dining: 100 m^2 Lodging: 300 m^2 Public area: 200 m^2
	A-2	Vegetarian dining hall	Dining	All demographics	1500	1800	2	Detached buildings	Ming-style imitation architecture	Dining: 700 m^2 Logistics: 700 m^2 Public area: 600 m^2 Administration: 100 m^2
	A-3	Heart-cleansing Pool	Outdoor yoga	Business professionals seeking stress relief	–	700	–	–	–	–
Wellness zone	B-1	Mind-nourishing Bath	Herbal Bath & Massage	Moderately sub-healthy individuals	1300	2000	1–2	Modular combination	Modern timber architecture	Hot spring Pool: 400 m^2 Massage rooms: 300 m^2 Rest area: 250 m^2 Supporting facilities: 350 m^2
	B-2	Blossom workshop	SPA	Urban women	500	700	1	Detached buildings	Modern timber architecture	SPA: 200 m^2 Massage area: 150 m^2 Rest area: 50 m^2 Public area: 100 m^2

(continued)

Table 6.2 (continued)

Item type	ID	Name	Project positioning		Construction indicators				Architectural style	Functional zoning allocation
			Function	Target audience	Building area (m^2)	Land area (m^2)	Number of floors	Building form		
Happy zone	C-1	Joyful garden	Recreational activities	Young people and children	5000	8000	3	Detached buildings	Modern architecture	Recreational area: 2000 m^2 Rest area: 300 m^2 Cinema: 600 m^2 Light dining: 500 m^2 Logistics: 400 m^2 Public area: 1000 m^2 Administration: 200 m^2
	C-2	Fragrant tea house	Tea tasting & DIY tea blending	All demographics	700	3000	1–2	Courtyard-style	Modern architecture with Zen elements	Tea art exhibition: 100 m^2 Tea tasting area: 300 m^2 Tea processing workshop: 200 m^2 Public area: 100 m^2

the landscape environment, with site selection and foundation suitability aligning with and utilizing existing terrain. There should be clear zoning and grading control measures for the nature and function, content and scale, standards and grades, location and height, volume and form, color and style of various buildings [3]. The determination of construction indicators requires a comprehensive consideration of zoning control requirements and judgments based on integrated coordination.

6.4 Sub-model II: Project Site Selection Model

The Project Site Selection Model is one of the sub-models of the Project Location Model and is closely linked to the Project Suitability Evaluation Model. This model uses the quantitative conditions derived from running the Project Suitability Evaluation Model as part of its parameters, coupling them with the site, ultimately determining the specific location for the project.

6.4.1 Building the Project Site Selection Model

The core of project site selection lies in establishing the coupling association between site characteristics and project design key points to filter out the site selection area. The mutual suitability of project site selection requirements and site conditions is reflected in the coupling process. Besides the construction indicators provided during the Project Suitability Evaluation process, project site selection requirements include specific criteria for relative elevation, slope, aspect, geomorphological resources, transportation and other supporting resources, as well as water bodies, vegetation, and other environmental resources. Table 6.3 lists the typical project site selection requirements, divided into three types and sixteen requirements.

(1) Spatial Requirements
Spatial requirements include relative elevation, slope, and aspect, which are the project's demands for the spatial position within the site. For instance, regarding slope, generally, construction activities are not suitable in regions with significant terrain fluctuations and steep slopes. However, certain special scenic park buildings may need to be constructed on steep cliffs, and the project can therefore limit the site selection range by specifying slope requirements.
(2) Functional Requirements
Functional requirements cover four aspects: land area, building area, building height, and transportation, reflecting the project's requirements for the construction plot. Among them, land area and building area were already derived in the previous research stage. Building height applies to building height-restricted areas within the scenic environment. For example, if the height limit for a cer-

Table 6.3 Statistical table of project site selection requirements

			Spatial requirements			Functional requirements				Special requirements								
										Water resources			Vegetation resources		Cultural resources			
Item type	ID	Name	Elevation (m)	Slope (%)	Aspect	Building area (m^2)	Land area (m^2)	Building height (m)	Transportation	Type	Area (m^2)	Distance (m)	Type	Vegetation Density	Type	Distance (m)	Privacy	View corridor
Zen retreat zone	A-1	Jingyi Mountain lodge	20 ~ 50	10 ~ 30	South, southeast	600	1500	< 6	Distance from general roads: Greater than 200 m	Ordinary Lake or pond	> 500	< 10	Bamboo Forest	Average	–	–	Relatively good	Good
	A-2	Vegetarian dining hall	< 10	< 5	–	1500	1800	< 9	Distance from urban arterial roads: Less than 200 m	–	–	–	–	–	–	–	Average	Average
	A-3	Heart-cleansing Pool	50~ 100	< 5	South, southeast	–	700	–	Distance from general roads: Greater than 200 m	Ordinary Lake or pond	500-1000	< 50	Mixed Forest	Relatively good	–	–	Good	Good
Wellness zone	B-1	Mind-nourishing Bath	< 50	< 10	–	1300	2000	< 6	Distance from general roads: Greater than 100 m	Hot spring	–	< 10	Mixed Forest	Average	Hot spring culture	< 100	Good	Average
	B-2	Blossom workshop	–	< 5	–	500	700	< 6	Distance from urban arterial roads: Less than 100 m	Hot spring	–	< 11	–	–	–	–	Good	Average
Happy zone	C-1	Joyful garden	< 10	< 6	–	5000	8000	< 12	Distance from urban arterial roads: Less than 50 m	–	–	–	–	–	–	–	Average	Average
	C-2	Fragrant tea house	< 50	< 10	South, south east	700	3000	< 6	Distance from general roads: Less than 100 m	–	–	–	Tea plantation	Good	Tea culture	< 50	Average	Relatively good

tain area is 6 meters, but a project intended for viewing purposes is expected to be 18 meters high, the area with a 6-meter height limit will be excluded from the selectable plots.

(3) Special Requirements
Special requirements reflect the characteristics of scenic park construction projects and are subdivided into water resource, vegetation resource, cultural resource, visual domain, and privacy. In scenic environments, waterfront buildings are common types that require proximity to water bodies for construction, thus generating waterfront distance site selection requirements.

The aforementioned requirements serve as parameters within the site selection model. By adjusting these parameters according to different project characteristics, different site selection results can be obtained. The project site selection model relies on the ArcGIS software platform, with site selection requirements used as parameter inputs, and all site selection operations completed within this platform.

The buildable areas derived from Ecological Sensitivity Evaluation and Land Use Suitability Evaluation serve as the basic regions for site selection operations. Table 6.3 provides detailed project site selection requirements. The running of the project site selection model filters out preliminary site selection areas that meet special project requirements, based on these requirements, using the buildable areas as the foundation and progressively deducting restrictive areas. It should be noted that site selection requirements are restrictive conditions; the more conditions added, the fewer corresponding plots available. Therefore, there may be instances where initial site selection fails to meet the requirements. If the plot area obtained from preliminary site selection does not satisfy the project's site selection requirements, feedback will be given to the Project Suitability Evaluation Model for site selection condition optimization, and the site selection will be reprocessed until compliant plots are obtained. Projects within landscape planning and design require overall consistency, thus necessitating coordination between various projects within the site and unified optimization of site selection, eventually deriving a preliminary general planning map and corresponding functional zoning (Fig. 6.4).

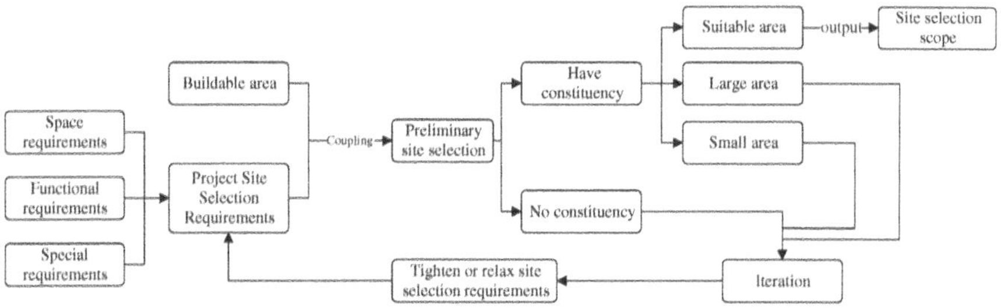

Fig. 6.4 Parameterized project site selection model of landscape architecture planning and design

6.4.2 *Application of the Project Site Selection Model*

The project site selection model is highly operable and practical. The author takes the northern area of the Niushou Mountain Scenic Area in Nanjing as an example and selects the "Jingyi Shan Room" construction project for a case study to discuss the application of the project site selection model.

6.4.2.1 Defining Site Selection Criteria

The case study "Jingyi Mountain Retreat" is situated within a meditation and wellness zone, primarily serving functions related to health preservation and Zen practice. It targets a modern demographic seeking high-quality experiences and spiritual tranquility. Through the project suitability analysis, preliminary estimates for both building and land area requirements have been established. Based on this, further refinement of site selection criteria is necessary to provide more detailed spatial constraints for siting decisions. Table 6.4 presents the preliminary site selection criteria for the "Jingyi Mountain Retreat." First, the site's elevation should not be too low in order to align with the spatial identity of a "mountain retreat." The slope gradient requirements may be moderately relaxed; however, since the project is intended for meditative use, a southern or southwestern slope aspect is preferred. Second, given that the project does not prioritize views or sightseeing, and that the built structures should not dominate the landscape, the maximum allowable building height is limited to 6 m. Third, the retreat primarily serves urban dwellers seeking stress relief and inner peace, necessitating a tranquil environment. As such, no stringent requirements are placed on accessibility; a distance of more than 200 m from secondary roads and over 500 from urban arterial roads is considered acceptable. Finally, the presence of open water surfaces can significantly enhance the meditative atmosphere, thus a moderate requirement for nearby water resources is specified.

6.4.2.2 Preliminary Site Identification and Feedback

Based on the preliminary site selection criteria for the "Jingyi Mountain Retreat," base maps were prepared using ArcGIS software, followed by parcel screening according to each individual criterion. Figure 6.5 comprises five thematic maps corresponding to elevation, slope gradient, slope aspect, proximity to water bodies, and transportation access—each illustrating the spatial extent of areas that meet the respective criteria within the site.

By overlaying these five thematic maps with equal weighting, it is possible to identify the parcels that simultaneously fulfill all five site selection criteria. As shown in Fig. 6.6, only a very limited number of areas within the site fully meet the initial requirements for the "Jingyi Mountain Retreat." Table 6.5 presents the area statistics for the resulting selected parcels. A total of 17 parcels were identified in

Table 6.4 Preliminary site selection requirements for "Jingyi Mountain Lodge"

Item type	ID	Name	Spatial requirements			Functional requirements				Special requirements		
										Water Resources		
			Elevation (m)	Slope (%)	Aspect	Building area (m^2)	Land area (m^2)	Building height (m)	Transportation	Type	Area (m^2)	Distance (m)
Zen practice	A-1	Jingyi Mountain lodge	20–50	10–30	South, southwest	600	1500	< 6	Distance from general roads: Greater than 200 mdistance from urban arterial roads: Greater than 500 m	Ordinary Lake or pond	> 1000	< 50

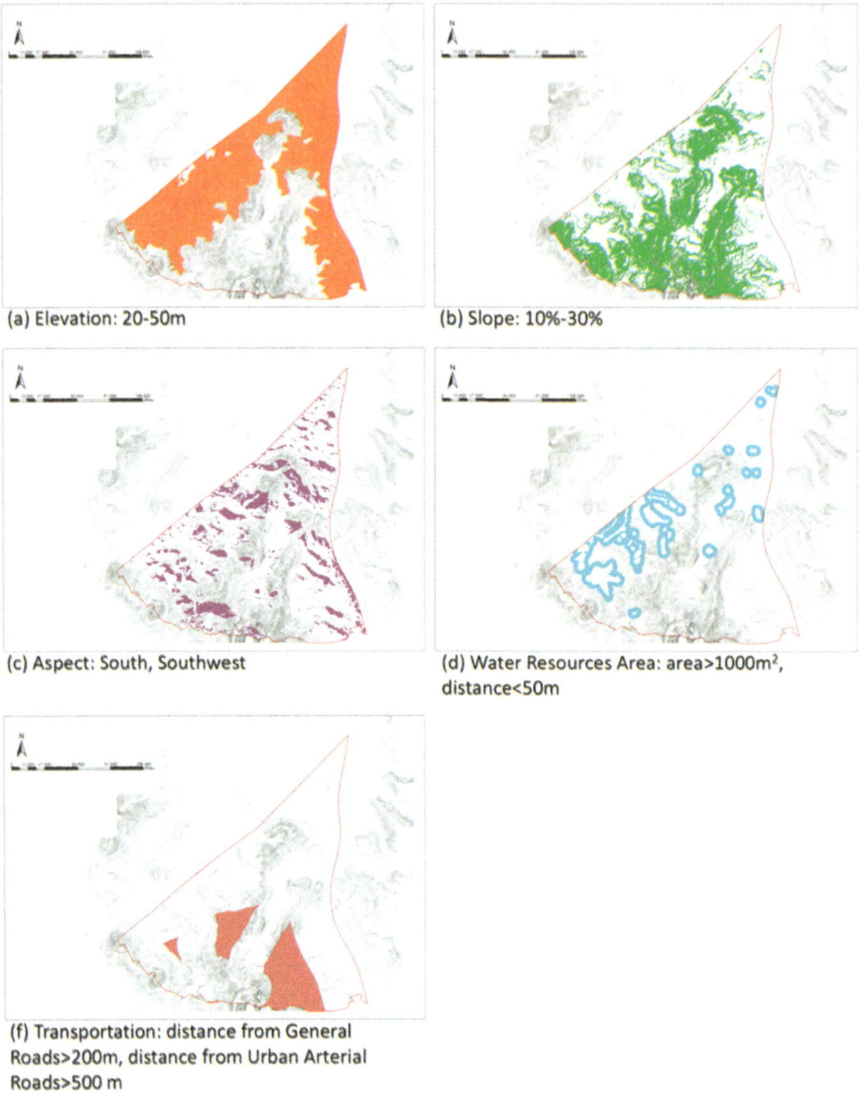

Fig. 6.5 Preliminary site selection process of "Jingyi Mountain Lodge"

the preliminary selection, with the largest measuring approximately 896 m^2 and the smallest only 0.68 m^2. When these results are compared with the land area requirements specified in the site selection criteria, it becomes evident that none of the parcels satisfy the size requirement. Since the selected parcels are generally much smaller than the desired land area, it is necessary to relax some of the selection criteria and conduct a second round of site screening.

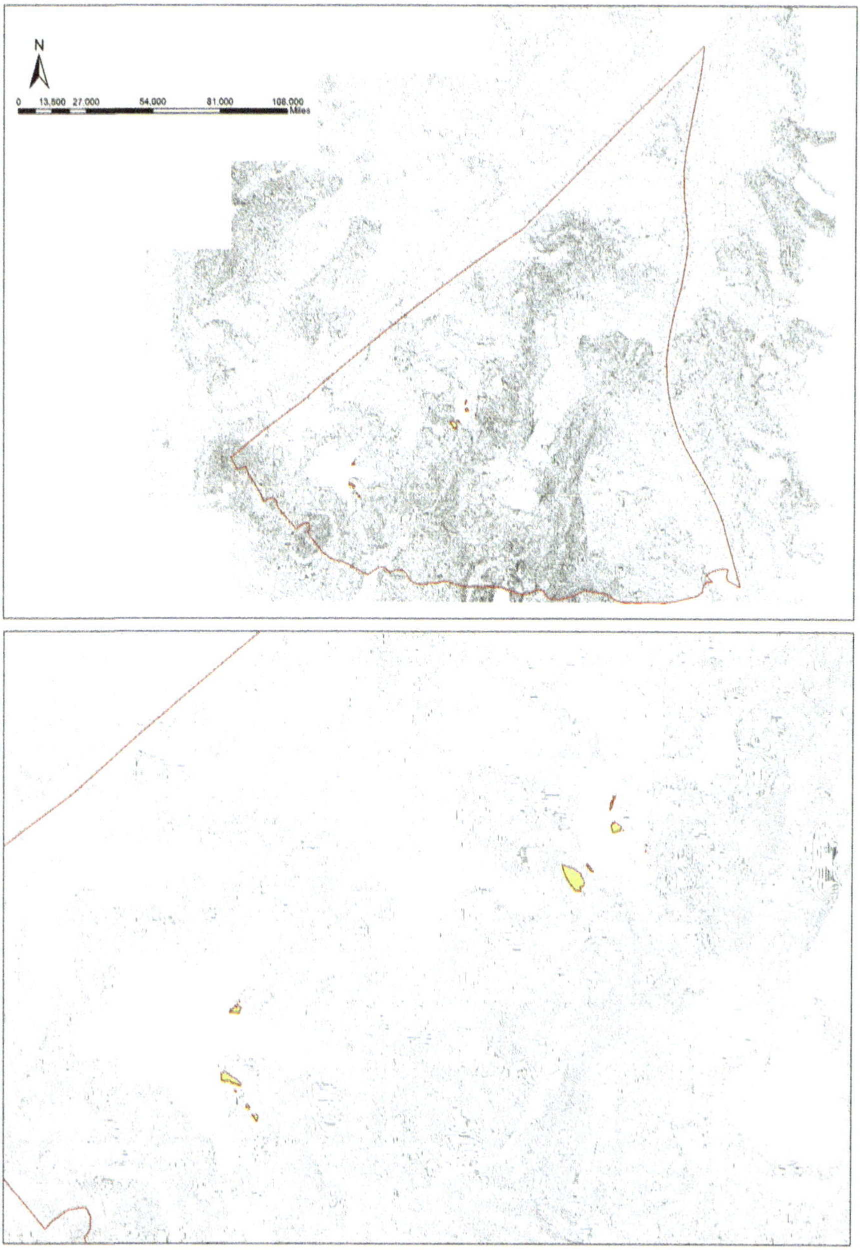

Fig. 6.6 Preliminary site selection results of "Jingyi Mountain Lodge"

Table 6.5 Preliminary site area statistics for "Jingyi Mountain Lodge"

OBJECTID	Shape_Leng (m)	Shape_Area (m^2)
1	62.23659542780	66.50884323500
2	3.77287329356	0.68466384569
3	60.19178289540	202.61268553000
4	3.77287329356	0.68466384569
5	9.14679273883	3.80629817210
6	38.36872562000	42.54413876030
7	3.77287329356	0.68466384569
8	134.24290191800	895.65831140300
9	7.58702791007	2.06419514352
10	3.77288662329	0.68466906925
11	62.56724130750	153.76639577200
12	96.54591709220	351.50597191000
13	11.92319454530	7.01682037348
14	23.19073553740	28.29490757870
15	3.77287329356	0.68466384569
16	4.00000000000	1.00000000000
17	32.78068986300	59.60169944630

6.4.2.3 Generation of the Second-Round Site Selection

The process of site selection is akin to searching for a target with a map—adding more constraints reduces the number of potential results and increases the likelihood that the outcome will not meet the project's requirements. Conversely, fewer constraints generally yield more results. When the selection results fail to meet the site requirements, it becomes necessary to either relax or tighten the selection criteria and re-screen for suitable parcels. Table 6.6 presents the second-round site selection criteria for the "Jingyi Mountain Retreat," in which requirements related to elevation range, slope aspect, transportation, and water resources have been relaxed.

The second-round selection process follows the same approach as the preliminary round, using ArcGIS software to produce thematic maps that reflect each individual criterion. As shown in Fig. 6.7, the relaxation of selection requirements led to a noticeable expansion in the acceptable ranges for elevation, water buffer zones, and proximity to roads compared to the preliminary selection. By overlaying these maps without weighting, the second-round site selection results, as shown in Fig. 6.8, were obtained. The number of suitable parcels increased significantly, mainly concentrated in the western part of the site where the existing water system is more developed.

The second-round site selection results for "Jingyi Mountain Retreat" represent potential ideal plots, and still need to be intersected with the buildable areas identified in the previous phase to yield a final set of parcels that meet actual construction conditions. Figure 6.9(c) shows the filtered parcels. A rough review of the area statistics table reveals that many of the selected parcels meet the land area

Table 6.6 Secondary site selection requirements for "Jingyi Mountain Lodge"

			Spatial requirements			Functional requirements				Special requirements		
										Water resources		
Item type	ID	Name	Elevation (m)	Slope (%)	Aspect	Building area (m^2)	Land area (m^2)	Building height (m)	Transportation	Type	Area(m^2)	Distance (m)
Zen practice	A-1	Jingyi Mountain lodge	20–50	10–30	South, southwest, southeast	600	1500	< 6	Distance from general roads: Greater than 100 mdistance from urban arterial roads: Greater than 400 m	Ordinary Lake or pond	> 1000	< 100

Fig. 6.7 Secondary site selection parameters of "Jingyi Mountain Lodge"

requirements of "Jingyi Mountain Retreat," indicating that this round of site selection yields feasible results. By using ArcGIS to filter for parcels larger than 1500 m^2, the final results are shown in Fig. 6.10. A total of six parcels meet the project's site selection criteria in the second round, with the largest parcel approximately 6342 m^2 in area and the smallest around 1674 m^2 (Table 6.7).

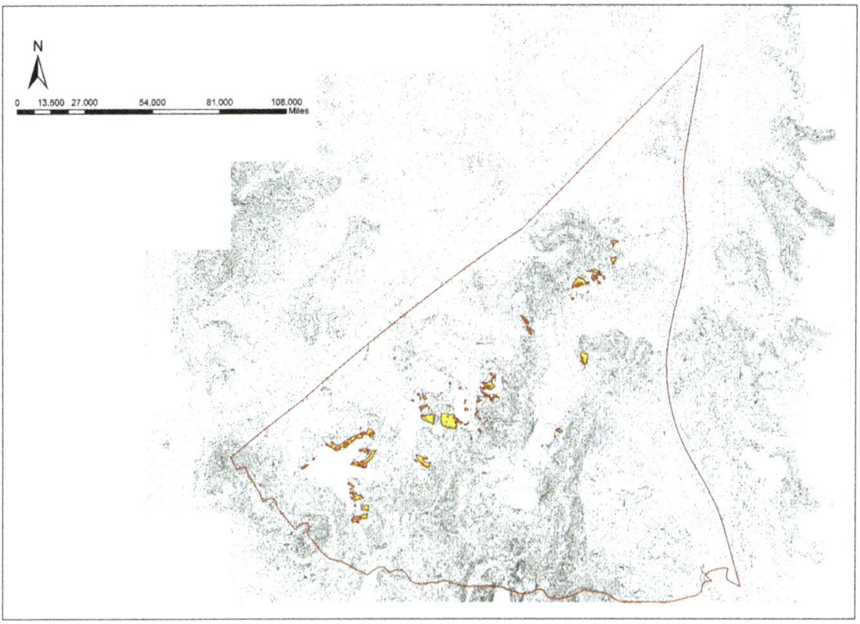

Fig. 6.8 Secondary site selection results of "Jingyi Mountain Lodge"

6.4.2.4 Formation of Functional Zoning

The final determination of a project's site requires comprehensive coordination and integration, for the following reasons: First, after the second-round site selection, multiple parcels may meet the site requirements. Second, due to the similarity of site selection criteria, there may be overlaps or partial overlaps between the selected sites of two or more projects. Third, if projects that share a common theme are located far apart, it may result in thematic fragmentation, which is detrimental to the formation of coherent functional zones. Therefore, coordination among projects is a necessary step for finalizing site selections and must be carried out after all individual project site selections are completed. Figure 6.11 illustrates the process of functional zoning formation. Once the site selection work for projects under the same theme has been completed, designers need to overlay, screen, and comprehensively assess them to determine the final construction parcels for each project. Fourth, the site selection for all themed project types should then be carried out in sequence. Fifth, all resulting zones are overlaid, and unified adjustments and optimizations are performed to generate the final functional zoning.

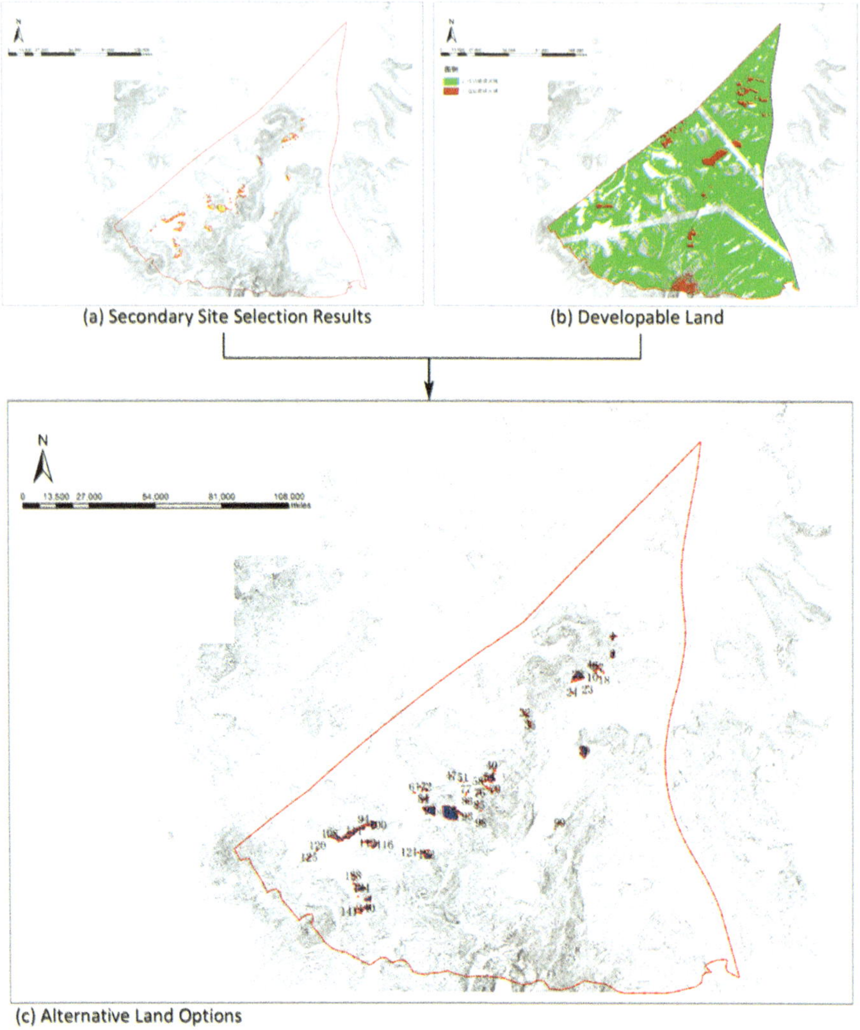

Fig. 6.9 Generation of alternative land options of "Jingyi Mountain Lodge" (For area statistics of alternative land, Please refer to Table A1 in the appendix for details)

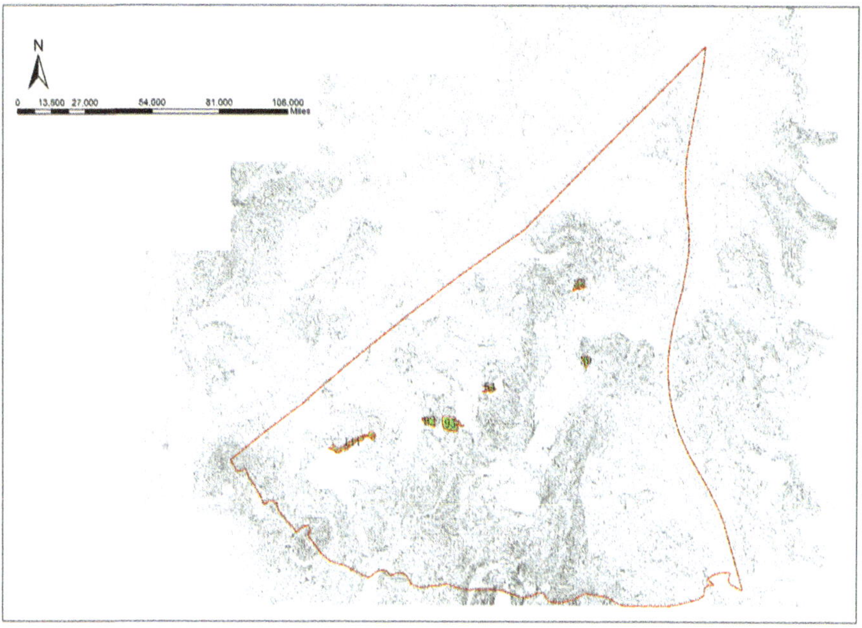

Fig. 6.10 Land screening results (Area > 1500 m^2)

Table 6.7 Statistical summary of alternative land parcels

OBJECTID	ID	Gridcode	Shape_Leng (m)	Shape_Area (m^2)
1	22	10	348.93327290900	2027.20060802000
2	39	10	194.32297990200	1674.51898669 000
3	54	10	388.24733228600	1710.68169177 000
4	92	10	279.48312638400	2516.83282095 000
5	95	10	548.35573322900	6342.33266278 000
6	111	10	1142.43395376000	5439.58287053000

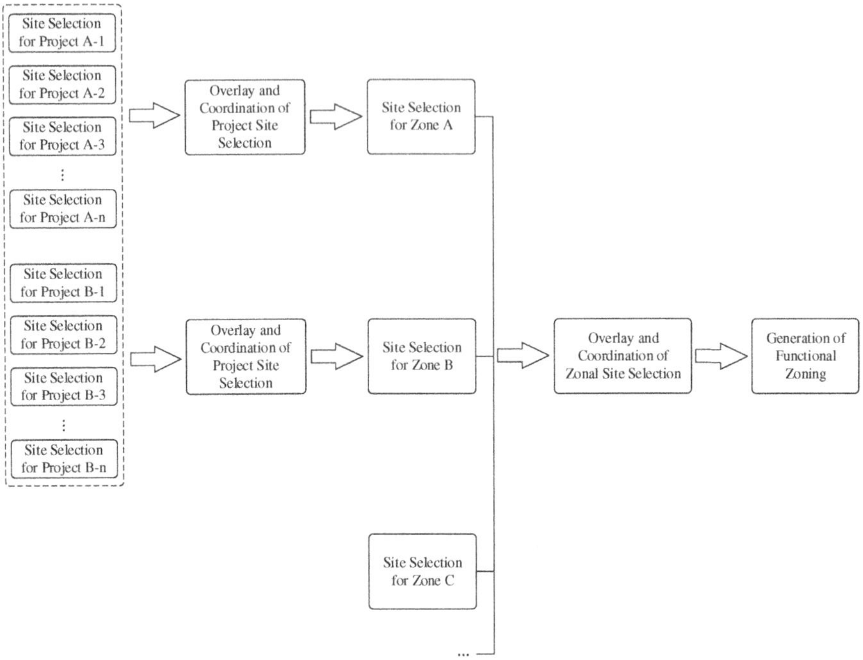

Fig. 6.11 Process of functional zoning based on project site selection

References

1. Agency for Cultural Affairs (2023) Bunkacho no kino kyouka: Kyoto iten [Reinforcement of the Agency for Cultural Affairs and relocation to Kyoto]. https://www.bunka.go.jp/seisaku/bunka_gyosei/kino_kyoka/index.html. Accessed 5 Apr 2025
2. National Administration for Quality Supervision, Inspection and Quarantine, Ministry of Housing and Urban-Rural Development of the People's Republic of China (2018) Standard for master planning of scenic and historic areas (GB/T 50298-2018). China Architecture & Building Press, Beijing, p 24
3. National Administration for Quality Supervision, Inspection and Quarantine, Ministry of Housing and Urban-Rural Development of the People's Republic of China (2018) Standard for master planning of scenic and historic areas (GB/T 50298-2018). China Architecture & Building Press, Beijing, p 22

Chapter 7
Road Route Selection Model

Roads are an inevitable component of the landscape environment, serving functions such as transportation. Route selection of roads involves comprehensively considering various factors including topography, hydrology, vegetation, soil, and earthworks, to determine a reasonable road alignment within the planning area through comparative evaluation. Due to the particularity of the landscape environment and the roads within the region, general requirements and methods for highway and urban road route selection cannot be directly applied. This chapter, based on a discussion of the characteristics and significance of road route selection in landscape architecture planning and design, reviews the development of route selection methods; elaborates on route selection algorithms based on the ArcGIS software platform; constructs a parameterized route selection model for landscape architecture planning and design roads; and, through comparison of cost-distance and path-distance algorithms and their parameters, explores the algorithms and parameters suitable for landscape architecture planning and design road route selection, further supported by illustrative examples.

7.1 Road Route Selection

7.1.1 *The Road System in Landscape Environments*

Road traffic in landscape environments is divided into external traffic and internal traffic. External traffic serves to connect the scenic environment with the surrounding environment and also functions to transport visitors into the scenic area. Internal traffic links various functional zones within the scenic environment, not only providing space for visitors but also guiding and distributing pedestrian flows. The "General Planning Standards for Scenic Spots" (GB/T 50298–2018) puts forward requirements for the planning of internal traffic and roads in scenic spots: make

Y. Yuan, *Parametric Landscape Architecture Planning and Design*,
https://doi.org/10.1007/978-981-95-8011-8_7

reasonable use of the terrain, select routes according to local conditions, and integrate with the surrounding landscape environment; organize scenic tours reasonably, facilitating the guidance and dispersal of visitors; avoid scenic and ecologically sensitive areas, and where avoidance is difficult, take effective protective and shielding measures; avoid deep excavation and high fill. Internal traffic in scenic environments does not pursue shortcuts, but needs to form good connections with each scenic node. It serves as both a link between scenic nodes and an important clue highlighting the landscape environment. Roads in scenic environments take scenic nodes as route control points, combining terrain and geomorphological characteristics as well as traffic mode requirements, and through comprehensive comparison, form a reasonable network system.

7.1.2 Summary of Road Route Selection in Landscape Architecture Planning and Design

Since landscape environments are often located in areas with abundant natural resources such as mountains, hills, and lakes, and generally feature complex terrain and geomorphology as well as ecologically sensitive zones, the overall road layout should conform to the natural terrain and landforms. The alignment should be flexible and varied, allowing appropriate curves and undulations according to site conditions. Large-scale earthworks should be avoided as much as possible to reduce human disturbance to the scenic environment, especially interference with ecologically sensitive areas. Roads designed according to local conditions often create good spatial landscape effects for visitors, provide favorable viewing perspectives, adapt to the landscape characteristics of the site, and blend harmoniously with the surrounding environment, becoming an integral part of the scenic environment. Article 6.1.4 of the "Park Design Code" (GB 51192–2016) specifies the following requirements for park road alignment design:

① Park roads should integrate with terrain, water bodies, vegetation, buildings, paved areas, and other facilities to meet traffic and touring needs and form a complete scenic composition;
② Create routes that orderly display landscapes or provide perspective lines for viewing forward scenery;
③ Road turns and connections should be smooth;
④ For main roads allowing motor vehicle traffic, the minimum radius of horizontal curves should be greater than 12 meters.

In view of the above, road alignment in landscape environments requires the integration of multiple factors and fulfillment of various requirements. In the parametric landscape planning and design process based on coupling principles, road alignment follows site selection. The landscape nodes identified in the previous "Site Selection" (Chap. 7) serve as control points determining the road direction.

Ecological sensitivity assessment results from the site suitability analysis provide constraints for road alignment. Factors such as terrain, geomorphology, and hydrological conditions also influence the alignment. Therefore, road alignment must be carried out based on prior research.

7.2 Progress in Road Route Selection Methods for Roads in Landscape Environments

7.2.1 Traditional Road Route Selection Methods

Traditional route selection methods rely on manual planning. Route designers conduct field surveys and collect and analyze data of the route selection area, then manually select points and draw connecting lines on large-scale topographic maps to predefine several potential route options. These plans are then subjected to on-site surveys and staked out section by section. After repeated comparisons, a relatively economical and reasonable route is determined. This manual method depends heavily on the designers' practical experience and technical proficiency, is time-consuming and labor-intensive, and can only form vague judgments about the economy and rationality of the route based on subjective experience. It not only relies heavily on the individual designer but also contains a degree of uncertainty, which makes it difficult to meet the demands of route planning for large-scale landscape environments. In terms of procedure, route selection in landscape planning and design is no different from that in traditional highway design. Designers often rely on topographic maps and their own experience to sketch out basic route paths, and then manually check basic requirements such as road gradients. The entire route selection process relies on the designer's intuition and experiential judgment. Due to the lack of a systematic and regularized operational platform, this method is overly dependent on experience. Moreover, route selection must consider numerous factors, and these factors vary greatly from site to site. Under the same conditions, the final design can differ depending on the designer's experience, skill level, and approach. Route selection methods can only be qualitatively summarized based on practical experience, drafting general principles to be followed during the process, but lack technical support and scientifically rational procedures.

7.2.2 Road Route Selection Methods Based on the ArcGIS Platform

Route selection is a complex and systematic task that requires the coordination of multiple objectives. The development of scientific technologies such as photogrammetry, computer technology, and information technology has provided more effective means than traditional methods, bringing route selection into its third developmental stage—computer-aided route selection—following the stages of field-based and photogrammetric route selection. This approach has been widely applied in fields such as transportation and surveying. In 1958, Professor Charles Leslie Miller of the Massachusetts Institute of Technology proposed the concept of automating the route selection process based on digital terrain models (DTMs). ArcGIS software possesses powerful spatial processing capabilities. It can import data from various sources in problem analysis, integrates spatial and attribute data, enables convenient construction of spatial analysis models, and supports comprehensive spatial qualitative, quantitative, and locational analysis. Accordingly, the ArcGIS software platform provides an excellent operational environment for route selection.

The methods of route selection using ArcGIS technology are flexible and diverse. From the perspective of basic spatial analysis methods, they can be categorized into route design methods based on digital elevation models, route optimization selection methods based on overlay analysis, and integrated route selection methods based on spatial analysis and simulation[1]. From the standpoint of the role ArcGIS plays in route selection, the methods can be divided into scenario-based selection and model-based selection. Among these, model-based selection includes both non-optimized and optimized route selection models[2]. The author's route selection model draws upon the integrated cost shortest path algorithm model from the non-optimized category. This model utilizes ArcGIS raster data and their adjacency relationships to perform spatial optimal path simulation and computation. It studies constraint conditions such as slope classification costs, maximum slope length, maximum longitudinal slope, and selection range, and uses the cost distance matrix method and shortest path analysis tools to calculate the optimal path. Based on a comprehensive cost surface, this method can effectively satisfy basic requirements for earthwork, longitudinal slope, and alignment in landscape road design, while also allowing effective avoidance of ecologically sensitive areas, regions with dense runoff, and other constrained zones. Another feature of this method is the conversion of various route selection factors into fundamental "cost" maps through evaluation, which allows for full utilization of the outcomes from the preliminary site analysis in landscape planning and design, transforming them into foundational maps of influencing factors for route selection. At the same time, the method is very easy to operate and requires only simple tools.

7.3 Road Route Selection Algorithm Model Based on ArcGIS

As the saying goes, "Only by knowing what it is can we know why it is so." Only by fully understanding the algorithmic principles behind the tool can it be effectively applied to landscape planning and design, thereby exploring route selection methods truly suitable for landscape environments. Therefore, the author explains and conducts targeted analysis of the algorithmic models and relevant concepts underlying the construction of the route selection model. GIS spatial data analysis mainly relies on two types of representation models: raster and vector models. Route selection primarily makes use of the distance tools in ArcGIS, where the target of calculation and analysis is the raster model, involving three algorithms: cost distance, path distance, and cost path.

7.3.1 Cost Distance Algorithm

The term "distance" refers to the degree of proximity between two entities and is a broad concept in spatial analysis. It not only refers to the straight-line length between two points but is also endowed with richer meanings. Euclidean distance is the most commonly used concept of distance, representing the straight-line distance between two points and can be considered an idealized distance. In spatial analysis, distance includes not only the straight-line length between two points but also the "resistance" or "cost" encountered when moving from one point to another. Here, "resistance" or "cost" represents a functional relationship of distance between the two points. "Resistance" corresponds to construction difficulty, while "cost" can also be understood as construction expense. In the cost distance algorithm, both "resistance" and "cost" can be transformed into "cost" problems. In landscape route selection, areas with steep slopes are not suitable for road construction, building roads across water bodies is uneconomical, and high-grade roads should not pass through ecologically sensitive zones—all of which are expressed as "cost" problems. Thus, in the ArcGIS routing algorithm, the concept of "cost" differs from everyday usage and is not limited to "monetary value." "Cost distance" refers to the "cost" or "expense" incurred when moving from one point to another in space and is calculated using a cost distance weighting function. Taking walking as an example, it is easier to walk on a park trail than on a muddy rural path—even if the straight-line distances are the same, the resistance encountered is different. The cost distance weighting function can calculate the least cumulative cost from each raster cell to the nearest source with the lowest cost, yielding cost direction data and cost allocation data. The cost direction data indicate the specific route from each cell to the nearest source along the path of least cumulative cost.

In route selection for landscape environments, factors such as slope, undulation, ecological sensitivity, and catchment areas must be considered, and thus become the "cost" of road construction. Based on single-factor cost, a comprehensive cost must be determined by integrating all factors within a unified cost classification system. Therefore, the importance of each factor in the total cost is determined through weighting, followed by overlay analysis to obtain the comprehensive cost, which is then used in the cost distance weighting function to refine the route selection.

7.3.2 Path Distance Algorithm

Compared to the cost distance algorithm, both the path distance algorithm and the cost distance algorithm can compute the minimum cumulative travel cost from a source to each location on a raster surface. However, the path distance algorithm can account for more complex factors, most notably the actual travel distance as well as both horizontal and vertical components. When a car is moving, surface friction and wind resistance both contribute to travel resistance. Additionally, when driving over undulating terrain, the actual distance traveled is greater than the straight-line distance between two points. These are all factors that affect movement. Compared to the cost distance algorithm, the path distance algorithm incorporates terrain undulation and slope as parameters into the calculation process, making it more suitable for road selection in landscape environments.

The first important parameter is elevation raster data. By computing the surface raster, the actual surface distance traveled from one cell to the next can be determined. Compared to planar ("straight-line") distance, this better reflects human or vehicle movement in a landscape setting. Accordingly, the more uneven the surface is, the longer the travel distance becomes. A longer distance implies higher cost, meaning that terrain with excessive undulation is unsuitable for road construction. Another important parameter is the vertical factor. The vertical factor reflects the relationship between slope and movement resistance; that is, a larger vertical factor indicates a steeper slope and more difficult movement. According to road design requirements, landscape road grades can be slightly more relaxed than those for urban roads or highways. However, areas with slopes greater than 30% are still unsuitable for construction, and this parameter can be expressed through the vertical factor. Figure 7.1 illustrates a linear and symmetrical functional relationship between the vertical factor (VF) and the vertical relative movement angle (VRMA) when the cut angle is ±90° and SLOPE (corrected slope) is 1/90 (0.01111). For landscape planning and design, slope is expressed as the angle formed with the horizontal plane. Whether positively or negatively related, it exists as slope and requires correction of movement cost via the vertical factor. Thus, a linear symmetric function is chosen for computation. Since areas with slopes greater than 30% are unsuitable for construction, the corresponding cut angle is set to ±30°, meaning that movement difficulty tends toward infinity in such zones. The SLOPE value reflects the slope between VF and VRMA. In the following sections, the value of this slope suitable for road selection in landscape planning and design will be further discussed in detail.

7.3.3 Cost Path Algorithm

Both the cost distance algorithm and the path distance algorithm calculate the minimum cumulative travel cost from a certain "source" (corresponding to a "control point" in the landscape environment) to each position on the raster surface. The

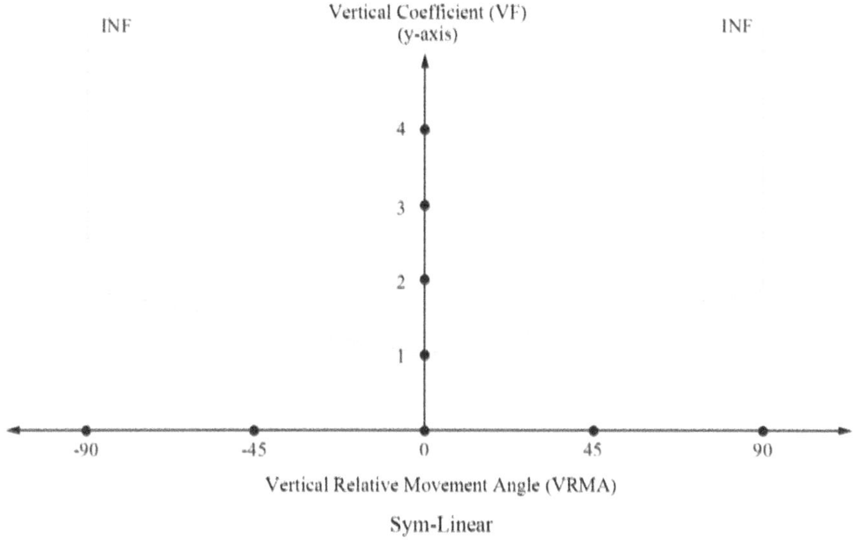

Fig. 7.1 Symmetrical Linear Vertical Factor Chart. *Source*: http://resources.arcgis.com/zh-cn/help/main/10.1/index.html#/na/009z000000z9000000/

calculation of the cost path is based on the above, obtaining the minimum cost path from one source or a group of sources to a target location or a group of target locations through the shortest path function. In ArcGIS software, the "shortest path" does not merely mean "the shortest straight line" in the usual sense; after incorporating "cost" into the calculation, the "shortest path" is the shortest under the condition of the lowest "cost." However, problems arise accordingly. According to the characteristics of the ArcGIS function calculations, the generated path is the shortest path under the premise of minimum cost. Due to limitations in the calculation method, the road alignment algorithm is difficult to fully meet the requirements of landscape architecture planning and design, that is, roads in landscape environments do not completely pursue the shortest path. Instead, winding and twisting roads can provide visitors with special visual and spatial experiences. Therefore, how to select paths based on existing algorithms while following the characteristics of road alignment in landscape architecture planning and design is the key focus of this research. The following will discuss in detail the targeted optimization of existing algorithms through parameter adjustment and comparison.

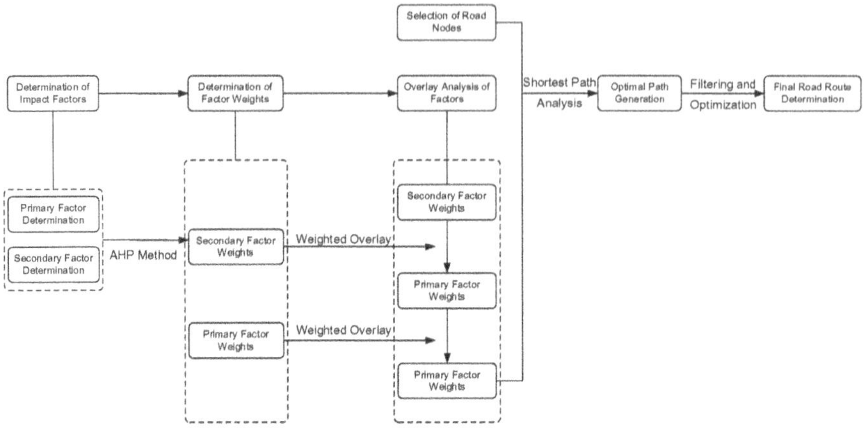

Fig. 7.2 Parameterized Road Alignment Model for landscape architecture planning and design

7.4 Parameterized Road Route Selection Model Construction for Landscape Architecture Planning and Design

Based on the cost distance or path distance algorithms and the cost path algorithm, a parameterized road alignment model for landscape architecture planning and design is constructed (see Fig. 7.2). Consistent with the sequence of operations involving the two algorithms, the model is divided into two parameterized calculation parts followed by a human-computer interaction part: comprehensive cost calculation, optimal path generation, and path screening and optimization. First, the comprehensive cost calculation includes three processes in sequence: determination of influencing factors, determination of factor weights, and overlay analysis of factors. These three processes are completed within the ArcGIS software, resulting in a comprehensive cost map that serves as a parameter input to the next calculation phase. Second, in the cost path calculation phase, road nodes and the comprehensive cost map are input parameters; through shortest path analysis under given conditions, a preliminary optimal alignment is obtained. Finally, the preliminary alignment is screened and optimized via human-computer interaction, eliminating unreasonable paths to derive the final road alignment.

7.4.1 Determination of Influencing Factors

There are many and complex factors affecting the road alignment scheme. Different regions have varying conditions, and planning and design objectives differ, resulting in different emphases on road alignment requirements. Therefore, it is necessary to determine the influencing factors based on the characteristics of the site. In most cases, the conditions that road alignment needs to satisfy can be summarized into three categories:

The first category consists of control point factors, including the start point, end point, and mandatory intermediate points, which also determine the route length;

The second category relates to geographic spatial factors, such as terrain, geomorphology, geological conditions, and hydrological conditions;

The third category includes restrictive factors, among which ecological sensitive areas are the most important in landscape architecture planning and design. In addition, this category also includes buildings, various facilities, and vegetation that need to be preserved, which provide restricted areas for road alignment.

All these necessary condition factors participate in the road alignment process. The influencing factors involved in the parameterized road alignment model for landscape architecture planning and design mainly consist of the second and third categories of conditions. According to the classification of site elements, they can be divided into primary factors such as terrain and geomorphology, hydrology, and corresponding secondary factors. The detailed classification can be found in Table 7.1.

7.4.1.1 Topography and Landform

(1) Elevation

Elevation is one of the important factors in road alignment, significantly affecting the longitudinal slope of roads. The longitudinal slope refers to the uphill and downhill along the direction of road progression. China's road design standards impose strict requirements on the longitudinal slope. For landscape environments, the requirements for road longitudinal slope are relatively relaxed, which can actually provide visitors with a better spatial experience and enhance the landscape effect. In certain special geographic scenic areas, the

Table 7.1 Influencing factors and weights for road alignment in landscape architecture planning and design

Evaluation objective	Primary evaluation factors	Weight	Secondary evaluation factors	Weight
Road route selection factors in landscape planning	Topography and landform	$W1$	Elevation	$^{W}11$
			Slope	$^{W}12$
			Terrain ruggedness	$^{W}13$
	Hydrological conditions	$W2$	Water bodies	$^{W}21$
			Watershed	$^{W}22$
			Runoff area	$^{W}23$
	Ecologically sensitive areas	$W3$	–	–
	Land use	$W4$	Existing roads	$^{W}41$
			Reserved facilities	$^{W}42$

elevation that roads should not exceed must also be reflected in the evaluation and grading of influencing factors.

(2) Slope

To meet vehicle traffic and road grade requirements, road construction imposes corresponding regulations on the longitudinal slope of the route. When the terrain slope exceeds certain limits and cannot satisfy the longitudinal slope requirements, corresponding engineering measures such as earthworks (cut and fill), retaining walls, and slope protection must be taken. These engineering measures not only increase construction costs but also cause significant disturbance to the original site environment. At the same time, steep slopes pose geological hazard risks. Generally speaking, the longitudinal slope of roads in landscape environments can be slightly increased within a certain range. Therefore, following the principles of minimizing disturbance and reducing earthwork volumes, road alignment in landscape design should respect the original terrain and adapt accordingly.

(3) Ruggedness

Ruggedness refers to the maximum relative elevation difference within a unit area and reflects the relative height variation of the terrain. It is a quantitative indicator describing the landform morphology. A higher ruggedness in an area indicates larger elevation changes, which means constructing roads there will cause frequent ups and downs, reducing driving comfort and increasing construction costs. Therefore, when selecting road alignments in landscape environments, routes should preferably pass through areas with lower ruggedness.

7.4.1.2 Hydrological Conditions

(1) Water Bodies

Water bodies refer to areas with a certain water surface area within the landscape environment, such as rivers, ponds, and lakes. Constructing roads across water surfaces involves high costs and should be avoided as much as possible. The periodic and seasonal fluctuations of rivers, lakes, and other water bodies significantly affect nearby areas. This not only complicates the treatment of road foundations but also raises the risk of flooding. Therefore, road construction should maintain a certain distance from water bodies.

(2) Catchment Areas

A catchment area is the surface region through which runoff converges toward a common outlet. With proper design and management, catchment areas in landscape environments have the potential to become new water landscape features, representing valuable scenic resources. Thus, when selecting road alignments in landscape environments, catchment areas should be avoided where possible. However, if combined with water landscape creation, building roads adjacent to catchment areas can achieve excellent waterfront landscape effects.

(3) Runoff Zones

Runoff refers to the flow of water over the surface or underground caused by rainfall, snowmelt, or irrigation under the force of gravity. Runoff typically occurs rapidly and with a high instantaneous flow rate, which can cause erosion features like gullies in areas with poor geological conditions, threatening road safety. On the other hand, road construction impacts runoff by obstructing flow convergence and increasing infiltration, which can jeopardize the safety of road foundations. Therefore, it is essential to fully consider runoff's impact on roads, as runoff also influences road morphology—for example, through the installation of water-crossing pavements or interception ditches on water-facing road surfaces.

7.4.1.3 Ecologically Sensitive Areas

Ecologically sensitive areas are regions designated based on an ecological sensitivity assessment of the site. These areas have fragile and sensitive ecological environments that require comprehensive protection. Therefore, roads in landscape environments should not pass through these areas and must maintain an appropriate buffer distance.

7.4.1.4 Land Use

(1) Existing Roads

Road alignment in landscape environments should make full use of existing roads. This approach not only reduces environmental disturbance but also leverages already established roadbeds, significantly lowering construction costs.

(2) Preserved Facilities

Preserved facilities in landscape construction mainly include reserved farmland and various man-made structures. Road alignments need to bypass these preserved facilities.

7.4.2 Determination of Factor Weights

Table 7.1 presents the typical influencing factors involved in route selection for roads. Once these factors are identified, the Delphi method and Analytic Hierarchy Process (AHP) are employed to determine their respective weights. The influencing factors and their weights vary depending on the specific site. Generally, factors such as slope and topographic relief, which directly determine the feasibility of road construction, are assigned higher weights. In landscape environments, partial road crossings over water surfaces are permitted and can enhance spatial layering through landscape bridges, offering visitors diverse visual experiences. Therefore, the weight assigned to the waterbody factor should not be too high. Utilizing existing roads and preserved facilities for new road construction undoubtedly reduces costs and minimizes environmental disturbance. However, overly high weight values for these factors may compromise the accuracy of route selection. This issue will be further discussed in the section on comparison and optimization. The factor weights serve as critical parameters in generating the comprehensive cost map, directly affecting the rationality of the route selection. Improperly assigned weights can result in flawed routes. Thus, it is essential to determine the weights based on accurate evaluations. Moreover, preliminary route results should be analyzed; if found unreasonable, the comprehensive cost map must be recalculated. This feedback process involves reassigning the weights to ensure improved route optimization.

7.4.3 Generation of Comprehensive Cost

Based on the determined factor weights, the various influencing factors are reclassified using ArcGIS software. The purpose of this process is to normalize different types of factors and assign them unified classification values, thereby incorporating them into a common evaluation system in preparation for overlay analysis. Since the classification of the comprehensive cost map generated by overlaying the factors corresponds to the classification of each individual factor, the way each factor is classified and valued directly affects the outcome. Classification determines the hierarchical structure of the comprehensive cost values. A richer hierarchy offers a broader range of cost values, increasing the complexity and accuracy of computation under given algorithms, ultimately improving the reliability of the selected route. Both classification and value assignment significantly impact route selection results and will be discussed in more detail in the following sections. The overlay of reclassified maps results in a comprehensive cost map, which serves as the base map for route selection. As such, the choice of influencing factors, the determination of their weights, and the classification of individual factors all affect the identification of the optimal path. These three elements must therefore be carefully addressed to enhance the rationality of route selection.

7.4.4 Selection of Road Nodes

Road nodes refer to starting points, endpoints, and mandatory intermediate points along a route. These nodes, together with the comprehensive cost map generated in the previous stage, serve as input parameters for the computation of cost distance, path distance, and cost paths. Landscape planning and design typically address medium- to large-scale regions, where road systems often form networks and multiple access points are set up to support different service functions. Given these characteristics, node selection in landscape environments differs significantly from that of urban roads or highways.

Determining starting and ending points. According to the algorithms for cost or path distance, a starting point must be provided as an input parameter. In landscape planning, all regional access points can generally serve as road starting points, functioning as connectors between external and internal transportation systems. Since there are usually multiple access points, the main entrance, being an essential node, must be included as a road starting point in the route selection process. Auxiliary entrances act as secondary starting points that contribute to forming the overall road network. Due to the networked nature of roads in landscape environments, there is no fixed "endpoint" in the conventional sense. Each designated project location determined in the site selection process must be connected by a road, and thus can be considered an endpoint relative to the starting point. The cost path tools support one or multiple source inputs, allowing all project locations to be entered as multiple sources for computation. However, due to algorithmic rules, the computed routes may bypass certain areas that designers intend the road to traverse. In such cases, additional points must be added to these areas as intermediate or required passage points and included as endpoints in the route selection process along with the project nodes. For example, if the route generated by the cost map does not pass a cliff area with distinctive views that the designer wishes to highlight, an intermediate node can be added to guide visitors through that area, thereby enhancing the variety of scenic experiences. The specific procedure and rationality of this approach will be detailed in the section "Comparison and Optimization of Multi-Point Route Selection Models." (Sect. 7.5.4).

7.4.5 Cost Distance Analysis

By inputting the comprehensive cost map and the road starting points into the cost distance tool, the cost-weighted distance and cost-weighted direction maps relative to the starting points can be generated. These maps respectively indicate the cumulative cost of each cell returning to the nearest source location, and the travel direction from each cell in the cost distance raster back to the source. Compared with the cost distance tool, the path distance tool requires more input parameters, such as the surface raster, angular limits related to the vertical factor computation, and SLOPE values. These additional inputs help further constrain the path selection process according to the requirements of landscape planning and design.

7.4.6 Optimal Path Generation

Depending on the number of input destination points, the generation of optimal paths can be divided into two-point and multi-point optimal path calculations. Construction projects within landscape environments—including landscape nodes and architectural nodes—must be connected to the main entrance, and in some cases to secondary entrances. Since there are usually multiple construction projects, if the sequence of nodes involved in path selection is artificially predetermined and two-point optimal paths are calculated sequentially, the resulting routes will reflect excessive human intervention and lack rationality. Moreover, generating a complete road network based on this method is extremely labor-intensive, especially in large-scale scenic area planning involving numerous road nodes, making it difficult to form a comprehensive network through pairwise path selection alone. Therefore, this study adopts a multi-point simultaneous calculation approach. This not only aligns with the logical traffic relationships between project nodes and entrances but also facilitates the formation of an internal road network within the planning area. Since each project node generates a path leading to the starting point, the number of project nodes determines the number of resulting paths, thereby affecting the density of the road network. It is thus evident that the number of control points—including both project nodes and supplementary nodes—is closely related to the network density, which will also be discussed in the following section on comparison and optimization. Due to the characteristics of the cost path algorithm, the calculated paths tend to form a dendritic (branch-like) structure, which is insufficient for satisfying the requirement of looped or circular paths in landscape planning and design. How to generate a network-like path structure through optimization methods is one of the key research issues addressed in the next section.

7.4.7 Path Selection and Optimization

Under the current computational methods and conditions, the paths generated through algorithms cannot fully meet the final network form required by planning and design. Therefore, manual selection and optimization are necessary. This work primarily involves two aspects. On one hand, it is necessary to eliminate redundant paths and retain only the required ones. To satisfy the requirement for looped roads, some nodes must be directly connected to two or more other nodes. However, each individual path computation only establishes a connection between two points, thus requiring multiple calculations. Some nodes, however, do not have high accessibility requirements and do not need to be connected via multiple paths. Addressing these conditions requires deliberate human selection and judgment. On the other hand, the alignment of the road paths needs to be optimized. Due to the characteristics of raster-based computation, the output paths from the cost path algorithm consist of a series of continuous raster cells. After exporting to vector format, manual

Table 7.2 Comparison of parameters involved in Cost Distance Method and Path Distance Method

Parameter	Influencing factor	Classification	Classification assignment	Road node			Surface raster	Vertical factor		
				Start point	End point	Intermediate point		Vertical raster	Vertical factor intersection angle	Slope value
Cost distance algorithm	✓	✓	✓	✓	✓	✓	×	×	×	×
Path distance algorithm	✓	✓	✓	✓	✓	✓	✓	✓	✓	✓

optimization of the path alignment is necessary in order to generate the final road layout.

7.5 Optimization of the Least-Cost Path Algorithm Model Based on Composite Cost

There are two approaches to path selection based on the least-cost path algorithm model: the Cost Path algorithm and the Path Distance algorithm. The difference between them lies in the specific computational tools used within ArcGIS software. In terms of parameter requirements, the Path Distance algorithm demands more input parameters than the Cost Path algorithm, thereby allowing more constraints to be incorporated into the computation (Table 7.2).

The following section provides a comparative analysis of the two methods in the context of road alignment in landscape environments, and proposes targeted optimization strategies. The research is divided into two parts:

1. In the case of two-point alignment, a comparison is conducted on the different parameter inputs between the Cost Path and Path Distance algorithms, as well as an evaluation of their relative rationality.
2. In the case of multi-point alignment, the comparison focuses on the number of control points and optimization methods, including a "1 + N" multi-point, multi-pass route selection strategy to support network formation.

The comparative and optimization research in this chapter is conducted using the northern area of the Niushou Mountain Scenic Area in Nanjing as a case study. This area is located in the southern part of the Nanjing urban area and is part of the Ning-Zhen mountain range, characterized primarily by mountainous terrain. The southeast and northwest borders of the area are adjacent to urban roads. Internally, the region contains existing roads, water bodies, current construction land, and forest conservation areas—conditions that meet the requirements for researching a parametric road alignment model for landscape planning and design.

7.5.1 Based on Cost Distance Algorithm: Comparative Analysis of Road Route Selection

Currently, road alignment methods based on the ArcGIS software platform are mostly applied in transportation engineering, especially in highway alignment studies. Most of these rely on Digital Elevation Models (DEM), using the cost distance algorithm tool combined with auxiliary optimization methods such as genetic algorithms. However, whether the cost distance algorithm is suitable for road alignment in landscape environments requires further exploration and research.

7.5.1.1 Characteristics and Problem Analysis of Road Route Selection Based on Cost Distance Algorithm

The first step in constructing a parameterized landscape architecture road alignment model based on the cost distance method is to identify the primary and secondary influencing factors affecting road alignment in the site, determine the costs for secondary factors, and assign weights to these factors through evaluation methods such as the Delphi method and Analytic Hierarchy Process (AHP). The second step is to normalize and assign values to these factors using ArcGIS's reclassification function. Third, overlay analysis is used to combine secondary and primary factor cost maps step-by-step to generate a comprehensive cost map. The final step is to use ArcGIS's shortest path analysis tool to calculate the optimal path. Since road alignment is a complex systemic problem, it cannot rely solely on computational methods to obtain the optimal road shape. After determining an initial road alignment scheme, it is necessary to optimize the route through human-computer interaction, which is a crucial part of the alignment process. According to the above procedure, the model variables include: influencing factors, classification levels, and classification values. The following discussion focuses on the alignment results generated under different parameter inputs of these three variables.

7.5.1.2 Comparison One: Selection of Influencing Factors

In landscape architecture planning and design, road alignment selection not only needs to consider general road construction requirements such as longitudinal slope, turning radius, and slope length, but should also emphasize the special demands of landscape environment roads in ecological protection, travel experience, and scenic effects. Among ArcGIS alignment selections based on DEM, slope is the easiest to obtain and also the most important basic influencing factor. A comprehensive cost map created based on multiple influencing factors can more accurately and reasonably reflect the site's foundational environment for route calculation. However, it should be noted that each factor influences the route results, and there may be mutual cost offsets among factors causing the comprehensive cost to be unreasonable—that is, an excessive or overly complex overlay of factors can invalidate the comprehensive cost map. The author conducted a comparative analysis of route results produced under different factor selections, which are: route results based on slope as a single factor (Fig. 7.3(a)), route results based on a composite of slope, water bodies, and undulation (Fig. 7.3(b)), and route results based on a composite of slope, water bodies, undulation, and existing roads (Fig. 7.3(c)).

In Fig. 7.3(a), the route bypasses areas with steep slopes, choosing flatter areas to travel through, but passes through existing water surfaces. In Fig. 7.3(b), the route changes little compared to Fig. 7.3(a) because the slope factor has a high weight of 0.6 in the comprehensive cost overlay, so slope still dominates the composite cost. Comparing the two paths, the path in Fig. 7.3(b) tends to lean toward the side with denser contour lines and avoids the water body. This is due to the addition of the

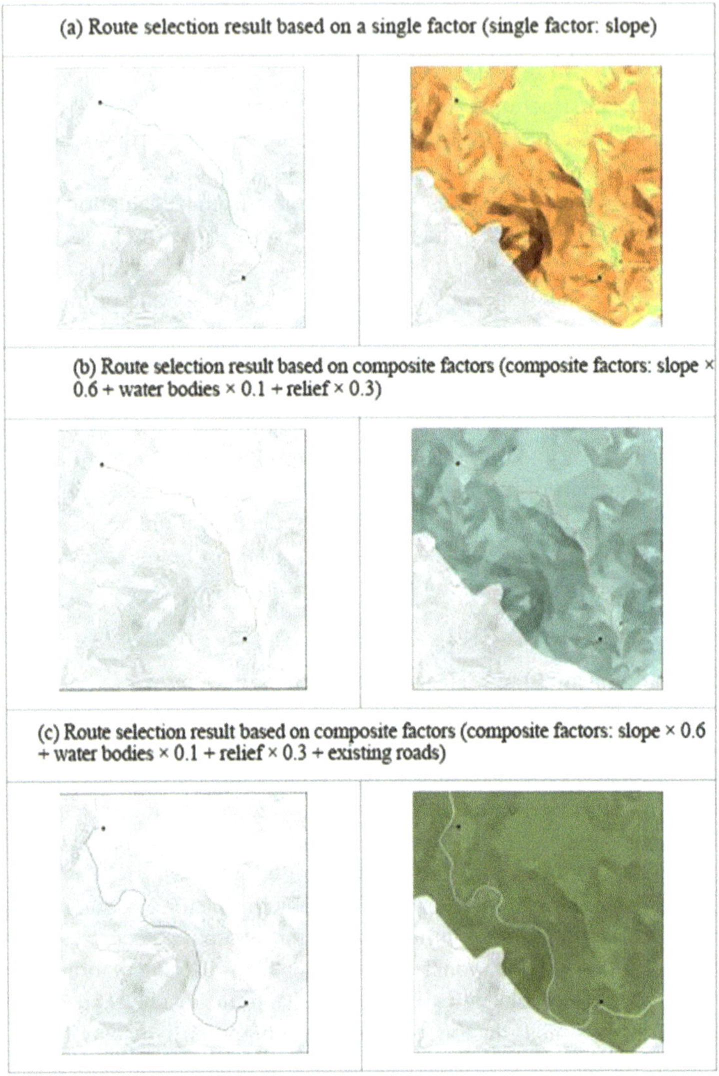

Fig. 7.3 Cost Distance Method—route selection results based on different factor cost maps

water body factor in the composite cost; since its weight is set relatively low, the deviation is not very pronounced. In Fig. 7.3(c), the existing roads are included as one of the factors in the composite cost. Because construction costs based on existing roads are lower than new routes, this area is assigned a value of 0, i.e., almost no cost. The calculation results clearly show that the route in Fig. 7.3(c) is completely different from those in Figs. 7.3(a) and (b) (Fig. 7.4). It crosses contour lines in a straight line and directly connects to existing roads, following those roads since the cumulative cost of this route is lowest. According to the rules of the cost path

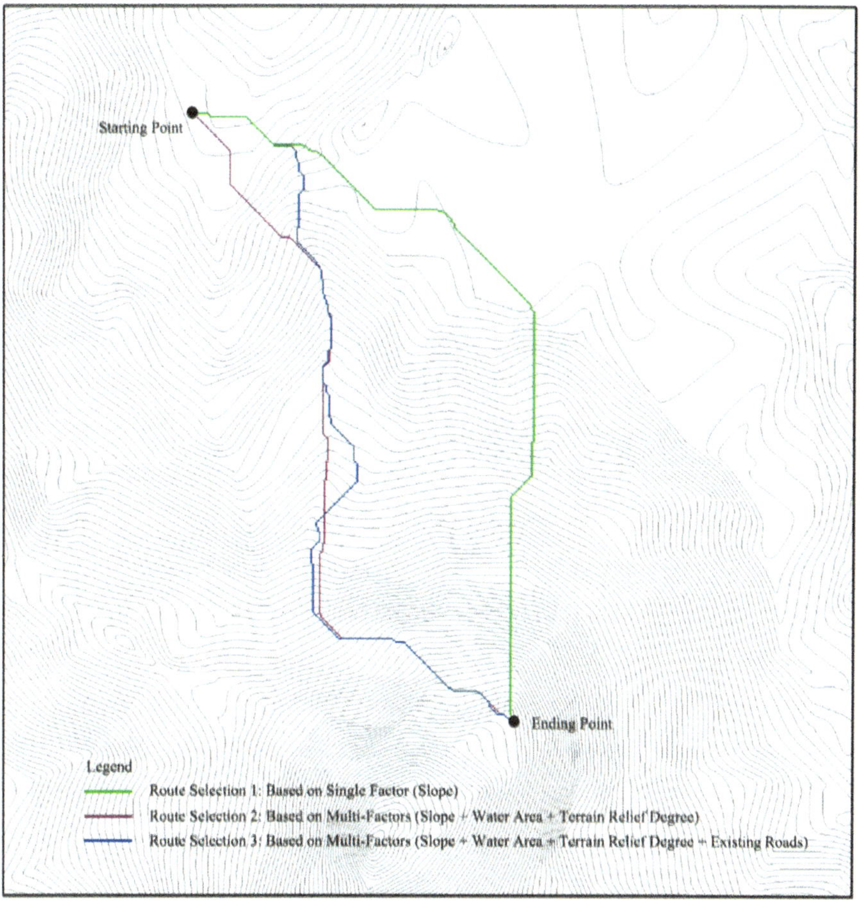

Fig. 7.4 Cost Distance Method—comparison of route selection results based on different factor cost maps

algorithm, whether including existing roads with extremely low costs interferes with the routing results will be specifically analyzed later.

7.5.1.3 Comparison Two: Classification of the Comprehensive Cost Map

The comprehensive cost map serves as the foundational map for cost distance calculations and therefore has a decisive influence on the route direction. Various types of factors must be normalized through dimensionless processing and incorporated into a unified evaluation system for overlay and assessment. An important step in achieving normalization using ArcGIS software's reclassification function is the reclassification and reassignment of values. Since the comprehensive cost map is composed of overlaid single-factor cost maps, the classification of the

Table 7.3 Discontinuity values and assigned values for different classification levels

Classification	9-level equal division	18-level equal division	32-level equal division	32-level unequal division	Assignment
1	10°	5°	3°	1°	1
2	20°	10°	6°	2°	2
3	30°	15°	9°	3°	3
4	40°	20°	12°	4°	4
5	50°	25°	15°	5°	5
6	60°	30°	18°	6°	6
7	70°	35°	21°	7°	7
8	80°	40°	24°	8°	8
9	90°	45°	27°	9°	9
10	–	50°	30°	10°	10
11	–	55°	33°	11°	11
12	–	60°	36°	12°	12
13	–	65°	39°	13°	13
14	–	70°	42°	14°	14
15	–	75°	45°	15°	15
16	–	80°	48°	16°	16
17	–	85°	51°	17°	17
18	–	90°	54°	18°	18
19	–	–	57°	19°	19
20	–	–	60°	20°	20
21	–	–	63°	21°	21
22	–	–	66°	22°	22
23	–	–	69°	23°	23
24	–	–	72°	24°	24
25	–	–	75°	25°	25
26	–	–	78°	26°	26
27	–	–	81°	27°	27
28	–	–	84°	28°	28
29	–	–	87°	29°	29
30	–	–	88°	30°	30
31	–	–	89°	31°	31
32	–	–	90°	90°	32

comprehensive cost aligns with that of the single-factor cost classifications. The more detailed the classification of costs, the more available values are provided for the calculation, thus increasing the possible route selections.

The ArcGIS software's reclassification function provides a maximum default classification of 32 levels, so the author uses 32 levels as the maximum number for classifying slope as a single-factor cost and compares it with classifications of 9 and 12 levels. Compared to highways, slope requirements for landscape environment roads can be somewhat relaxed, but construction on slopes greater than 30° is still inadvisable. Based on this, the study adds a comparison of equal and unequal divisions for the 32-level slope classification. The difference lies in the discontinuous

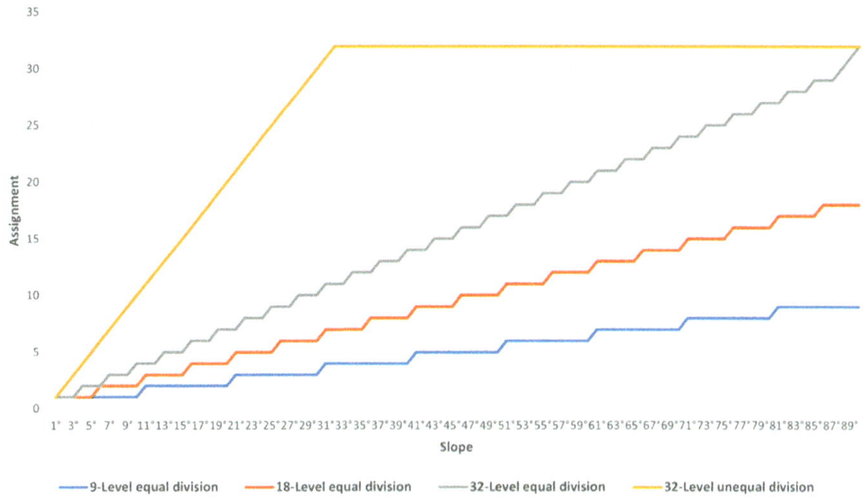

Fig. 7.5 Line chart of assigned value increase corresponding to different classification levels

value intervals, as shown in Table 7.3. Unequal classification means that the slopes suitable for construction are divided in more detail, effectively increasing the richness of classifications. For example, in the 9-level equal division, the range of 1°–10° is considered a single grade with the same cost of "1"; whereas in the 18-level equal division, this same slope range is split into two grades, 1°–5° and 5°–10°, with corresponding costs of "1" and "2"; the 32-level equal division has more than 3 different cost values, and the 32-level unequal division contains 10 cost values. Thus, the more levels, the finer the slope classification, providing more detailed cost options for the calculation. In summary, the author compares the comprehensive cost map classifications based on slope single-factor costs at 9-level equal divisions, 18-level equal divisions, 32-level equal divisions, and 32-level unequal divisions (Fig. 7.5).

According to Fig. 7.6, the single-factor slope cost maps display different levels of color richness depending on the classification levels, representing the number of cost categories. Figure 7.7 shows that the routes selected based on 9-level and 18-level equal interval classifications have similar travel directions, while the results from 32-level equal interval and 32-level unequal interval classifications are quite close. Specifically, the route from the 9-level equal interval classification travels along the gentler slopes of the valley in the first half, then ascends almost in a straight line cutting across contour lines to the endpoint. The 18-level equal interval route follows the gentler ridge slopes throughout, climbing in a generally winding manner. The 32-level equal interval route chooses very gentle slope areas in the first half, resulting in a longer path; it then climbs along relatively gentler slopes in the middle, and finally ascends at a smaller angle cutting across contour lines to the endpoint. The 32-level unequal interval route resembles the 32-level equal interval one in the first half, and travels almost in a straight line at a small angle toward the endpoint in the middle and latter sections.

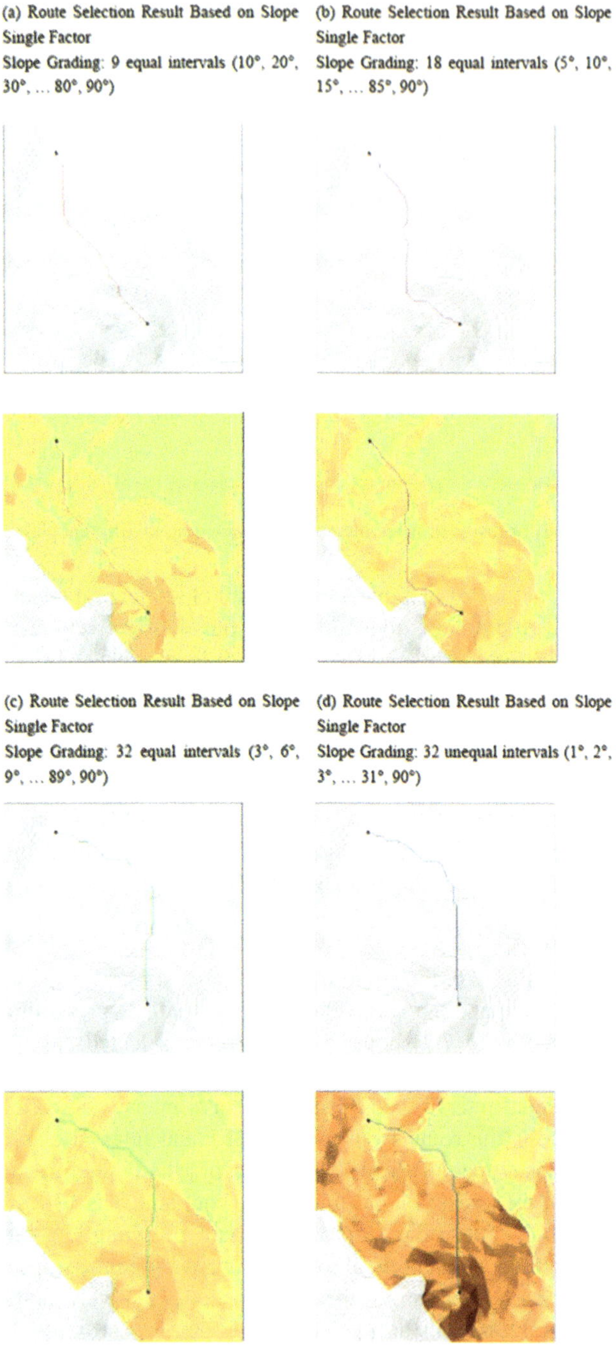

Fig. 7.6 Cost Distance Method—route selection results based on different classification levels

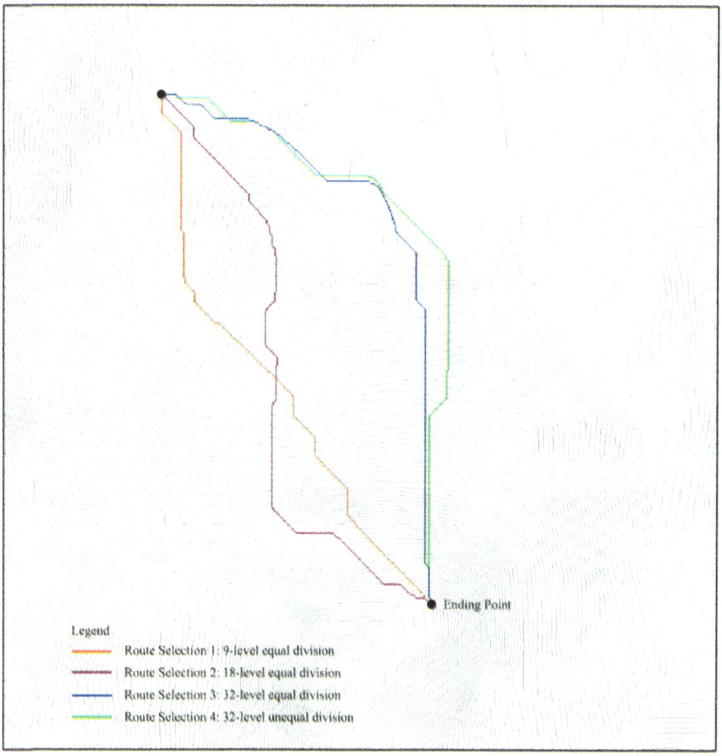

Fig. 7.7 Cost Distance Method—comparison of route selection results based on different classification levels

The author compared the routing results of different classification levels over longer distances. Figure 7.8 shows that regardless of the classification level, the routes all approach the endpoint from the northwest side. This is because the contour lines on the southeast side are denser, making crossing costs higher. Since the overall slope of the northwest route is gentler and slope values involved are concentrated in a smaller range, the assigned values for the 32-level equal and unequal interval classifications show little difference, resulting in completely overlapping routes. A closer look at the red box area in Fig. 7.8 reveals that compared to the 9-level equal interval route, the other three routes detour around small peaks, choosing longer paths to avoid crossing too many contour lines. Among these, the paths generated by the 32-level equal and unequal interval classifications show better avoidance of areas with dense contour lines than the 18-level equal interval route.

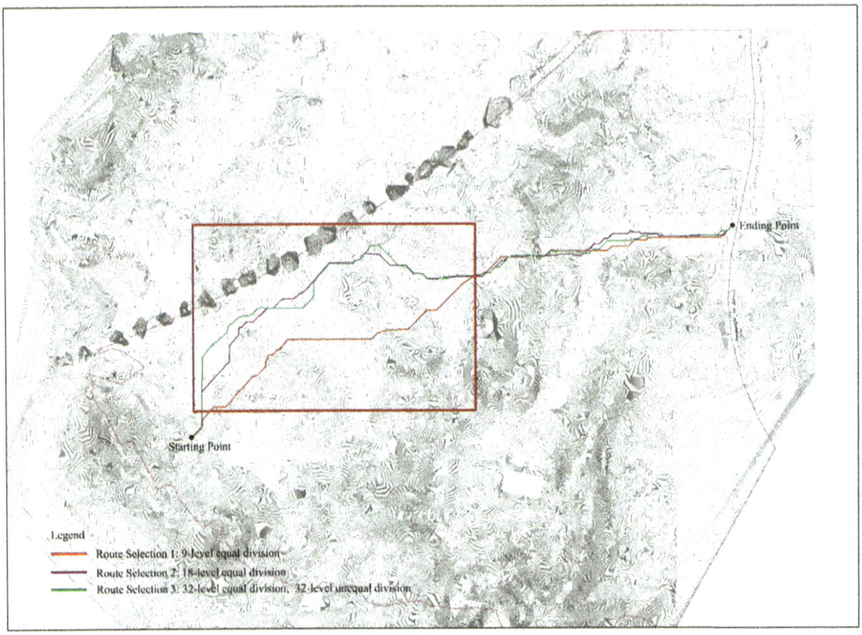

Fig. 7.8 Long-Distance Route Selection—comparison of results based on different classification levels

7.5.1.4 Comparison Three: Graded Value Assignment of the Comprehensive Cost Map

According to the cost path algorithm, the number of grading levels provides different cost values for computation, thereby increasing the range of possible selections. The value attribute of cost is reflected in the assignment of grading values. In practice, the following problem often arises: the cost value assigned to the slope range of 25°–30° is 6, while that for the 30°–35° range is 7, with only a difference of "1" between them. This can easily cause the calculated path to pass through the 30°–35° slope area rather than the 25°–30° area. Such issues occur because the assigned values of adjacent grades are too close, making the cost of steeper slopes relatively lower under the shortest distance calculation. Therefore, it is necessary to use methods to enlarge the graded value assignment, by increasing the "cost" difference between levels, so as to constrain the path selection tendency towards areas with gentler slopes.

The author chose to apply an exponential assignment method, comparing the path selection results generated by the original values, exponential values with base $^{1}/2e$, and exponential values with base 2. As shown in Table 7.4, the exponential assignment results in geometrically increasing graded values, with the base 2 exponential showing a faster growth rate and the largest value gap between grades (see Fig. 7.9).

Table 7.4 Different grading valuation table

Classification	32-level equal division	Non-exponential assignment	Exponential assignment with base ¹/2e	Exponential assignment with base 2
1	3°	1	1.36	2
2	6°	2	1.8496	4
3	9°	3	2.515456	8
4	12°	4	3.42102016	16
5	15°	5	4.652587418	32
6	18°	6	6.327518888	64
7	21°	7	8.605425688	128
8	24°	8	11.70337894	256
9	27°	9	15.91659535	512
10	30°	10	21.64656968	1024
11	33°	11	29.43933476	2048
12	36°	12	40.03749528	4096
13	39°	13	54.45099358	8192
14	42°	14	74.05335126	16,384
15	45°	15	100.7125577	32,768
16	48°	16	136.9690785	65,536
17	51°	17	186.2779468	131,072
18	54°	18	253.3380076	262,144
19	57°	19	344.5396903	524,288
20	60°	20	468.5739788	1,048,576
21	63°	21	637.26-06112	2,097,152
22	66°	22	866.6744313	4,194,304
23	69°	23	1178.677227	8,388,608
24	72°	24	1603.001028	16,777,216
25	75°	25	2180.081398	33,554,432
26	78°	26	2964.910702	67,108,864
27	81°	27	4032.278554	134,217,728
28	84°	28	5483.898834	268,435,456
29	87°	29	7458.102414	536,870,912
30	88°	30	10143.01928	1,073,741,824
31	89°	31	13794.50622	2,147,483,648
32	90°	32	18760.52846	4,294,967,296

Figure 7.10 shows the route selection results corresponding to three different valuation methods. The route selected with the non-exponential valuation chooses the area with relatively sparse contour lines in the first half of the path, but cuts straight through the densely contoured area in the second half. The routes selected with the ¹/2e exponential valuation and the 2-exponential valuation are similar in direction. Among them, the ¹/2e exponential valuation route follows a gentler ridge, while the 2-exponential valuation route takes a more

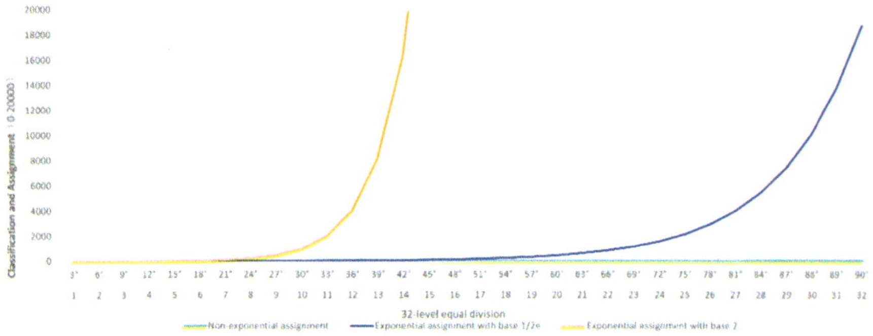

Fig. 7.9 Growth curves of different grading valuations

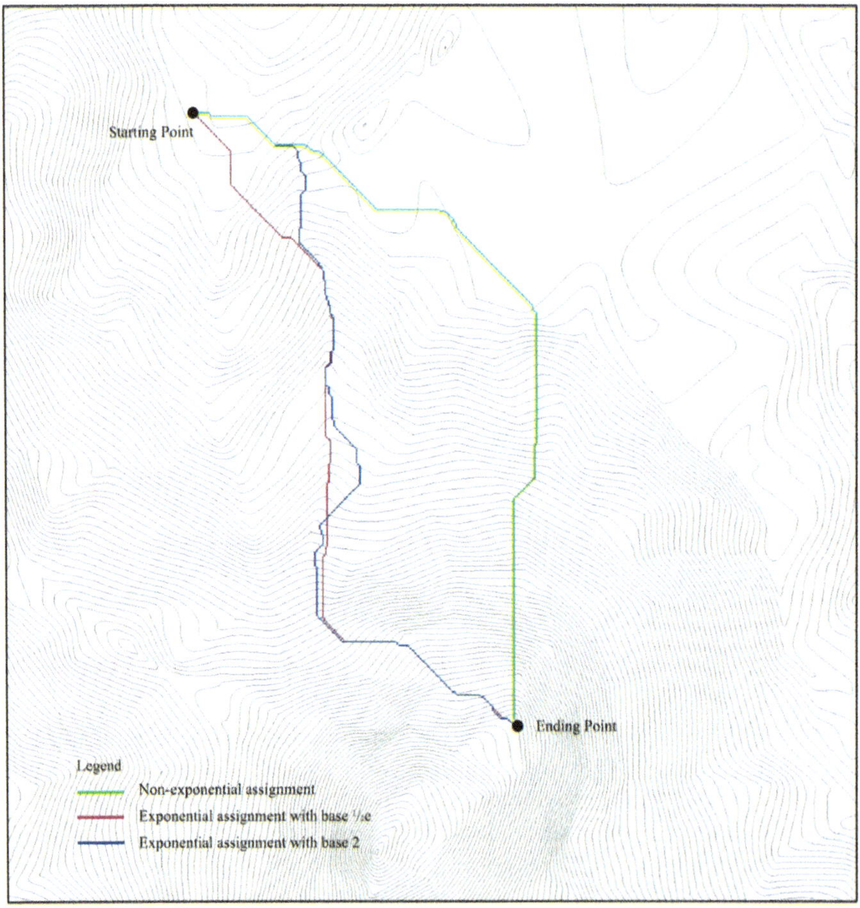

Fig. 7.10 Cost Distance Method—comparison of route selection results based on different grading valuations

winding and detoured path. The reason for the route chosen under the 2-exponential valuation is that the higher cost assigned to steep slopes makes the cumulative cost of a longer detoured path lower than that of a straight crossing. This selection outcome better aligns with the ideal form for route planning in landscape architecture environments.

7.5.1.5 Analysis and Summary

From the above analysis, it can be seen that:

> Compared to single-factor cost maps, comprehensive factor cost maps can more comprehensively reflect the site conditions for route selection. By controlling costs to impose constraints on the route selection, it helps to avoid ecologically sensitive areas, water bodies, and other regions that roads should not cross. Whether existing roads should be included as part of the comprehensive cost calculation remains to be further explored, because excessively low costs can greatly reduce the cumulative cost of the route. As a result, the path tends to seek the nearest point to access the existing road from the starting point, causing direct crossing of steep slope areas and seriously interfering with the route selection results.

Regarding the classification levels of the comprehensive cost map, having more levels does not necessarily mean the route selection result is more reasonable. In terms of route selection, the 18-level equal interval classification performs best, while the first half of the 32-level equal interval classification performs relatively well. The reason is that a larger number of levels, represented by the 32-level unequal interval classification, causes the cost values to increase too rapidly with slope, excessively raising the route cost corresponding to steep slopes. This leads the route to preferentially choose the gentlest slopes, but then to climb steep slopes in the latter part to reach the endpoint, resulting in an overall less reasonable route. Analysis of long-distance route selections shows that the number of classification levels has a more significant impact on long-distance routing. The computed route generally follows the gentlest slope area. Comparatively, the 32-level equal and unequal interval classifications perform better in such cases.

From the comparison of route selection results based on different classification value assignments, the 1/2e exponential assignment and the 2-exponential assignment yield better results. Among them, the 2-exponential assignment effectively avoids long straight crossings of densely contoured areas, producing more reasonable outcomes. This shows that enlarging the cost difference between slope levels can effectively optimize the route, generating road routes more consistent with landscape planning design. The drawback of the 2-exponential assignment is obvious: as slope increases, the classification values grow exponentially, causing the assigned data to rapidly increase and adding computational burden. Whether there are better methods than simply enlarging classification values to control route selection will be further discussed in the next section of this study.

7.5.2 *Comparative Analysis of Road Routing Based on the Path Distance Algorithm*

In the previous section, road routing results using the cost distance algorithm under different parameter inputs for various variables were compared and analyzed. The path distance algorithm is one of the key algorithms within ArcGIS's distance tools. Compared to the cost distance algorithm, it incorporates more parameters into the computation process, including surface raster, horizontal factor, and vertical factor. Regarding road routing in landscape environments, the path distance algorithm better satisfies the characteristics of route selection compared to the cost distance algorithm.

7.5.2.1 Characteristics and Analysis of the Road Route Selection Based on the Path Distance Algorithm

The Road Route Selection model of the path distance algorithm is fundamentally similar to that of the cost distance algorithm. However, the biggest difference in applying the path distance algorithm to road routing in landscape planning lies in the inclusion of two important variables: one is the surface raster, which incorporates the actual movement distance of objects across undulating terrain into the calculation; the other is the vertical factor, which defines the critical element of slope in routing and controls the impact slope has on the route selection. Compared to the cost distance algorithm, which guides routing by enlarging the assigned values between classification levels, the path distance algorithm offers a more reasonable route guidance and is easier to operate. The vertical factor variable includes two parameters: the vertical factor angle and the vertical parameter SLOPE value. Determining which of these variables meaningfully contribute to road routing in landscape environments is one of the issues discussed in this section. Additionally, this section will analyze how the choice of influencing factors from the previous section affects the routing results when using the path distance algorithm. By controlling different variable parameters, the study explores the optimal routing model for road selection in landscape planning design using the path distance algorithm.

7.5.2.2 Comparison One: Selection of Influencing Factors

The only parameter differing between routing calculations based on comprehensive factors and those based on slope as a single factor is the comprehensive cost map; all other parameters remain consistent. Figure 7.11 shows routing results generated by the path distance algorithm for both cases under a 32-level equal interval cost classification. It is evident that the routing results differ significantly. The routing based on comprehensive factors tends to select relatively gentle valleys and ridges, traversing contour lines in a winding manner and approaching the

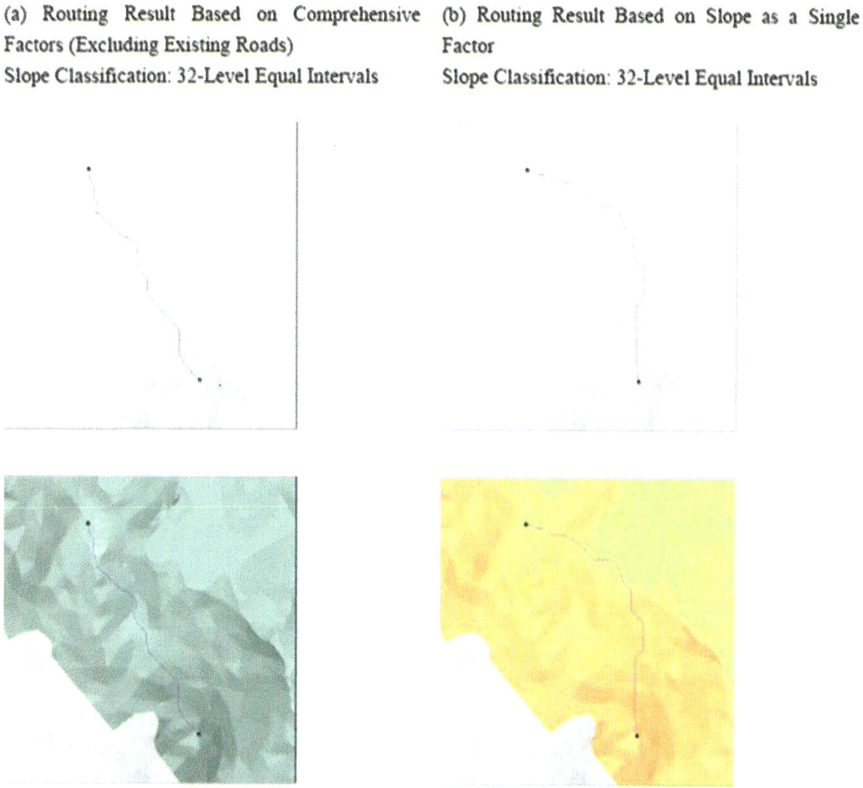

Fig. 7.11 Path Distance Method—routing results based on different cost factor maps

destination; whereas the routing based solely on slope tends to avoid densely contoured areas as much as possible, traveling through the gentlest slopes in the first half, but in the latter half features some direct crossings of contour lines (see Fig. 7.12).

7.5.2.3 Comparison II: Vertical Factor Angle

The vertical factor angle refers to the vertical relative movement angle when moving from one raster cell to the next in the calculation. It can be understood as the vertical angle between the start point (rear foot) and the end point (front foot) of a step during walking, with each step representing one unit of distance. The significance of setting a limit for the vertical factor angle lies in treating movement over a certain slope angle as having an infinite cost—thus controlling the routing calculation to avoid areas with steep slopes.

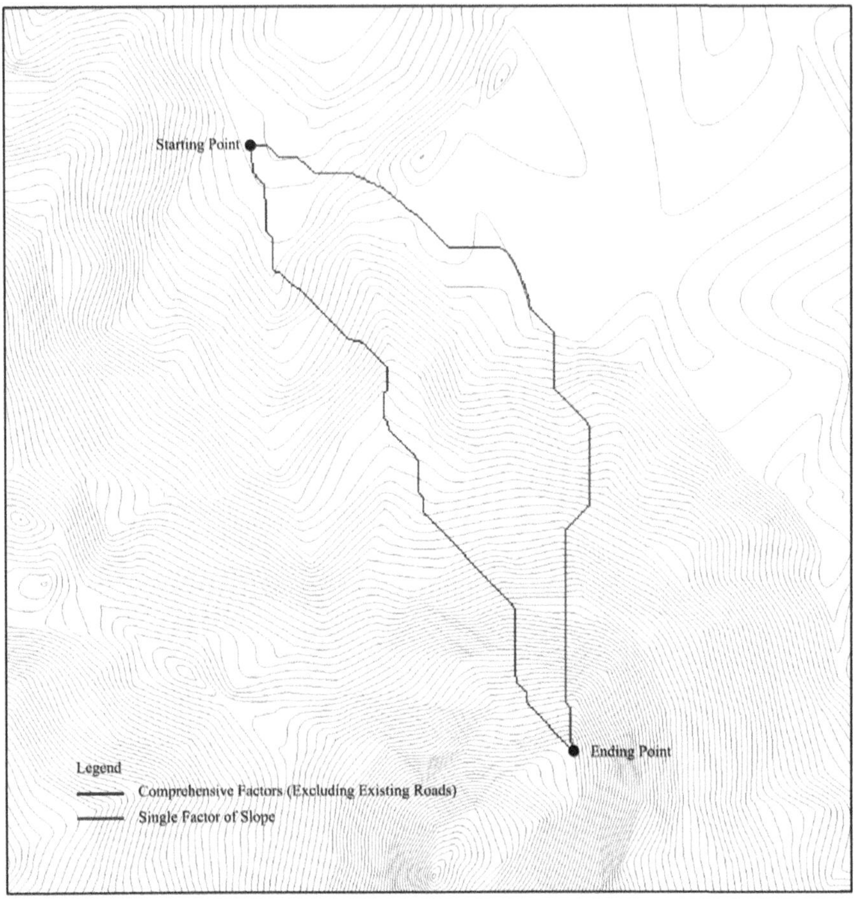

Fig. 7.12 Path Distance Method—comparison of routing results based on different cost factor maps

Generally speaking, in a landscape environment, areas suitable for road construction should not have slopes greater than 30°. Based on this, the vertical factor angle limit can be set at ±30°, meaning that whether ascending or descending along the terrain surface, the angle should not exceed 30°, otherwise the cost is considered to approach infinity.

As shown in Fig. 7.13, the two routing results correspond to input parameters for the 32-class equal-interval composite cost map with vertical angle limits set at ±15°, ±30°, and ± 90°. From Fig. 7.14, we can see that the routing results for the ±30° and ± 90° limits are completely overlapping, while the route under the ±15° constraint takes a more winding path through gentler slope areas to reach the destination and differs significantly from the other two.

The reason lies in the fact that the slope values in the area between the start and end points mostly do not exceed 30°. Therefore, setting the limit at ±30° and ± 90°

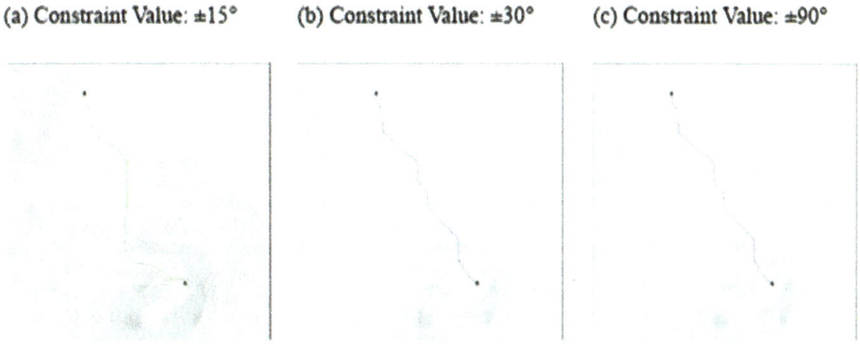

Fig. 7.13 Path Distance Method—routing results under different constraint values

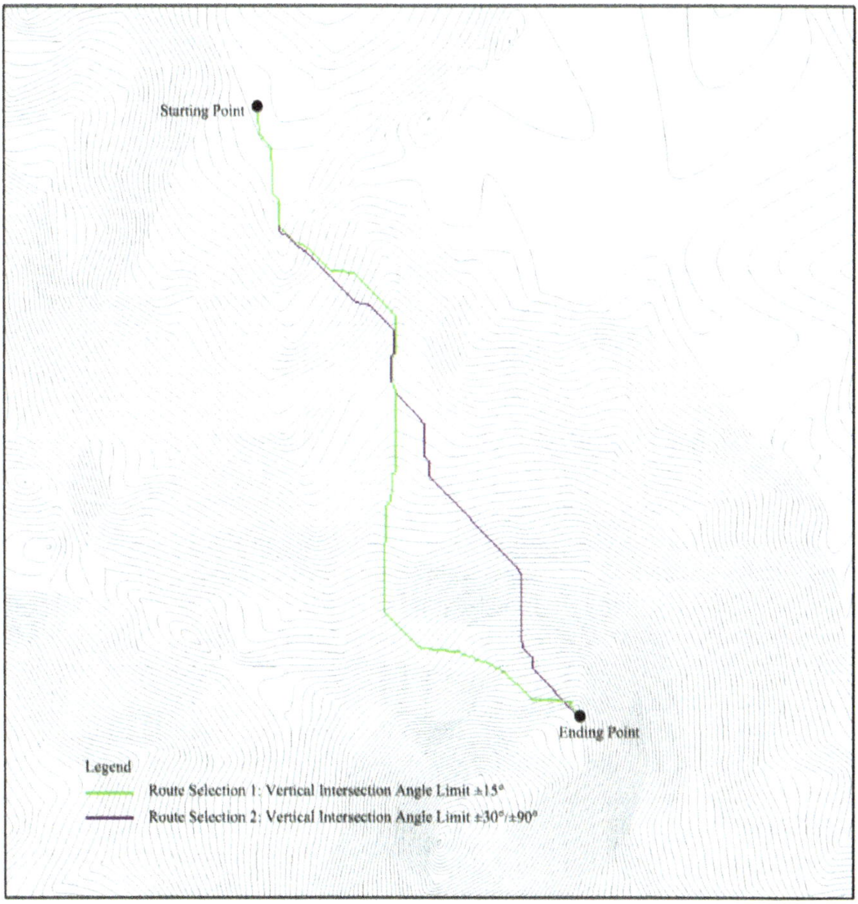

Fig. 7.14 Path Distance Method—comparison of routing results under different constraint values

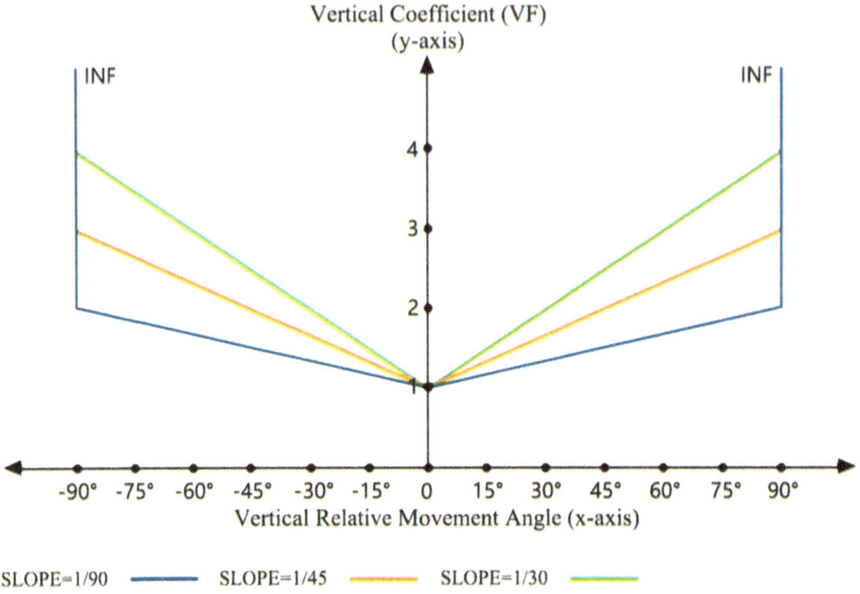

Fig. 7.15 Symmetric linear function vertical coefficient curves corresponding to different SLOPE values (vertical coefficient angle = ±90°)

yields effectively the same routing constraints, resulting in identical routing outcomes. The route generated under the ±15° constraint stands out more in this particular case.

7.5.2.4 Comparison Three: Vertical Parameter SLOPE Value

The SLOPE value is a modifying coefficient that reflects the functional relationship between the vertical coefficient and the vertical relative movement angle. According to the characteristics of road alignment planning in landscape architecture, a symmetrical linear function is selected during calculation, meaning that whether the vertical relative movement angle is positive or negative, the vertical coefficient always follows a linear function, and these two linear functions are symmetric about the vertical coefficient (y-axis).

As shown in Fig. 7.15, as the angle decreases, the corresponding SLOPE value increases, and the slope correspondingly becomes steeper. The slopes in the figure decrease in order as follows: SLOPE = 1/30 > SLOPE = 1/45 > SLO PE = 1/90.

Figure 7.16 presents the routing results corresponding to four different SLOPE values. The results for SLOPE values of 1/15, 1/30, and 1/45 are quite similar, with the paths for 1/15 and 1/30 nearly overlapping. The path for SLOPE = 1/90 shows similarity with the others only in the first half but deviates significantly in the latter

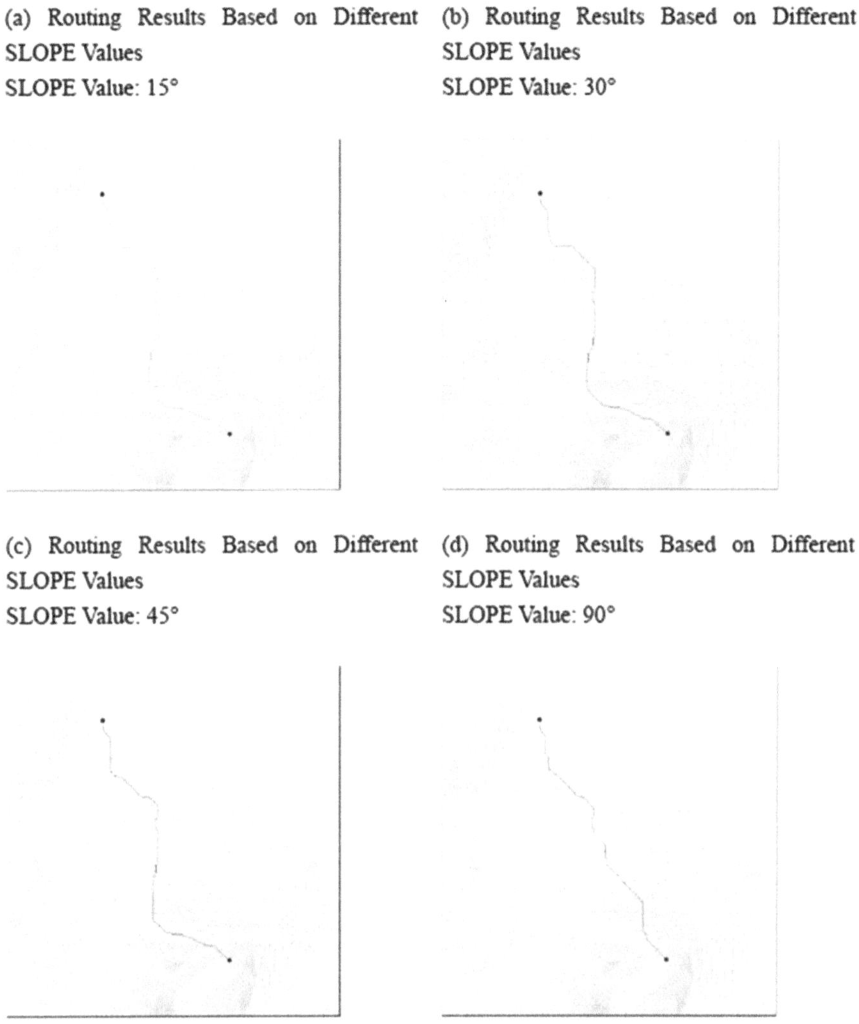

Fig. 7.16 Path Distance Method—routing results based on different SLOPE values

half. The 1/90 path tends to cut across contour lines almost in a straight line, which is more pronounced compared to the others. The routing results for SLOPE values of 1/15, 1/30, and 1/45 effectively avoid areas with dense contour lines and choose to travel through gentler slopes (Fig. 7.17).

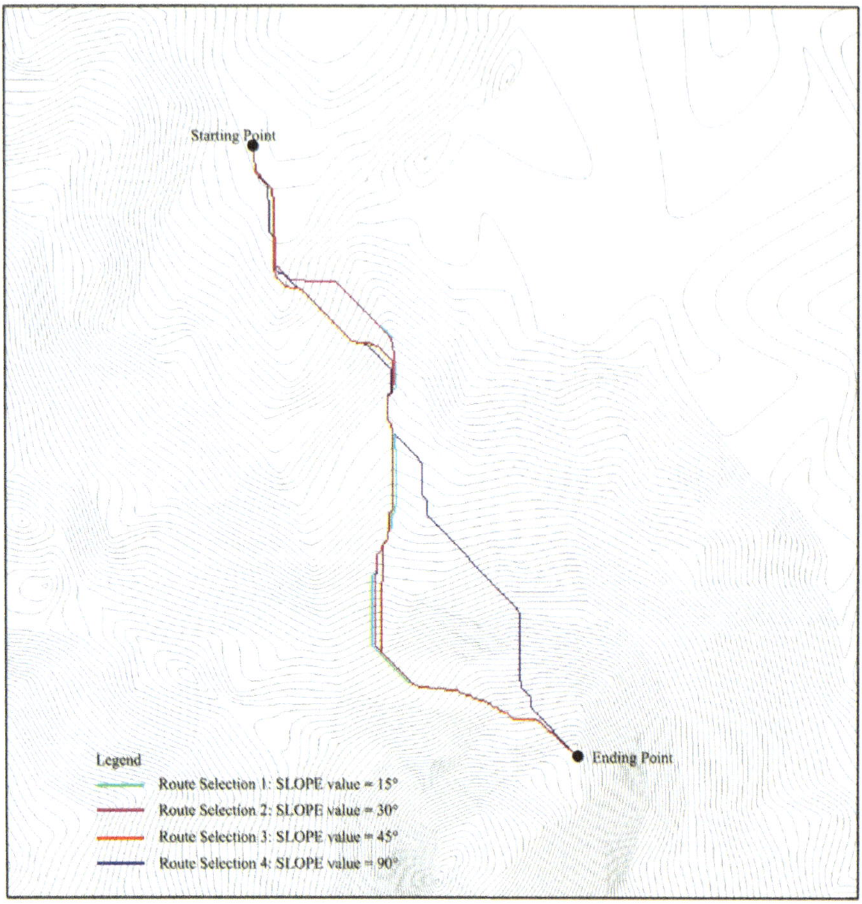

Fig. 7.17 Path Distance Method—comparison of routing results based on different SLOPE values

7.5.2.5 Analysis and Summary

From the above comparisons, it can be seen that, similar to the cost distance algorithm, the comprehensive factor cost map can more comprehensively reflect the site conditions for route selection, while the slope single-factor cost map emphasizes the impact of slope on the route selection. According to the comparison results, the slope single-factor cost routing chooses gentler slope areas in the first half of the path, showing better performance; however, in the latter half, in order to reach the destination, it has to cross areas with dense contour lines in an almost straight manner, resulting in a less ideal route. The comprehensive factor cost routing does not detour around areas with dense contour lines at the starting stage, but instead advances in a winding way toward the destination. Overall, the routing based on the comprehensive factor cost avoids crossing water bodies in the site, and the path is relatively reasonable. It should also be noted that the path distance algorithm, due to its multiple parameter inputs as route selection controls, can better satisfy the slope requirements for routing through the setting of these parameters, thus eliminating the need to strengthen slope costs to control route slope.

The limitation of the vertical factor angle is a highly significant parameter for road routing in landscape architecture environments. This is because setting this parameter can enforce reasonable slope constraints on the route selection, thereby avoiding road construction in excessively steep areas. The comparison shows that for different landscape environments, the vertical angle limit should be chosen according to the slope conditions, so that the path distance algorithm can more reasonably compute the routing path and produce results that fit the site conditions. For example, when routing roads in environments with large slope fluctuations and steep slopes, the vertical angle limit can be set higher, such as greater than ±30°, whereas for environments with gentle slopes, the limit should be set to a smaller range, such as around ±15°.

The SLOPE value, as a slope parameter, determines the magnitude of the vertical factor for different slopes through a linear symmetric function. This coefficient adjusts the difficulty of climbing, where a larger slope value indicates that the climbing difficulty for a given slope is greater compared to the same slope with a smaller slope value. Based on the comparison of routing results under different SLOPE values, in landscape environments, a SLOPE value of 1/45 can basically meet routing requirements, while SLOPE values of 1/15 and 1/30 can better satisfy the routing demands.

7.5.3 Comparison of Cost Distance Algorithm and Path Distance Algorithm for Road Route Selection Models

By individually comparing different parameter inputs for the variables of the cost distance algorithm and the path distance algorithm, preliminary parameter selections suitable for road routing in landscape architecture planning and design have been established. Given the characteristics of these two algorithms, this section focuses on discussing which algorithm is more appropriate for road routing in

Table 7.5 Comparison of parameters for road routing models: Cost Distance Algorithm vs. Path Distance Algorithm

Parameters		Cost Distance Algorithm	Path Distance Algorithm
Influencing factors		Composite factors (excluding existing roads)	Composite factors (excluding existing roads)
Cost classification		32-class equal interval division	32-class equal interval division
Class value assignment		Without exponentiation	Without exponentiation
Vertical parameters	Vertical factor angle	–	±30
	SLOPE value	–	30°

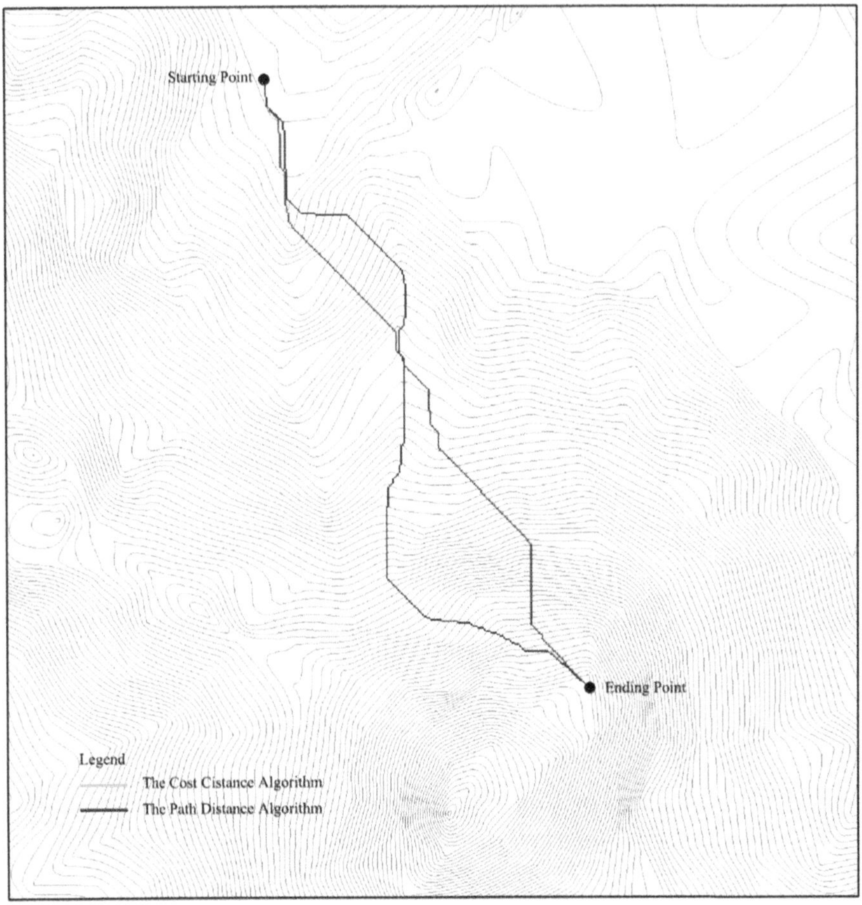

Fig. 7.18 Comparison of routing results between Cost Distance Algorithm and Path Distance Algorithm

landscape architecture environments. Table 7.5 lists the parameters required by both algorithm models and the selected parameters used in the comparative analysis of the routing results. The parameters chosen are those identified in previous discussions as suitable for road routing in landscape architecture design, and the common parameters are kept consistent between the two algorithms to ensure a more reasonable comparison.

Figure 7.18 compares the routing results of the two algorithm models. The routing path based on the cost distance algorithm approaches the endpoint in a relatively linear manner with some straight diagonal crossings over contour lines, resulting in a less ideal routing outcome. In contrast, the routing path based on the path distance algorithm meanders from the starting point, following gentler slopes along valleys and ridgelines to reach the endpoint. By comparing the results of the two algorithm models, it is clear that the routing outcome based on the path distance algorithm is significantly better than that of the cost distance algorithm.

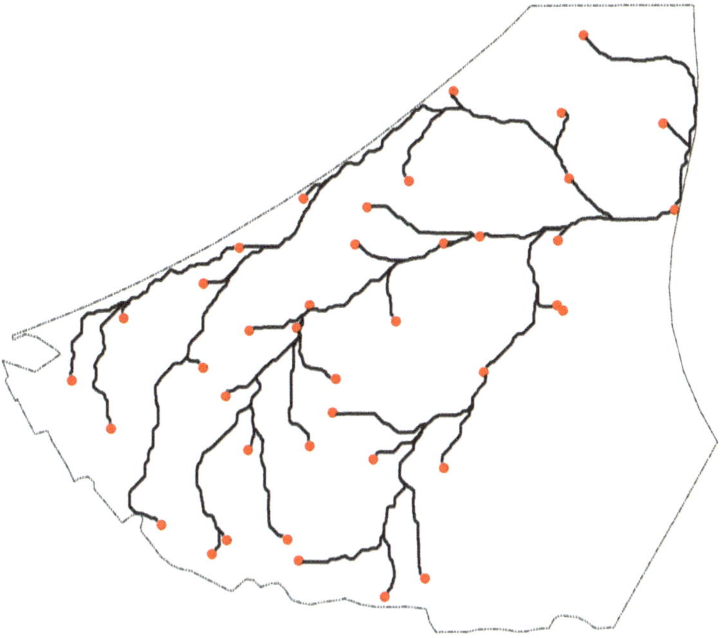

Fig. 7.19 Roads with a "branch-like" structure generated through multi-point path selection

7.5.4 Comparison and Optimization of Multi-point Road Route Selection Models

In landscape architecture planning and design, the construction areas determined by project site selection are future landscape nodes, including points for recreational facilities, scenic viewing spots, and so forth. These landscape nodes must be connected by a road network to facilitate visitor access. Therefore, these nodes become "control points" in the road routing process — serving as the start points, intermediate points, and end points of roads. Landscape architecture design often deals with medium- to large-scale areas, where multiple entrances and exits exist based on different service positioning within the region. Every landscape node needs smooth access from the main entrance of the area, making it necessary to establish connections with the main entrance. Consequently, the road routing model developed here takes the location of the main entrance as the starting point parameter input. Given the multitude of scenic spots, road routing in landscape environments inevitably faces a multi-point routing problem. To avoid excessive workload and minimize subjective human interference, this study treats multiple points as simultaneous endpoints, inputting them as parameters in one go for the routing calculation. This approach is called the "1+N multi-point multi-route method" (Fig. 7.19), proposed as an optimization of road routing based on the ArcGIS software platform. In this section, a detailed discussion will be provided on the comparison of the number of

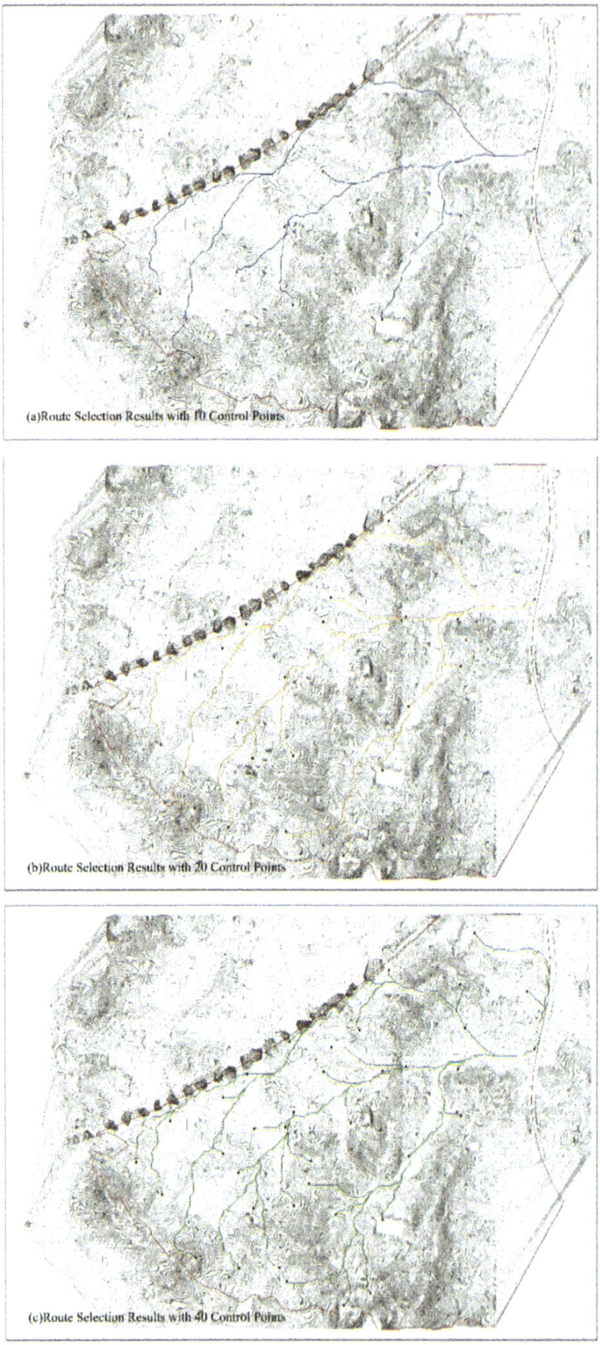

Fig. 7.20 Road alignment results based on different numbers of control points

control points and how to implement the "1 + N" multi-point multi-route method effectively.

7.5.4.1 Comparison: Control of the Number of Control Points

Figure 7.20 shows the road routing results using the path distance algorithm with 10, 20, and 40 control points respectively. The following characteristics can be observed:

First, the routing paths form a "branch-like" structure. Due to the nature of the path distance algorithm, all "end points" seek the minimum cost shortest path to reach the "start point," so every "end point" connects back to the "start point." At the same time, according to the calculation rules, the input cost parameters and the start point determine the minimum cost and backtracking direction from each grid cell to the next. Thus, the shortest path in a certain direction is established under these premises. Control points located near this direction will seek the shortest path to this route and eventually reach the "start point" along this confirmed shortest path.

Second, as the number of control points increases, the density of the "branches" also grows. This is intuitive: more control points mean more points connected to the start point, thus generating more routing paths. The number of control points is determined by project site selection, which varies continuously depending on different sites and projects.

Controlling the road direction through the "control points" derived from the site eventually produces the designed road network. This road routing process reflects the "coupling" principle and effect of planning and design starting from the site conditions.

7.5.4.2 Optimization: The "1 + N" Multi-point Multiple-Run Route Selection Method

In landscape environments, road network planning typically requires the formation of loops and meshes to ensure good accessibility to each scenic spot and to facilitate the orderly flow and dispersal of visitors. From the previous analysis, it can be seen that multi-point routing conducted in a single run produces road paths with a "tree-like" structure, which does not form a mesh network or looped roads. This is because each control point entered as an "endpoint" only connects to the "start point," and there is no connection between the endpoints themselves. However, in a road network, each node should be connected to at least two or more points. Therefore, the road paths obtained from a single routing run cannot meet the requirements for road routing in landscape environments. Due to the scale of landscape environments, there are usually multiple entrances and exits rather than just one. After the main entrance is used as the "start point" in the first routing run, the other entrances/exits also need to be connected to nodes inside the area. Hence, these entrances/exits can serve as "N" start points for routing computations. Generally, only one additional

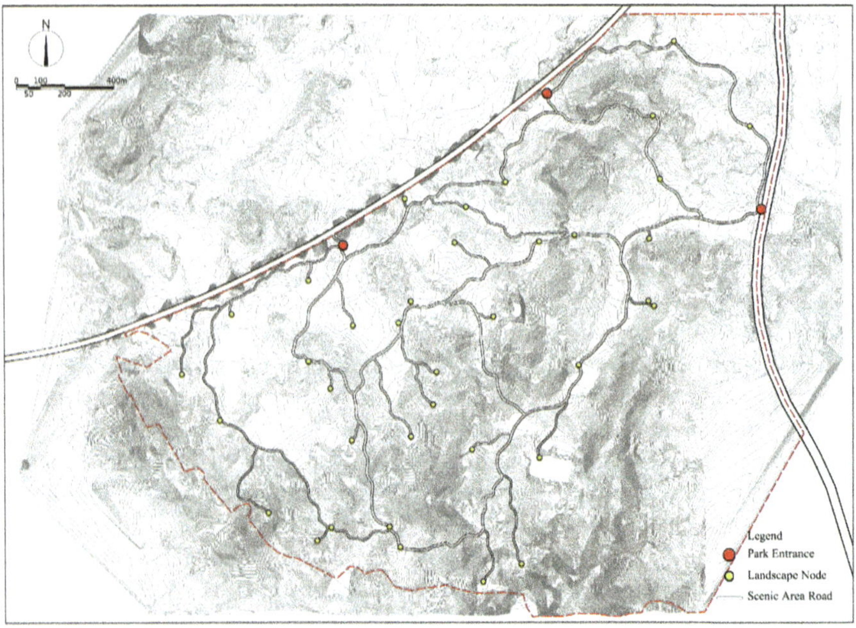

Fig. 7.23 Road network formed by optimized routing results

7.6 Alignment Optimization and Road Network Generation

The routing lines calculated based on raster maps are polylines and cannot be directly used as the final routing results. Therefore, these lines need to be smoothed into curves to meet road design requirements (Fig. 7.22). Using advanced editing tools in ArcGIS software, the routing paths are edited by inputting an appropriate "maximum allowable deviation" value to curve the paths. Unreasonable parts are then manually optimized to produce the final routing alignment. Based on the generation of all road network paths, primary and secondary roads are distinguished. The selected routing paths serve as the road centerlines, resulting in the final road network (Fig. 7.23).

Chapter 8
Simulating Natural Waterscape Construction Model

The creation of waterscapes holds positive significance both aesthetically and ecologically within landscaped environments. As one of the key themes in landscape planning and design, waterscapes—particularly those featuring natural elements—can take on countless forms through artificial intervention. However, their formation must respect the laws of nature and be based on a thorough understanding of site conditions. Waterscapes generated under limited human intervention, while appearing natural in form, also become integral components of the environment by adhering to objective principles. The construction of simulating natural waterscapes reflects a multi-objective and intensive design approach, emphasizing reduction and sustainability.

This chapter explores the parameterized methods for constructing simulating natural waterscapes in landscape environments. Starting from their significance in landscape planning and design, it elaborates on the scientific and engineering connotations of simulating natural waterscapes. Drawing extensively from engineering practice and based on the ArcGIS platform, this chapter establishes a constructive model for simulating natural waterscapes and illustrates its application through the example of simulating natural waterscape construction in the northern area of Niushou Mountain Scenic Area in Nanjing.

8.1 Waterscape Construction in Landscape Planning and Design

8.1.1 Waterscapes in Landscape Planning and Design

Humans have always had a unique emotional connection to water. Not only is it indispensable for production and daily life, but it has also been imbued with aesthetic values throughout human history. This has given rise to the saying, "There is

Y. Yuan, *Parametric Landscape Architecture Planning and Design*,
https://doi.org/10.1007/978-981-95-8011-8_8

no vitality in a landscape without water." Throughout history and across cultures, waterscape design has remained a vital part of landscape architecture. From ancient Egypt to Rome, through the Renaissance in Europe, and into the nature-revering English landscape gardens, waterscapes have consistently been a timeless theme. In traditional Chinese gardens, water is a critical component, with the art of stacking mountains and managing water being an essential element of garden design. As early as the Spring and Autumn and Warring States periods, King Wen of Zhou's Ling Pool marked the earliest model of architectural water management. From Emperor Qin Shi Huang diverting the Wei River into Lanchi Pool to the introduction of Yuquan Mountain springs into Kunming Lake at the Summer Palace, Chinese garden designers have persistently sought harmony between mountain and water landscapes by skillfully leveraging natural elements. "The lake reflects autumn moon in harmony; its surface undisturbed like an unpolished mirror."[1] Countless poems by Chinese poets throughout history have captured the beauty of waterscapes. Moreover, in traditional cultural frameworks, "water" carries moral and educational significance. The culture of water has deeply rooted itself in Chinese civilization and continues to shape aesthetic tastes.

Beyond its aesthetic and cultural meanings, the ecological significance of waterscapes in landscape planning and design should not be overlooked. "Water" in landscape environments has a dual meaning: on the one hand, it is a vital part of the landscape itself; on the other, mismanagement of surface runoff can cause flooding and pose threats to both the landscape and human settlements. Consequently, concepts such as Low-Impact Development (LID), Water-Sensitive Urban Design (WSUD), and Sponge Cities have emerged globally to address water management in urban environments, reflecting a growing emphasis on water ecology. While pristine natural environments have an inherent "sponge effect" due to undisturbed ground surfaces, the "sponge bodies" in urban built environments include rivers, lakes, ponds, and supplementary infrastructure such as green spaces, gardens, and permeable pavements. Landscape planning and design is not limited to urban built environments; it also covers vast peripheral and scenic areas with minimal artificial construction. These areas function as large-scale peripheral "sponge bodies," and water retention within them is equally important. Natural depressions in the terrain collect precipitation to form ponds, pools, and wetlands. Aside from groundwater recharge, these natural water bodies are primarily replenished by surface and subsurface runoff. Water bodies that temporarily or permanently collect rainfall in landscape environments hold immense ecological value. The presence of such water bodies—especially when connected into water systems—enables the collection of rainfall within catchment areas. On one hand, they slowly release water into surrounding soil during dry periods, alleviating water shortages and balancing water levels. On the other hand, they buffer the impact of sudden and heavy rainfall events, thereby reducing downstream erosion and aiding in soil and water conservation. Furthermore, the interface between land and water is among the richest habitats for

[1] From *Looking at Dongting Lake*, a poem by Liu Yuxi, a Tang Dynasty poet.

biodiversity. The existence of water systems fosters favorable ecological environments and enhances species diversity in the region.

In summary, the creation of waterscapes contributes positively to both the aesthetic and ecological aspects of landscape environments. How to rationally manage water storage while achieving both water resource conservation and scenic enhancement is a valuable research topic in contemporary landscape planning and design. Beyond traditional aesthetic and ecological concerns, the organic integration of water form and use—scientifically retaining, storing, and utilizing water resources—has become an indispensable area of study in modern landscape architecture.

8.1.2 Construction of Simulating Natural Waterscapes

Starting from the Tang Dynasty through the Northern Song and into the Ming Dynasty, Hangzhou's West Lake underwent four major phases of human-led development, ultimately becoming today's internationally renowned natural-style landscaped environment centered around natural waterscapes. Even more exemplary is the grand simulating natural waterscape system of the Old Summer Palace (Yuanmingyuan), which stands as a masterpiece of ancient waterscape construction. The nature-revering ethos of Chinese cultural traditions has long shaped a design philosophy centered on imitating nature, thereby fostering a tradition of simulating natural waterscape creation. In contemporary landscape planning and design, waterscape construction remains a key focus among designers. Landscape planning involves both natural and built environments, and the design of simulating natural waterscapes is applied to both types of settings. Within these two environmental contexts, the concept of "simulating natural" carries relatively broad implications, which can be summarized in two layers: First, in landscaped environments with manageable scale, the emphasis lies in mimicking natural forms—interpreting terrain and landforms to produce waterscapes with natural charm, where *ordering* (理) plays a central role. Second, in built environments with larger and more varied scales, "simulating natural" construction is more about human efforts to emulate natural laws. Though these two interpretations differ in scope, they share a common essence—understanding and adhering to natural principles.

8.1.2.1 The Connotation of "Nature"

The term "nature" is rich in meaning. In philosophy, the term "first nature" refers to environments unaltered by human intervention. Within the scope of this research, "nature" is used in contrast to "artificial," denoting environments untouched—or minimally affected—by human activity. It may also refer to *primitive* nature, manifesting as unspoiled landscapes [1]. "Simulating natural" literally means "like nature." Thus, the construction of simulating natural waterscapes refers to *artificially created water landscapes that resemble those found in nature*. "Natural waterscapes" are water bodies formed by natural processes, such as rivers, lakes, seas, and waterfalls—e.g., Poyang Lake, Huangguoshu Waterfall, and Sun Moon Lake—which are considered true natural landscapes (Fig. 8.1). In contrast, the formation of simulating natural waterscapes is distinct from symbolic, idealized representations such as the traditional "one pond and three hills" layout—these are considered imitative "false landscapes." What this research emphasizes is the creation of *simulating natural "true landscapes."* Unlike artificial waterscapes that mimic or copy the forms of natural water bodies, the focus of simulating natural waterscape construction in this study is on the principle of *ordering*. As Master Pan Guxi stated in his renowned work *The Art of Landscape Structuring in Jiangnan*, *"Natural mountain-and-water scenery is not the same as 'garden scenery'; it cannot be fabricated—it can only be utilized... 'Ordering' refers to governance. Ordering may involve creation, such as building gardens or bonsai; or adaptation, such as following the terrain and water, arranging elements skillfully so the beauty of mountain and water is fully expressed and utilized*" [2]. Thus, "ordering water" means utilizing the existing topography and hydrological conditions—working with the terrain and water dynamics—to reorganize or optimize them, forming new waterscapes. The "nature" in simulating natural waterscape construction bears a dual implication of *form* and *law*. From the perspective of *form*, the construction of simulating natural waterscapes involves artificial interventions that produce water features with natural appearances. It is important to note that *curved shapes alone do not equate to natural forms*. Natural water bodies are typically irregular and free-form, shaped through natural processes. On land, natural waterscapes may include rivers, lakes, streams, ponds, wetlands, waterfalls, springs, and more. These water bodies have extremely flexible boundaries and exist within open systems. Their forms vary based on environmental factors—e.g., water levels fluctuate seasonally, which in turn alters their shapes. From the perspective of *law*, constructing simulating natural waterscapes requires minimal human intervention to generate natural hydrological systems. A water system refers to the water network within a watershed, inherently forming a complete aquatic ecosystem. As the saying goes, *"Follow the laws of nature."* In constructing simulating natural waterscapes, not only must

Fig. 8.1 Natural waterscapes Sun Moon Lake and artificial waterscapes West Lake (Cited from the Internet)

the resulting forms appear natural and beautiful, but the creation process must also conform to natural laws. The ability to form a self-sustaining and ecologically complete system is a defining feature of simulating natural waterscapes. Designers must begin with existing site conditions and the *forces of nature*—analyzing current topography, hydrological patterns, and other environmental factors—while mastering ecological principles such as habitat requirements. Only then can construction proceed in harmony with nature's logic.

8.1.2.2 The Significance of Simulating Natural Waterscapes

In recent years, many countries around the world have begun re-evaluating river channel engineering projects that previously disrupted the natural environment. New development concepts such as "near-natural rivers" and "multi-natural river construction" have emerged, prompting efforts to modify artificial rivers and restore them to a more natural state. This global trend reflects a growing contemporary understanding of the ecological and hydraulic engineering significance of nature-based river systems. The same realization applies to landscaped environments. The construction of simulating natural waterscapes holds not only positive ecological value at a regional scale, but also plays a vital role in the regulation and optimization of regional water resources. Through the processes of retention and gradual release, such systems can alleviate both water shortages and the risks associated with excess water. Incorporating artificially constructed water features into a broader natural system helps maintain the integrity and systemic nature of the regional hydrological network. Additionally, the creation of water systems can optimize the spatial distribution of water bodies within a region and enable rational allocation of water resources. Beyond ecological and hydrological benefits, simulating natural waterscape construction also reflects the principle of intensification. It emphasizes a site-specific approach—beginning with detailed analysis of the site's topography, hydrology, and other existing conditions—to identify locations with natural potential for water feature development. By working with the terrain and hydrological flow, the design minimizes environmental disturbance. This "research – analysis – control – optimization" process allows for the most effective use of existing natural conditions to generate water features in a natural state. Such an approach not only ensures optimal landscape forms but also promotes ecological protection, satisfies aesthetic and recreational needs, and is economically viable. Thus, the construction of simulating natural waterscapes fulfills the objectives of multi-target intensification and embodies the principles of low-impact design. Waterscapes that align with natural laws also reflect sustainability—they require little to no long-term maintenance and, by establishing a complete aquatic ecosystem, are capable of self-sustaining development without human intervention.

One illustrative example is the Ziqing Lake Tourist Resort in Nanjing, located in Tangshan Town, within the western extension of the Ning-Zhen Mountain Range. The area is characterized by significant terrain variation and serves as a runoff corridor for mountain flows. Heavy rainfall generates large runoff volumes that pose safety risks to surrounding man-made structures. During the planning and design of the resort, the terrain and watershed within the planning area were first studied and analyzed to identify runoff convergence zones. Based on elevation, outlet positions, and runoff volumes, a system of graded dams was constructed along the mountain's flood pathways at different elevations, creating multi-level water surfaces to regulate water resources. This design, grounded in the site's natural conditions, not only produced simulating natural waterscapes for visitors to enjoy—meeting aesthetic demands—but also enabled scientific and rational water management. In dry periods, water is retained and stored; in wet periods, runoff is mitigated to reduce downstream impact. These strategies collectively meet the requirements of sustainable development (Fig. 8.2).

8.1.2.3 Key Principles in Simulating Natural Waterscape Construction

Natural water bodies form in naturally occurring depressions and rely on water sources to maintain their levels. The principles underlying the formation of simulating natural water bodies are essentially the same—they require depressions and stable water supplies. One of the key reasons water cannot be retained in artificial environments is the absence of closed surface depressions. Therefore, the creation of "basin-like depressions" is the primary prerequisite for waterscape construction. These "basin-like depressions" are not limited to naturally formed low-lying areas, but also include artificial depressions created through damming or excavation (Fig. 8.3). Since simulating natural waterscape construction seeks to minimize intervention, it does not advocate large-scale earth excavation to form depressions. Instead, water retention is achieved by selecting runoff areas and strategically constructing dams based on hydrological analysis—this approach forms waterscapes with the least environmental disturbance. Because simulating natural water bodies are not naturally formed and excavation is avoided, their main sources of water are surface runoff from precipitation, supplemented by shallow subsurface runoff, with little to no reliance on groundwater. Under the assumption of a fixed water-holding

(a) 3D Elevation Analysis

(b) Runoff Analysis and Water Body Generation

Fig. 8.2 Construction of mimic-natural water scapes in Nanjing Ziqing Lake Tourist Resort

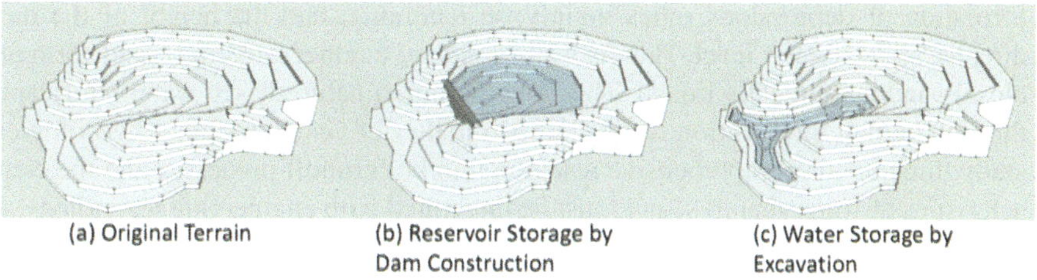

Fig. 8.3 Schematic diagram of pond basin depression formation

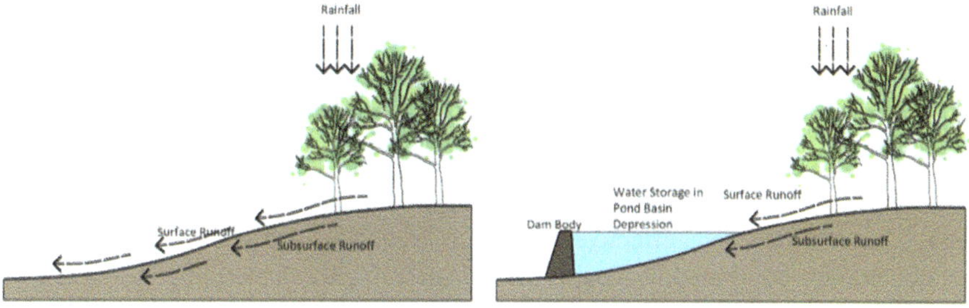

Fig. 8.4 Diagram of the principles of imitating natural waterscape construction

capacity in the basin, the volume of available water determines the water body's storage level. The aesthetic form of the water body is one of the most visible indicators of successful waterscape design. Since one of the primary functions of waterscapes is to serve people, their shape and form are key subjects of design. As the saying goes, "Water has no form—its shape is defined by its container." The shape of the water body, therefore, is determined by the depression in which it is held. In natural topography, the form of a water body is closely related to the elevation at which it is submerged—reflected in the intersection between water level and landform, as well as the spatial relationships among multiple water bodies. Due to the complex undulations of natural terrain, different water levels yield varied water body forms. Likewise, given the uniqueness of each site, no two simulating natural water systems will be exactly alike. Thus, simulating natural waterscape construction that is closely integrated with site-specific design is not only the most contextually appropriate, but also the most distinctive. In summary, there are three key factors to be considered in the construction of simulating natural waterscapes: water volume, engineering effort, and form (Fig. 8.4).

Whether in rivers, lakes, ponds, springs, or waterfalls, natural waterscapes do not exist in isolation; they are interlinked to form a hydrological network system. The construction of such systems requires grading based on runoff direction, with excess water from higher-level units flowing into lower ones—thus forming a complete system. As discussed earlier, water level and water volume jointly determine the shape of a water body within a specific depression. These three elements—water, engineering, and form—are therefore dynamically interrelated. The artificial formation of depressions relies mainly on damming, and the height of the dam directly affects water level. Hence, water volume, engineering effort, and form are closely and dynamically linked. The construction of nature-based water landscapes through human intervention is not simply a matter of visual design or hydrological study. It demands comprehensive research into both runoff processes and the characteristics of water features, and must be integrated with engineering technology. In this sense, simulating natural waterscape construction is a systematic project, as well as a comprehensive and dynamic process of balance and adjustment.

8.2 Construction of a Parametric Model for Simulating Natural Waterscape Formation

Simulating natural waterscapes are typically situated within landscaped environments and often span relatively large areas. The hydrological analysis tools available in ArcGIS can effectively support relevant analytical tasks, offering parameterized computation capabilities for specialized calculations. The creation of simulating natural waterscapes mirrors the formation processes of natural aquatic systems, necessitating the identification of runoff volume and water-retaining depressions. Based on these two key elements, the process can be broken down into the following steps: hydrological analysis—site selection for waterscape—water volume estimation—preliminary dam height assessment—waterscape morphology simulation. In the generation of simulating natural waterscapes, parameters such as precipitation, runoff volume, dam height, and waterbody morphology are incorporated throughout the computational process. These variables are interrelated and influence one another reciprocally. Figure 8.5 presents a parameterized model for simulating natural waterscape construction, illustrating the entire formation process and the logical relationships among the parameters.

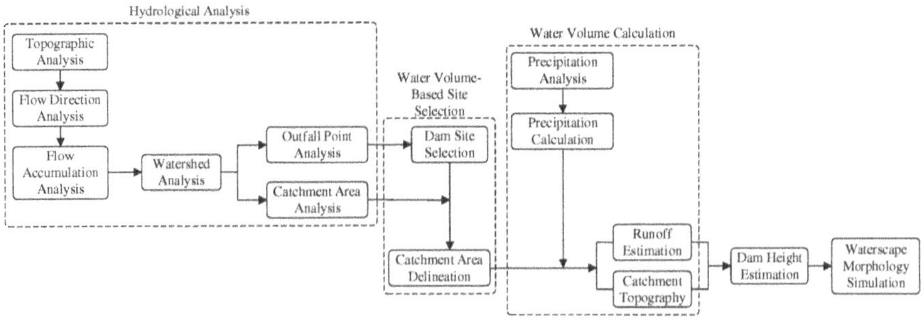

Fig. 8.5 Landscape architecture planning and design parameters-based simulated natural waterscape construction model

8.2.1 *Hydrological Analysis*

8.2.1.1 Runoff Analysis

The ancient Chinese landscape treatise Yuan Ye stated in the section on site selection: "When establishing a base, first investigate the water's source, trace the route of the flow, and observe the origin of the water" [3]. The first challenge in constructing a simulating natural waterscape is thus identifying an appropriate water source. In contrast to water features that require artificial supplementation, a key characteristic of simulating natural waterscapes is their reliance on natural water supply. Ideally, this source should be stable and capable of providing a continuous supply to sustain the waterbody. Historically, Chinese gardens were often located in regions with favorable natural conditions, drawing water either directly from surface sources or by well-digging to replenish garden waters. Notable examples include Emperor Qin Shi Huang channeling water from the Wei River into the Lanchi Pond, the Huayuan Garden of the Wei and Jin periods sourcing water from the Zhang River into Tianyuan Pool, and the Summer Palace diverting springs from Yuquan Mountain into Kunming Lake. These methods relied on either horizontal channels or vertical wells, which required large-scale engineering efforts, had poor cost-efficiency, and caused significant environmental disturbance—suitable only for imperial garden projects and not broadly applicable today. In contrast, simulating natural waterscape construction utilizes natural precipitation, capturing it through the creation of surface basins or depressions. The process begins with an analysis of the site's hydrological characteristics, particularly focusing on runoff, catchment areas, and outflow points (Fig. 8.6).

When rain, melting snow, or irrigation water moves under gravity—excluding portions lost to evaporation, plant interception, infiltration, or depression storage—it forms what is known as runoff. Runoff can be classified into surface runoff and

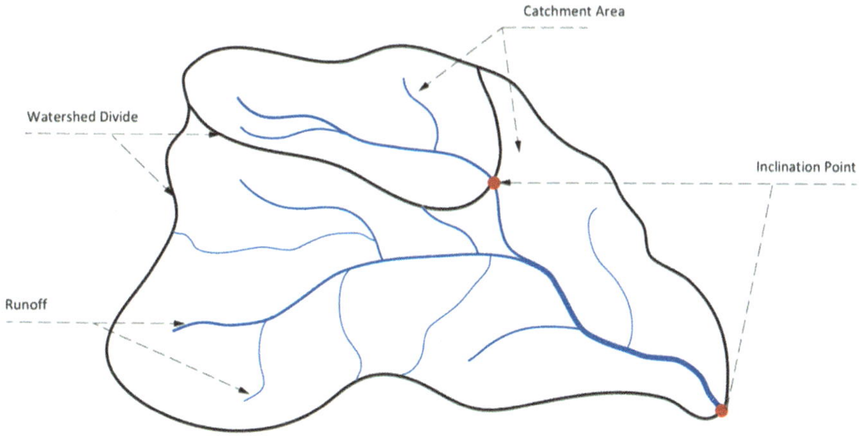

Fig. 8.6 Composition of catchment basin

subsurface runoff, with surface runoff further divided into overland flow and channel flow. Once water converges into a defined channel, it is considered channel flow. In the context of simulating natural waterscape design, the focus is on overland flow. Using a Digital Elevation Model (DEM), ArcGIS can analyze the surface runoff patterns across the landscape design site, identifying a hydrological "network" formed by converging flows. By adjusting threshold values, it is possible to generate runoff networks of varying densities. As runoff converges toward depressions, it accumulates, merges, and gradually forms a hierarchical system. River networks are commonly classified using the Strahler [4] and Shreve stream ordering methods [5]:

Strahler Method [6]:

1. Headwater streams are classified as first-order streams.
2. When two streams of the same order merge, the downstream stream is assigned an order one level higher.
3. When streams of different orders merge, the downstream stream takes the higher of the two orders.

Shreve Method:

1. All headwater streams are assigned an initial order of 1.
2. When two streams merge, the downstream stream's order is the sum of the upstream stream orders.

Compared to the Strahler method, the Shreve method more effectively reflects the fact that higher-order streams typically carry greater volumes of water and sediment (Fig. 8.7). Therefore, for the analysis of runoff in simulating natural waterscape construction, the Shreve method is adopted in this study. It more clearly captures the cumulative effect of flow convergence and better indicates runoff magnitude. According to the Shreve method: A first-order runoff segment has no tributaries; A second-order runoff must include at least two first-order tributaries; A third-order runoff must have at least two second-order tributaries, and so on. Through runoff

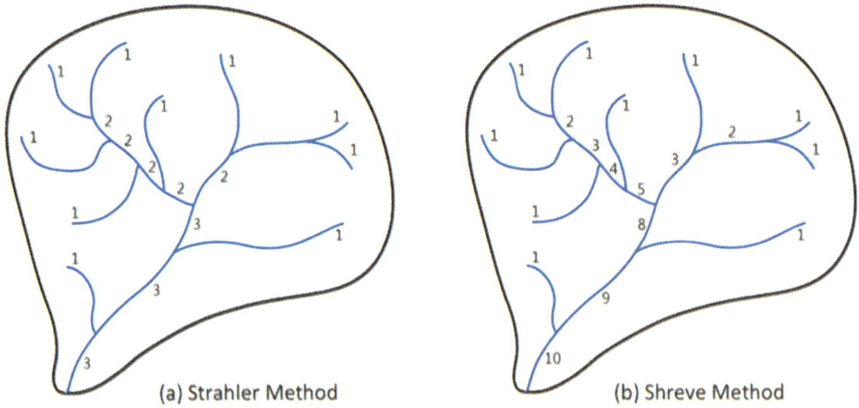

Fig. 8.7 Strahler method and Shreve method

analysis, one can determine how precipitation converges on the surface. The higher the runoff order, the greater the volume of water it carries. In any given site, the number of runoff paths decreases as their order increases. Consequently, first-order runoff segments are the most numerous in any runoff network system. Runoff analysis forms the foundation for generating simulating natural waterscapes and is the first stage of the construction process. It aims to delineate the site's hydrological network to support subsequent analysis and calculation. This network represents the site's catchment structure, and watershed lines reveal the flow paths of water supply. Hence, areas within watershed lines are likely to retain water, marking them as potential zones for simulating natural waterscape development.

8.2.1.2 Catchment Area Analysis

A catchment area, also referred to as a drainage basin or watershed, is the geographical area that collects and channels precipitation to a common outlet. Its boundaries are inherently shaped by the topography. Due to variations in terrain elevation, water from precipitation naturally flows downslope, concentrating in lower-lying areas. The highest points in the landscape, such as ridgelines or mountaintops, act as divides or watershed boundaries. Precipitation falling within adjacent catchments converges internally along its own drainage hierarchy before eventually merging into higher-order runoff outside the immediate basin. Thus, catchment areas function as independent hydrological units, yet are also hierarchically nested based on the classification of runoff order. Following the same logic as stream order: A first-order catchment corresponds to the drainage area traversed by a first-order runoff stream; A second-order catchment contains second-order streams as its highest stream order; A fifth-order catchment necessarily includes a fifth-order stream as its dominant drainage line. Just as higher-order streams are formed by the confluence of multiple lower-order streams, a higher-order catchment is fully composed of multiple lower-order catchments. This creates a nested hierarchical structure (Fig. 8.8), wherein the water volume a higher-order catchment can retain equals the total runoff volume of its sub-catchments, and likewise, its spatial area is the sum of all subordinate catchment areas. It is important to note that certain high-order drainage systems may form independently without being embedded within a broader nested structure. Through catchment analysis, the location and extent of the watershed that encompasses the proposed simulating natural waterscape can be determined. The area of the catchment serves as a critical basis for calculating runoff volume. Constructing waterscapes within higher-order catchments facilitates the development of multi-tiered water systems, where waterbodies in lower-order sub-catchments function as upstream sources to those in higher-order basins. These hierarchically connected water features together constitute a cohesive waterscape system.

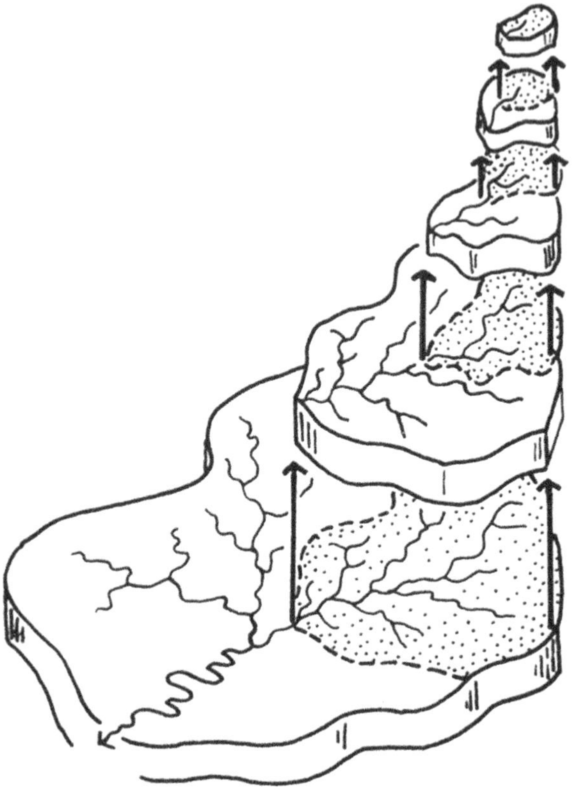

Fig. 8.8 Schematic diagram of the nested hierarchy system of catchment areas. Quoted from: Marsh, W. M. (2006). *An environmental approach to landscape planning* (Q. Zhu, L. Huang, K. Y. Yu, et al., Trans.). Beijing, China: China Architecture & Building Press. (Original work published 2006, p. 172).

8.2.1.3 Pour Point Analysis

A pour point is the outlet of a catchment area—the lowest point in the drainage network where runoff exits and transitions into a higher-order drainage system. Because runoff accumulates hierarchically by order, the resulting flow network resembles a branching tree structure, with the pour point situated at the terminus of each branch. Specifically, the pour point is located at the downstream end of the highest-order runoff segment within a catchment, making it the outlet through which the entire basin's water volume is discharged. As such, the pour point carries the maximum flow volume within the catchment, aggregating all upstream runoff. Due to its unique status as both the lowest elevation point and the maximum discharge point, identifying and analyzing pour points is essential for simulating natural waterscape planning. Accurate localization of pour points supports the effective design of water outlet features, ensures realistic simulation of hydrological behavior, and enhances the ecological performance of the constructed waterscape.

8.2.2 *Waterscape Site Selection*

Runoff networks are unevenly distributed across the landscape of a site designated for landscape planning and design. Although areas along watershed lines offer potential for water supply, the specific location for waterscape construction must still be determined by integrating both design objectives and existing topographic conditions. For instance, a certain area may require a water feature, yet the current site may lack an existing water system or only contain fragmented water bodies. In such cases, the feasibility of creating a new waterscape can be evaluated using runoff network maps. If the analysis indicates potential, the topography can be leveraged to identify suitable dam construction locations. This process involves two key technical considerations within the framework of the parametric simulating natural waterscape construction model for landscape planning: determining the dam location and delineating the catchment area.

8.2.2.1 Determining the Dam Location

Two critical conditions for water retention are water source availability and the existence of basin-like terrain. Runoff analysis reveals the water supply routes, and the classification of runoff order provides insight into the quantity of available water. As such, areas along watershed lines meet the conditions for supplying water and thus for constructing a simulating natural water body. A defining principle of simulating natural waterscape design is to minimize artificial interventions. Therefore, dam construction should align with natural terrain features. As noted in the Yuan Ye (The Craft of Gardens), in the section "Mountain Forest Sites," it is stated: "Excavate the water by following the source into the depths, dig where the land lies low [7]." This principle aligns with site selection strategies used for large-scale hydraulic projects such as the Three Gorges Dam—identifying pour points within runoff networks as optimal dam sites. When selecting suitable basin terrain for damming, existing topographic conditions must be carefully considered. The shape and enclosure of terrain surrounding a potential dam site significantly affect the scale of the required engineering. If the surrounding terrain is well-enclosed, the "mouth" of the terrain at the pour point is narrower, thereby reducing the required construction volume for a dam of the same height. Damming along higher-order runoff streams is advantageous because water inflow is positively correlated with stream order—the higher the runoff order, the greater the inflow and thus the greater the potential volume available for retention. However, it is important to note that in undulating terrain, the volume of retained water is not necessarily proportional to the surface area of the water body. In waterscape design, dam locations often correspond to hydraulic elevation differences, marking points of water level transition. These become critical nodes for features such as cascades or waterfalls. Therefore, dam location and elevation have a direct impact on the morphology and aesthetics of the waterscape.

8.2.2.2 Catchment Delineation

A catchment area, bounded by surface and subsurface watershed divides, is defined as the drainage basin that collects precipitation and channels it to an outlet. Conventionally, the term refers to the surface water drainage area [7]. The size of a catchment represents the total contributing area from which rainfall is collected—after losses due to evaporation, infiltration, and vegetation interception—and directed via runoff to the catchment's pour point and then into a higher-order catchment. For a dammed waterbody, the corresponding catchment area is the direct supply zone, and the volume of rainfall it captures constitutes the primary water source for the waterscape. Given a known rainfall amount, a larger catchment area yields a higher volume of water, thereby increasing the water available for landscape use. Consequently, water volume calculations require the delineation and measurement of the catchment area. Once the dam location is established, ArcGIS hydrological tools can be employed to extract the associated catchment area for the pour point and to compute its total area. This becomes a fundamental input for determining the feasible water volume for the intended waterscape.

8.2.3 *Estimation of Water Volume*

Atmospheric precipitation is the primary source of surface freshwater. Variations in geographic conditions lead to differences in annual precipitation across regions, and both interannual and intra-annual fluctuations can occur within the same region. Precipitation also serves as the main natural source of water supply in landscape environments. In hydrology, rainfall over a watershed, after deducting various losses, contributes to runoff through both surface and subsurface pathways, eventually converging into river networks. This flow measured at the outlet cross-section of a watershed is referred to as streamflow or runoff [6]. The transformation from rainfall to runoff is typically divided into two processes: runoff generation and runoff concentration. Once precipitation reaches the land surface, it sequentially contributes to direct rainfall onto water surfaces, surface runoff, subsurface runoff, and groundwater recharge. The portion of rainfall that remains after losses—known as effective rainfall—is equal in volume to total runoff, though the timing of surface and subsurface runoff generation may differ. Given this, during the formation of runoff, interception and evaporation result in the loss of a portion of the total precipitation, making it impossible to convert all rainfall into available water for simulating natural waterscape construction. Therefore, the amount of water available for waterscape construction must be estimated in two parts: The total precipitation received by the catchment area; The losses incurred, to derive the actual amount of water that can be used.

8.2.3.1 Estimating Precipitation Volume

Precipitation volume refers to the depth (in millimeters) of water that would accumulate on a flat horizontal surface over a given period, assuming no infiltration, no runoff, and no evaporation. The annual precipitation is the sum of the average monthly precipitation over a year. The long-term average annual rainfall of a location is often referred to as its "mean annual precipitation."[2] In natural sciences, precipitation data is typically categorized by year, season, month, ten-day period, or growing season. Given the implementation timeline of landscape planning and the expected duration of landscape effects, this study adopts the annual cycle for calculations, using long-term average annual precipitation data. Since simulating natural waterscapes do not require extensive excavation of existing terrain, the main sources of water replenishment include direct rainfall on the water surface, surface runoff, and subsurface runoff, among which surface runoff contributes the most. As the waterscapes are newly created rather than existing water bodies, inter-basin water transfer can be excluded. The water volume available for waterscape construction can be calculated using the following formula [8]:

$$W = P \times A \times 10^3 \tag{8.1}$$

Where:

- W = total precipitation volume within the catchment (m^3)
- P = long-term average precipitation depth for the region (mm)
- A = area of the catchment basin for the waterscape (km^2)

8.2.3.2 Estimating Effective Runoff Volume

A portion of rainfall is lost due to evapotranspiration by vegetation and soil, and therefore cannot be fully utilized for waterscape construction. The annual evapotranspiration within the catchment is the primary loss that must be deducted from total precipitation. The influence of annual precipitation and evapotranspiration on total runoff varies by climatic region. In humid regions, precipitation is the dominant factor affecting runoff, with evapotranspiration playing a relatively minor role [9]. Conversely, in arid regions, most precipitation is lost to evaporation, making both precipitation and evapotranspiration equally important in determining runoff. After precipitation reaches the ground, it follows four primary pathways: Interception by plant canopies (loss); Infiltration into the soil (loss); Depression storage in surface micro-topographies (loss); Surface runoff, which ultimately flows into channels, rivers, or ponds (Fig. 8.9) Factors such as land cover (land use), soil composition and structure, and surface slope significantly affect infiltration and runoff. In areas with dense vegetation, interception and infiltration are substantial, often rendering

[2] http://www.weather.com.cn/static/html/knowledge/20080507/721.shtml.

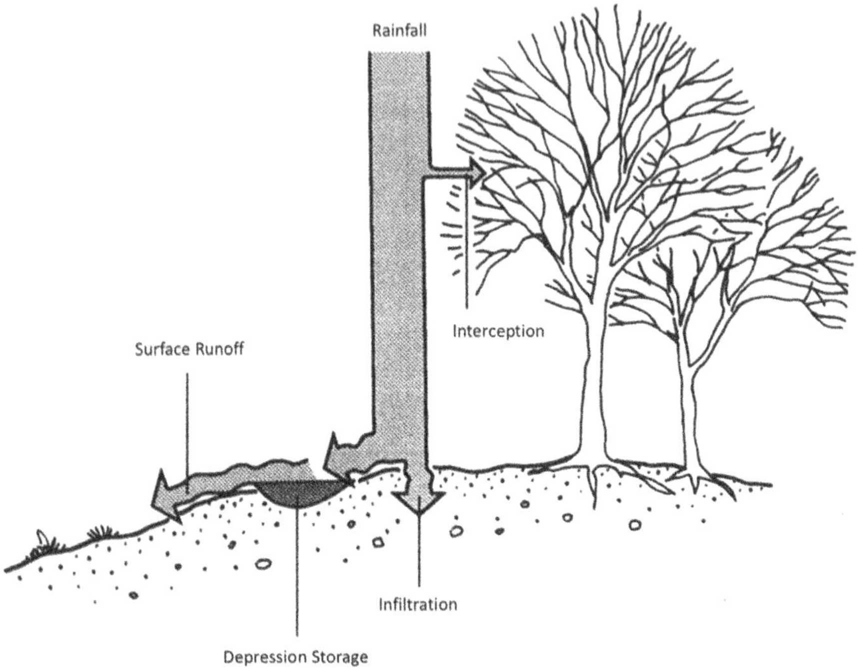

Fig. 8.9 Four components of precipitation after reaching the ground: interception, infiltration, depression storage, surface runoff. Quoted from: Marsh, W. M. (2006). *An environmental approach to landscape planning* (Q. Zhu, L. Huang, K. Y. Yu, et al., Trans.). Beijing, China: China Architecture & Building Press. (Original work published 2006, p. 172)

surface runoff negligible. In contrast, arid zones or urbanized landscapes with sparse vegetation and hardened surfaces exhibit much higher runoff ratios. Generally: Runoff increases with slope; Runoff decreases with higher soil organic content and coarser particle size; Runoff increases with impervious surfaces; Runoff decreases with increased vegetation [9]. Since rainfall intercepted by vegetation is considered a loss, it must be deducted when calculating effective runoff. Once soil becomes saturated, any additional water contributes to subsurface runoff, which—like surface runoff—moves downslope through aquifers and recharges water bodies. However, groundwater formation via infiltration is a long-term process (decades to centuries), with relatively minimal short-term loss. Therefore, it is excluded from this estimation. Due to storage and retention effects caused by terrain and infiltration, a reduction coefficient must be applied to adjust for actual usable water. Hydrological cycles are complex nonlinear systems, and the conversion of rainfall into runoff is influenced by a wide range of variables. While detailed hydrological modeling can offer precise simulations, such methods fall beyond the scope of this study. Instead, a simplified approach is adopted here to estimate the runoff volume in the catchment. By subtracting vegetation interception and infiltration losses using coefficients, the following formula is applied [8]:

$$W = \sum_{1}^{n} k_i \cdot m_i \cdot A_i \cdot P \cdot 10^3 \tag{8.2}$$

Where:

- W = effective water volume available for waterscape construction
- *ki* = runoff coefficient for land use type *i*
- *mi* = reduction coefficient for land use type *i*
- A *i* = catchment area under land use type *i* (m^2)
- P = long-term average precipitation depth (mm)

The runoff coefficient is the ratio of total runoff to precipitation within a defined area, ranging from 0 to 1. For example, a coefficient of 0.7 implies that 70% of rainfall contributes to surface runoff, while 30% is lost to infiltration. Surface cover types and soil properties directly affect runoff coefficients. Compared to rural areas, urban areas with higher proportions of impervious surfaces tend to have lower infiltration and higher runoff coefficients.

8.2.4 Estimating Dam Height

In the artificial construction of simulating natural waterscapes, it is necessary to form ponded depressions for water storage through dam-building. The crest elevation of the dam corresponds to the designed normal water level of the future waterbody. When there is excess inflow, overflow will be directed downstream, creating scenic features such as waterfalls and cascades. For safety, ecological health, aesthetics, and the requirements of aquatic plant growth, the water depth should not be excessively deep. Based on calculations in the preceding step, the volume of water that can be collected from the catchment area—i.e., the maximum available water volume for the waterscape—is determined. The dam crest elevation corresponding to this volume is defined as the maximum possible water level. To ensure perennial water presence, the dam crest elevation must not exceed this maximum level. The volume formula $V = S \cdot h$ indicates that, given a fixed volume, the water surface area and dam height are interrelated. ArcGIS software provides a "Surface Volume" tool that can calculate the corresponding waterbody volume using topographic data and dam elevation as input parameters. The author constructed a storage–elevation curve for the proposed waterbody and used an approximation method to establish a functional relationship among surface area, dam height, and water volume. The concept of "storage capacity" originates from hydraulic engineering and refers to the volume of water stored below a given reservoir water level or between two levels. Here, the term is borrowed to describe the volume that can be held within the constructed ponded depression. The "storage–elevation curve" illustrates the relationship between water level and corresponding storage capacity, typically with water level on the vertical axis and storage volume on the horizontal axis. In this study, the curve expresses the relationship between dam crest elevation and stored water volume in the ponded area. Conventional methods for calculating reservoir storage include the contour method, cross-section method, grid method, and triangulation method. Using ArcGIS, a Digital Elevation Model (DEM) based on an irregular Triangulated Irregular Network (TIN) can be constructed, allowing for volume calculations using predefined functional models. However, due to terrain undulations and spatial heterogeneity, inverse calculations—determining the precise dam height for a given water volume—are challenging. Therefore, the author employed an approximation method to reverse-engineer the storage–elevation curve and estimate the threshold dam height. Different dam elevations result in distinct water surface morphologies, and the aesthetic quality of the waterbody's form is a key criterion for the success of the landscape waterscape. Thus, under the constraints of available water volume, the final dam height must be determined based on an evaluation of the waterbody's form. Dam height, water volume, and waterbody morphology together constitute a parametrically dynamic relationship.

8.2.5 Simulation and Quantitative Evaluation of Waterbody Form

"Water has no inherent shape—it takes the form of its container." The morphology of a waterbody is directly related to the shape of the ponded terrain. Given the heterogeneity of topography, different water levels inevitably result in varying surface morphologies. Simulating natural waterscape design emphasizes minimizing human intervention. Thus, within the range of maximum water levels, the dam crest elevation corresponding to the most aesthetically pleasing naturally formed water surface should be selected as the final design elevation. This natural form can then be further refined and optimized to derive the final design outcome. Quantitatively describing the quality of waterbody morphology is therefore a critical aspect of simulating natural waterscape studies.

8.2.5.1 Quantitative Representation of Waterbody Morphology

The study of waterbody form involves the comparison and analysis of freeform geometries. Compared with qualitative descriptions, quantitative analysis focuses on revealing underlying patterns. Quantitative analysis aids in understanding the logic and inherent rules behind morphological formation. It allows for the scientific description of shoreline features and supports the derivation of design guidelines based on formal characteristics and regularities.

Waterbody form is essentially a geometric shape enclosed by a shoreline. In limnology, lake morphometry refers to the numerical characterization of the horizontal and volumetric geometry of lakes. Indicators include lake length, maximum and average width, length and orientation of the major and minor axes, shoreline length, shoreline development index, surface area, island ratio, water surface area, maximum and average depth, volume, and catchment area. These metrics can be directly or indirectly extracted from satellite imagery and geographical datasets, and are easily translated into design constraints and visualized representations. Accordingly, the author employs a set of quantitative geometric indices to describe waterbody morphology.

Natural forms can be classified into two types: those with integer dimensions that can be described by traditional Euclidean geometry, and those—such as coastlines, rivers, and mountain ridges—that exhibit irregularity and self-similarity and are more accurately characterized by fractional dimensions. Compared to traditional Euclidean indices such as shoreline development index, circularity ratio, shape index, and compactness [10],[3] fractal theory offers a more effective approach for capturing the complexity of lake shoreline variation. Previous studies have introduced fractal theory into lake ecology and morphology, using fractal dimensions to reveal hydrological characteristics of lakes and river systems [11, 12]. In this study, the author incorporates fractal theory into waterbody form analysis and employs fractal dimension as a quantitative indicator to reflect shoreline complexity. The evaluation system for waterbody morphology integrates basic geometric indices, fractal geometry indices, and Euclidean geometry indices (Table 8.1). Among the basic geometric indices, surface area (A) and perimeter (P) are directly measurable, while the maximum axis length (L) requires calculation. The fractal dimension describes the complexity of an individual shoreline. Since lake shorelines are two-dimensional curves, the fractal dimension of a waterbody typically falls between 1 and 2. A value closer to 1 indicates stronger self-similarity and a more regular shoreline form; a higher value implies greater complexity [13]. From the perspective of landscape ecology, longer shorelines offer significant edge effects. Higher shoreline complexity contributes not only to ecological benefits but also to richer spatial experiences. However, overly complex shoreline forms can appear fragmented, diminishing aesthetic appeal. Therefore, the author compares the effectiveness of using both the shoreline development index and the fractal dimension for evaluating waterbody morphology. By analyzing and comparing different waterbody forms resulting from various dam heights, the study identifies the most desirable form and determines the corresponding dam location and elevation. The Euclidean geometry indices used include the shoreline development index (D_L), shape index (F_R), circularity ratio (C_R), compactness (C), and the spatial enclosure area of the waterbody (ΔA).

[3] "Circularity index" and "shape ratio" are shape indices in geographical application research, used to briefly describe the spatial shape characteristics in three-dimensional conditions.

Table 8.1 Quantitative evaluation indicators for water body morphology

Indicator category	Indicator name	Calculation method	Representation meaning
Basic geometric morphological indicators	Area (A)	Measurable	Using the boundary formed by the interface between water and land, the area within the water surface boundary is measured to describe the size of the water body
	Perimeter (P)	Measurable	Represents the total length of the water body shoreline
	Long axis (L)	Exhaustive algorithm	Distance between the two farthest points along the water body shoreline
Fractal geometric morphological indicators	Fractal dimension (FD)	$FD = \frac{2\ln P/4}{\ln A}$	The fractal dimension index indicates the complexity of the water body shape. The fractal dimension ranges between [1, 2]. The closer the value is to 1, the stronger the shape similarity of the water body, the more regular the shape, and the simpler the geometric morphology; conversely, the closer the fractal dimension is to 2, the poorer the shape similarity, the more irregular the shape, and the more complex the geometric morphology
Euclidean geometric morphological indicators	Shoreline development coefficient (D_L)	$DL = \frac{S_L}{\sqrt{\pi A}}$	The more irregular the shoreline, the more tortuous and varied the water body shoreline is, resulting in a larger shoreline development coefficient
	Form ratio (F_R)	$F_R = \frac{A}{L^2}$	Indicates the openness of the water surface and the development of horizontal water circulation. Water bodies with smaller shape index values exhibit large variations in water surface width, high shoreline curvature, and relatively enclosed local shapes
	Circularity ratio (C_R)	$CR = \frac{4\pi A}{P^2}$	Reflects spatial dispersion degree; the smaller the compactness ratio, the greater the shoreline curvature and the longer the shape perimeter
	Compactness ratio (C)	$C = \frac{A}{A_o}$	Represents a shape index of the water body reflecting water surface openness; the larger the value, the more open the water surface; conversely, it indicates a narrower and elongated water surface
	Water body spatial enclosure area ($\triangle A$)	$\triangle A = \tau(R_0\text{-}R^2)$	Nder constant water area conditions, changes in water surface spatial structure cause variations in the controlled land area. Water bodies with equal area but different shapes control different extents of adjacent land

Note: A = area; P = perimeter; L = longest axis length; R = standard circle with equivalent area; R_0 = radius of maximum circumscribed circle; S_L = lake shoreline length; A_0 = minimum circumscribed area

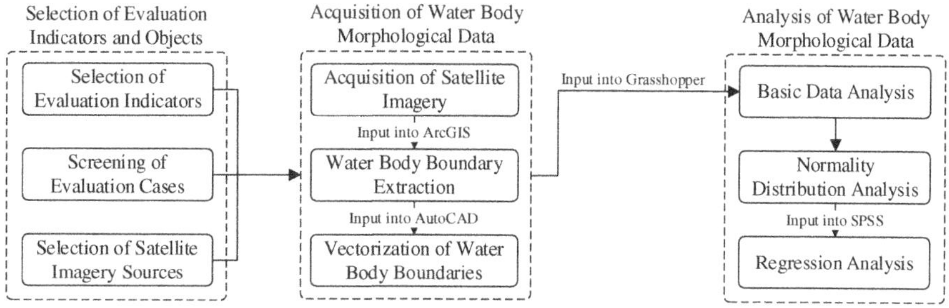

Fig. 8.10 Technical approach for quantitative evaluation of landscape water body morphology

8.2.5.2 Technical Framework and Methodology for the Quantitative Evaluation of Waterbody Morphology

This study explores the characteristics and underlying patterns of waterbody morphology through a quantitative evaluation and analysis approach. The technical framework comprises three main components: (1) selection of evaluation indicators and case studies, (2) acquisition of waterbody morphological data, and (3) analysis of morphological data. First, based on the characteristics of landscape waterscapes, appropriate quantitative indicators for waterbody morphology are identified, representative case studies are selected, and suitable sources of satellite imagery are chosen. Second, satellite images are mosaicked, georeferenced, extracted, and processed using software such as ArcGIS, AutoCAD, and Grasshopper, resulting in a dataset of waterbody morphological indices. Third, statistical software (SPSS 19.0) is employed to perform distribution analysis, correlation analysis, and regression analysis on the collected indices, thereby deriving key quantitative parameters for the case studies, such as value ranges of specific metrics (Fig. 8.10).

The simulating natural waterbodies featured in landscape waterscapes are typically irregular in form, organically shaped, and highly variable. The quantification of such morphologies presents several challenges: ① Calculating values such as the maximum axis length and the minimum circumscribed circle area for irregular shapes is inherently complex; ② Due to the varying computational approaches for different types of indicators, it is often necessary to work across multiple software platforms, resulting in difficulties with data interoperability; ③ Some stages of data calculation and aggregation must be performed manually, which is labor-intensive and prone to error.

To address these challenges, the author developed a computational workflow using Grasshopper, a visual programming platform operating within the Rhino environment (Fig. 8.11). On one hand, Grasshopper enables precise computation and output of multi-type data, ensuring seamless data integration while avoiding errors associated with manual processing. On the other hand, by scripting algorithms within Grasshopper, the workflow allows for accurate and efficient batch calculations (Fig. 8.12). Furthermore, by leveraging Rhino's capabilities, the

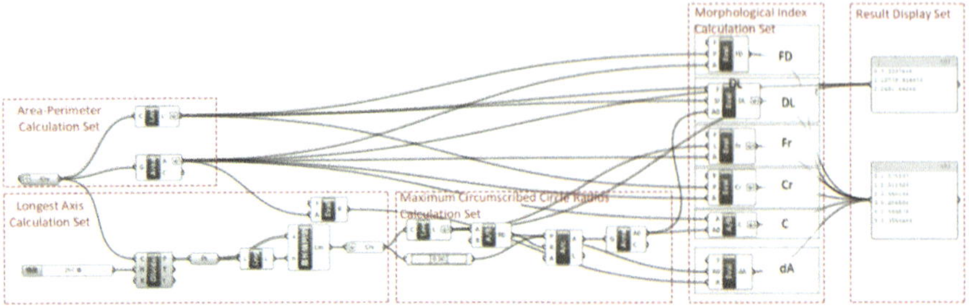

Fig. 8.11 Grasshopper calculation set

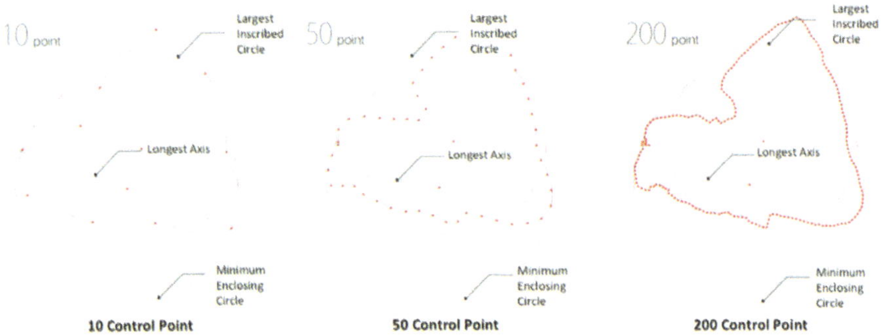

Fig. 8.12 Using grasshopper to solve the longest axis and minimum enclosing circle of water body morphology

computational results generated in Grasshopper can be visualized and updated in real time, thereby facilitating iterative adjustment and optimization.

8.3 Application of the Parametric Simulating Natural Waterscape Construction Model

This section elaborates on the operational process of the parametric simulating natural waterscape construction model through a practical case: the design of the *Zen-seeking Path* in the northern area of the Niushou Mountain Scenic Zone.

8.3.1 Site Hydrological Analysis

8.3.1.1 Runoff Analysis and Runoff Hierarchy Classification

Understanding the hydrological conditions of the design area begins with a runoff analysis, which is based on topographic raster data using ArcGIS software. To avoid unrealistic flow directions, the existing Digital Elevation Model (DEM) is processed to generate a depressionless DEM. Flow direction and flow accumulation rasters are then generated, forming the foundational dataset for all subsequent analysis. Topographical analysis shows that the site's terrain is higher in the south and lower in the north, with the highest elevation located in the southeastern corner. Two distinct and parallel ridgelines extend northward from this high point. The northwest part of the site is relatively flat and at a lower elevation. The flow direction algorithm in ArcGIS adopts the D8 single-flow direction model [14, 15]. Unlike multiple-flow models, the D8 model assumes that water from a given raster cell flows into only one neighboring cell—the one with the steepest downslope—which is a standard method for river network extraction. The D8 algorithm, introduced by Jenson and Domingue in 1988, determines water flow direction by identifying the maximum slope between a raster cell and its eight adjacent cells. Figure 8.13(b) shows the resulting flow directions across the site, categorized into eight types. Based on the flow direction data, a flow accumulation raster is produced. Each raster cell in this dataset represents the total upstream contributing flow volume. Cells with high accumulation values indicate potential runoff-generating points. The flow accumulation threshold is a critical parameter in runoff analysis: a lower threshold results in a denser runoff network. For instance, when the threshold is set to 500, all cells with accumulation values greater than 500 are considered part of the runoff network. Since more cells exceed 500 than 1000, the runoff network is denser at the 500 threshold. In hydrological studies, careful threshold selection and further calibration (e.g., using stream length thresholds) are essential for producing realistic representations of surface water networks. For simulating natural waterscape design, however, the threshold requirements are comparatively relaxed. Precise simulation of natural hydrological systems is not necessary; rather, what matters is clearly illustrating the relationships between surface runoff paths. Overly dense

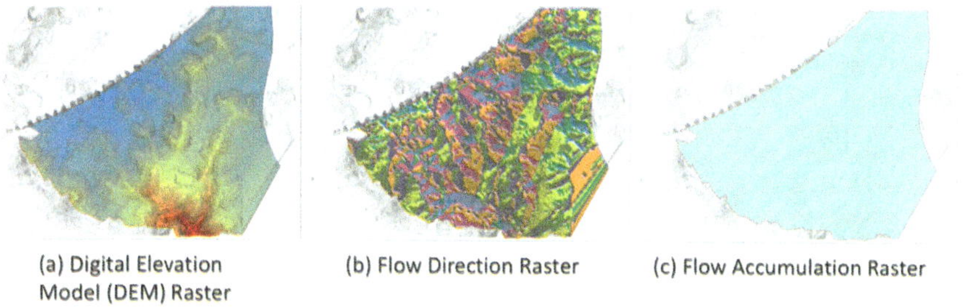

Fig. 8.13 Terrain, flow direction and flow accumulation analysis map

networks may hinder a designer's ability to interpret runoff flows and increase the workload for subsequent analysis. Figure 8.14 displays the generated runoff networks using thresholds of 1000, 2000, and 3000. As seen in the figure, the networks take on a tree-like structure, with the densest network occurring at the 1000 threshold. As the threshold increases, the number of terminal branches decreases and the runoff paths become more distinct. Since waterscape design typically occurs at smaller spatial scales compared to hydraulic engineering (e.g., reservoirs), overly simplified runoff networks may omit smaller regions suitable for waterscape construction, thereby narrowing the range of possible design sites. Therefore, careful threshold selection is necessary. Based on a comparison of network density and structure, a threshold of 2000 was ultimately chosen for this project.

Once the runoff network is determined, the next step in the parametric simulating natural waterscape construction model involves classifying the hierarchy of individual runoff segments. This is essential for identifying dam sites and calculating contributing runoff volumes. The *Strahler Stream Order* method is used for this classification. Figure 8.15 shows the results, with runoff hierarchy represented by a color gradient from cool to warm, and numerical labels indicating stream order. Analysis reveals that first-order streams (in blue) dominate the network, consistent with the common pattern in most stream networks where the lowest order occupies the largest proportion (see *Appendix Table 2: Runoff Hierarchy Statistics*). Warmer colors represent higher-order streams, which carry the cumulative runoff of all contributing upstream segments. The main branches transition from cool to warm colors, indicating a gradual increase in runoff volume due to upstream convergence.

8.3.1.2 Basin Analysis and Selection

A basin refers to a catchment area enclosed by watershed boundaries. By delineating basins within the site, one can clearly distinguish between runoff catchments, thereby informing site selection and layout for waterscape construction. Catchment areas exhibit a hierarchical nesting structure. In this study, large-scale areas defined by the entire runoff network are referred to as basins, while the drainage area corresponding to each individual runoff stream is termed a sub-catchment. Using ArcGIS software, the topographic raster, flow direction raster, and flow

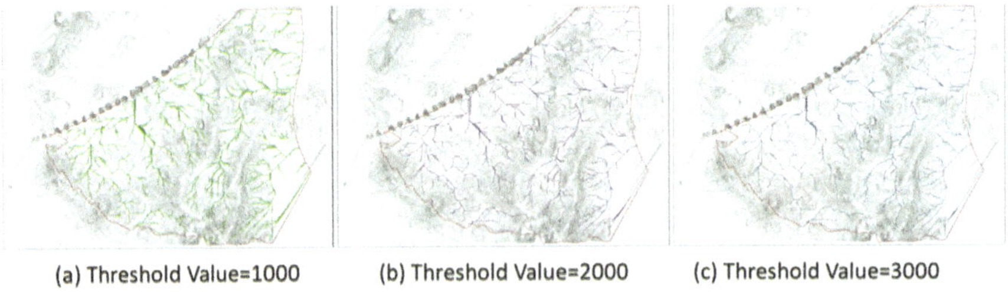

Fig. 8.14 Channel network density corresponding to different threshold values

Fig. 8.15 Runoff classification map

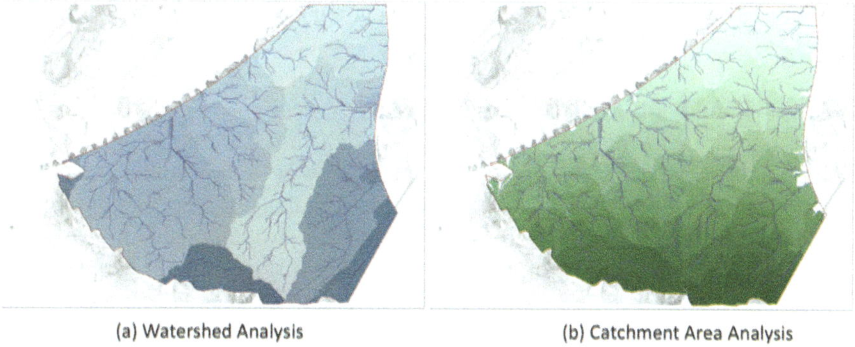

Fig. 8.16 Runoff watershed and catchment area analysis map

accumulation raster are input as parameters to generate a basin and sub-catchment analysis map, as shown in Fig. 8.16. This map reveals the division of the site into basins and illustrates the sub-catchment area associated with each stream, providing a spatial basis for locating waterscape features. Thirteen primary basins (numbered 1 to 13) were identified within the study area, as illustrated in Fig. 8.17(a). Among these, Basins 2, 3, 4, and 5 are the most prominent. From Fig. 8.17, it can be seen that these four basins have relatively complete runoff networks and large catchment areas, which suggests a correspondingly high volume of runoff—ideal conditions

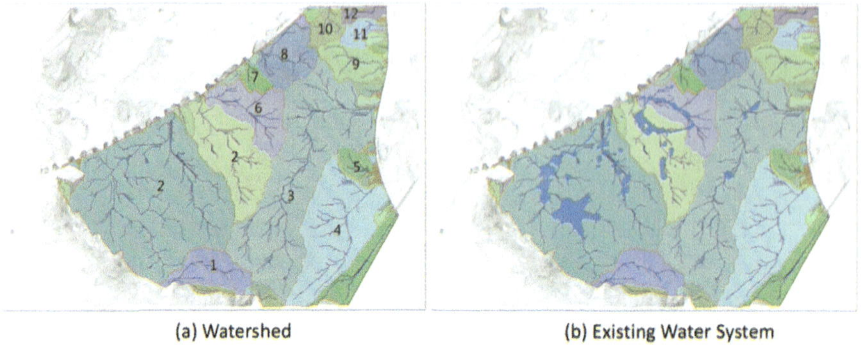

Fig. 8.17 Watersheds and existing water systems

for constructing landscape water systems. According to the current hydrological conditions, Basins 2 and 3 already support relatively well-developed water systems, including substantial surface water bodies. Basin 4, located in a mountain valley, has a well-structured runoff network but currently contains only scattered small water bodies, suggesting potential for further development into a full-fledged landscape water system. In addition, the overall site planning positions this area as a Zen culture thematic zone. Through the construction of water features, the design aims to evoke the aesthetic of gently flowing mountain streams. Based on the aforementioned analysis and its integration with the design vision, Basin 4 was ultimately selected as the site for implementing the simulating natural waterscape design.

8.3.1.3 Pour Point Analysis and Extraction

Pour points—also known as outlets—are the exit points of flow within a catchment, corresponding to the lowest elevation point in a drainage basin. Identifying pour points provides key references for the placement of artificial dams in waterscape construction. Each segment of a runoff stream has a starting point and an ending point, with the latter often serving as a junction node between two or more stream segments. By analyzing the connectivity between runoff segments, it is possible to identify all starting points, ending points, and pour points (i.e., nodes) within the entire runoff network, as illustrated in Fig. 8.18(a). A total of 1175 nodes were generated through this analysis. Appendix Table 3: Runoff Node Statistics Table lists details for runoff nodes numbered 1 to 80, including the number of connected runoff segments per node, the hierarchical level of each node, and the stream order of the connecting segments. From this comprehensive set of nodes, the pour points for the entire site were extracted based on their attribute values, as shown in Fig. 8.18(b). These points represent critical reference locations for the implementation of the simulating natural waterscape construction model.

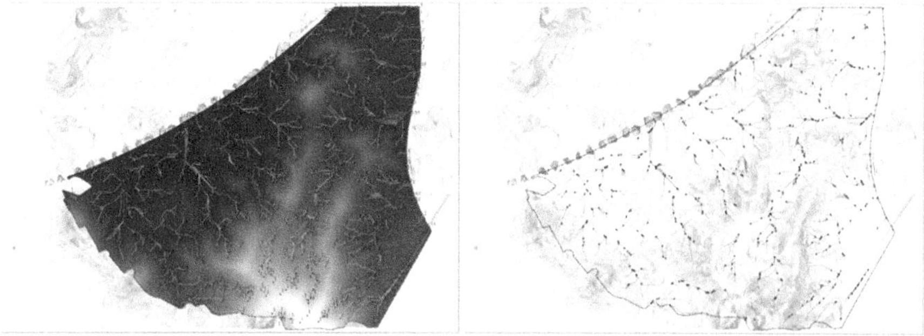

(a) Runoff Network Nodes Analysis (b) Inclination Point Analysis

Fig. 8.18 Runoff network nodes and inclination points analysis map

8.3.2 Determination of Waterscape Location

8.3.2.1 Extraction of the Waterscape Basin

After identifying the basin where the waterscape is to be located, it is necessary to extract the relevant data for this basin to support subsequent computational procedures. By querying the attribute information of the selected basin and using the "Extract by Attributes" tool in ArcGIS, the basin can be isolated from the raster dataset of all basins in the planning area to generate an independent raster dataset (Fig. 8.19).

8.3.2.2 Comparison and Selection of Dam Locations

By interpreting the positions of runoff lines in relation to contour lines within the selected basin, areas with potential for surface water accumulation can be preliminarily identified. In Fig. 8.20, the yellow-circled area highlights one such potential site: a valley located between two ridgelines, which is relatively enclosed and naturally forms an open-ended basin. The terrain is higher on the southwest side and lower on the northeast side, making the outlet naturally located toward the northeast. Based on pour point analysis, four potential dam locations—A, B, C, and D—are identified in the diagram. Each location is situated along high-order runoff streams, ensuring sufficient water supply. Location A lies at the edge of the basin. If dammed, the resulting water surface would extend toward the southwest, but the terrain in that direction is highly variable, which would likely result in a constrained water surface. Location B is situated in a relatively flat central area of the basin but lacks sufficient topographic enclosure, which would increase engineering demands. Location C shares similar disadvantages with B and is thus excluded. Location D, located in the lower-elevation northeast portion of the basin, has favorable surrounding topography and, compared with the other three, receives a greater volume

Fig. 8.19 Extracted watershed

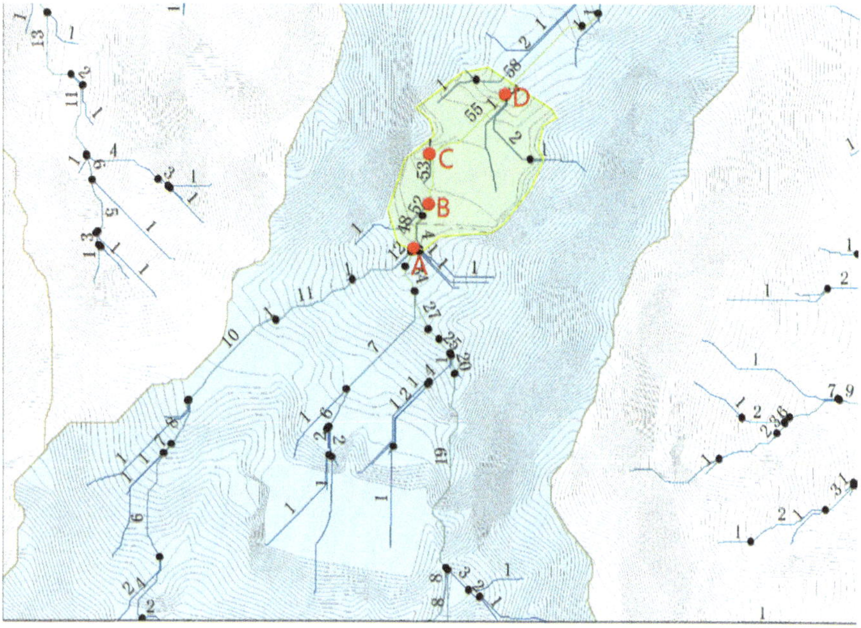

Fig. 8.20 Site selection for dam construction

of runoff. This location can take full advantage of the terrain to form a water surface and is therefore selected as the most suitable dam site.

8.3.2.3 Delineation of the Catchment Area

Following the selection of the dam site, the corresponding catchment area D—i.e., the runoff sub-basin that contributes directly to the water supply—is extracted. Since this is a standalone water body with no upstream reservoirs or diversion systems, the amount of water it receives corresponds directly to the runoff volume of the stream at the dam site. According to the principles of runoff network hierarchy, the water volume of the selected stream includes the cumulative contribution from all upstream streams. Therefore, the delineated catchment area D is defined as the union of all sub-catchments associated with the upstream runoff (Fig. 8.21).

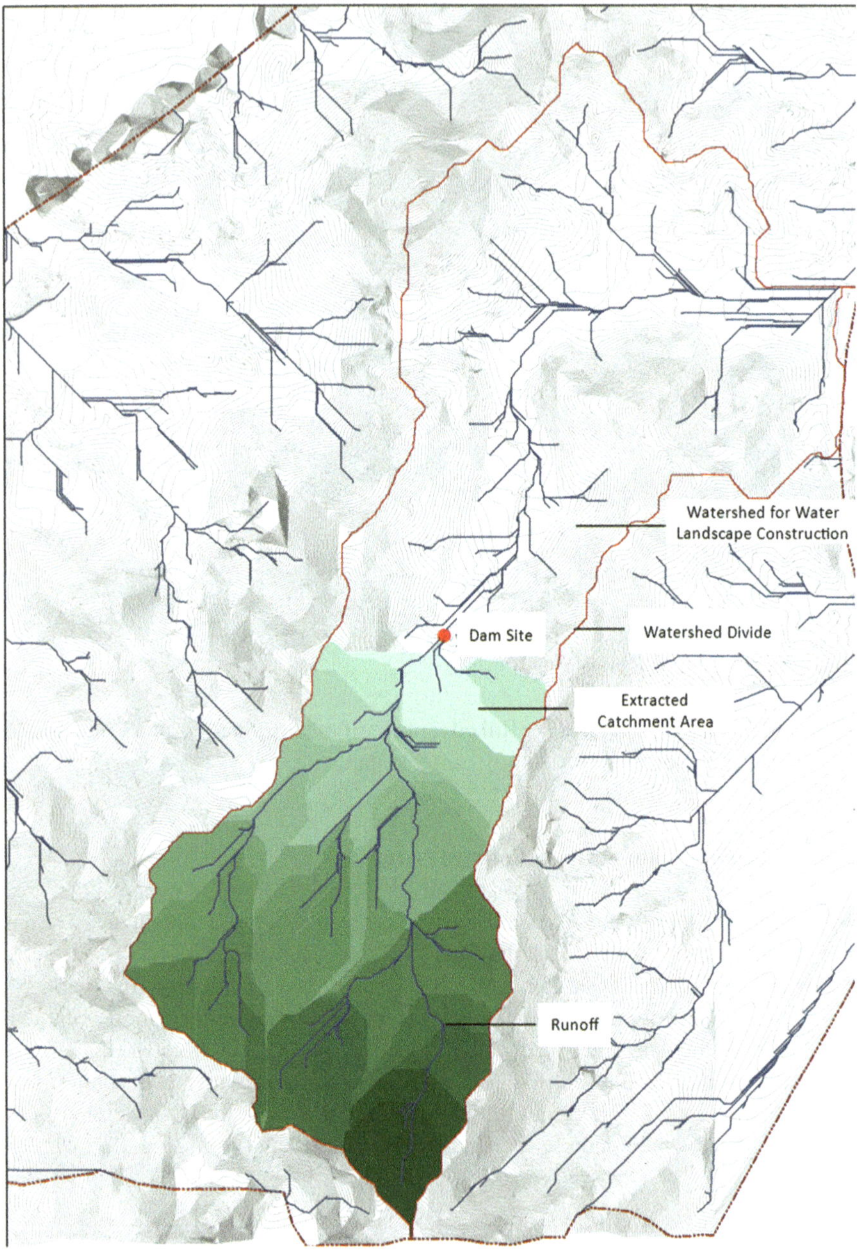

Fig. 8.21 Catchment area extraction

8.3.3 Calculation of Water Volume

8.3.3.1 Calculation of Precipitation Volume

Table 8.2 lists the monthly and annual precipitation data for Nanjing from 2002 to 2013. Based on these data, the average annual precipitation over the 12-year period is calculated to be 1117.3 mm,[4] which is adopted in this study as the representative annual precipitation for Nanjing. This value represents the gross precipitation, without accounting for losses such as vegetation interception, soil infiltration, and evapotranspiration during the rainfall process. By querying the attribute data of the extracted catchment area D, its area is found to be 406,910 m^2. Applying Eq. (8.1) provided earlier, the total precipitation volume for catchment area D (corresponding to dam site D) is calculated to be: 454,640.5 m^3.

8.3.3.2 Calculation of Runoff Volume

Within catchment area D, three main land use types are identified: forest land, sparse grassland, and construction land. Their respective areas and proportions of the total catchment area are as follows: Forest land: 392,145.69 m^2 (96.4%); Grassland: 5353.468 m^2 (1.3%); Construction land: 9410.842 m^2 (2.3%). Referring to the study "Potential Analysis and Calculation of Rainwater Resource Utilization in Small Watersheds of the Central Sichuan Hills" [8], and based on observation data and actual site conditions in the Niushou Mountain area, the runoff coefficients and reduction coefficients for different land use types are listed in Table 8.3. According to the water volume Eq. (8.2) mentioned earlier, the total runoff volume at dam site D is calculated as follows:

$$
\begin{aligned}
W = & 0.35 \times 0.45 \times 392145.69 \times 1.1173 + 0.36 \times 0.65 \times 5353.468 \times 1.1173 \\
& + 0.7 \times 0.45 \times 9410.842 \times 1.1173 \approx 73719.5\,\text{m}^3
\end{aligned}
$$

The result calculated by the formula represents an estimated volume of rainfall that can be effectively retained after deducting losses due to interception, infiltration, and evapotranspiration. Compared with the gross precipitation in the region, it is evident that a significant portion of the water is lost, primarily because the dominant land use type in the area is forest land. As a result, the underlying surface has a strong capacity for rainwater retention, and the area exhibits good soil and water conservation conditions.

[4] http://www.jssb.gov.cn/.

Table 8.2 Precipitation statistics for Nanjing City (2002–2013)

	(Unit: mm)													
Year	January	February	March	April	May	June	July	August	September	October	November	December	Annual Total	Annual Average
2013	17.8	69.9	42.9	22.8	110.1	172.6	229.6	115.5	67.0	22.4	17.2	10.6	898.4	1117.3
2012	21.0	73.3	79.3	56.2	62.4	17.8	176.4	198.3	68.7	54.6	42.8	66.4	917.2	
2011	10.8	17.2	43.2	11.6	40.6	312.9	278.0	284.3	12.6	28.7	21.3	15.8	1077.0	
2010	18.8	115.6	117.8	197.9	56.1	62.1	343.2	142.1	181.1	32.1	7.6	24.0	1298.4	
2009	32.2	112.8	48.3	59.2	56.0	168.7	485.4	102.7	102.9	3.2	113.7	78.4	1363.5	
2008	110.1	18.9	32.2	90.0	81.4	131.7	193.3	191.0	42.4	38.4	27.5	18.1	975.0	
2007	14.7	41.1	89.7	50.6	38.3	94.8	428.5	81.7	127.4	39.7	23.7	40.7	1070.9	
2006	107.1	61.2	12.1	132.2	90.9	155.7	248.2	106.7	72.4	9.3	101.2	9.8	1106.8	
2005	25.5	71.1	43.8	80.9	58.6	63.6	235.9	214.9	85.8	37.9	62.9	11.4	992.3	
2004	53.3	28.7	44.7	106.3	114.6	276.7	83.1	116.5	37.8	1.2	77.7	34.5	975.1	
2003	39.0	68.2	122.5	175.9	32.6	162.0	555.0	128.4	190.5	90.0	72.2	22.0	1658.3	
2002	31.1	40.7	120.0	89.8	153.5	177.9	79.3	172.6	34.5	67.2	4.7	103.3	1074.6	

Compiled from: Jiangsu Statistical Yearbook 2002–2013, Jiangsu Provincial Bureau of Statistics

Table 8.3 Runoff coefficients and reduction coefficients for different land use types

Land use type	Runoff coefficient	Reduction coefficient
Paddy field	0	0.05
Sloping cultivated land	0.6	0.55
Forest land	0.35	0.45
Wasteland/grassland	0.36	0.65
Water body	0.7	0.45
Bare land	0	0.05
Sloping cultivated land	0.65	0.65

Table 8.4 Reservoir capacity and water surface area corresponding to different water levels

Water level (m)	Reservoir capacity (m^3)	Water surface area (m^2)
60.00	279.20	520.16
61.00	1120.37	1253.02
62.00	2974.82	3646.18
63.00	10757.68	12400.25
64.00	25224.86	16681.27
65.00	42942.15	18811.41
66.00	63002.13	21760.74
67.00	86110.88	24507.22

8.3.4 Water Level Calculation

Using ArcGIS software, the storage capacity and water surface area corresponding to various water levels can be derived, as shown in Table 8.4. When the water level is 60.00 m, the storage capacity is only 279.20 m^3, and the resulting water surface area is relatively small, failing to meet the design requirements. As the simulated water level gradually rises, both the storage capacity and water surface area increase accordingly. When the water level reaches 67.00 m, the storage capacity is 86,110.88 m^3, and the water surface area is 24,507.22 m^2. According to previous calculations, the maximum runoff volume of Catchment Area D is 80,423.1 m^3, corresponding to a water level of 66.8 m, which is taken as the maximum water level. Consequently, a storage capacity curve for Catchment Area D can be drawn (Fig. 8.22), reflecting changes in reservoir capacity and aiding in the evaluation of water system morphology and the selection of water level. The curve for Catchment Area D is convex: the growth in storage capacity is relatively slow when the water level is between 60.00–62.00 m; it becomes more uniform between 62.00–64.00 m; and beyond 64.00 m, the rate of increase in storage capacity significantly surpasses the rate of water level rise. As shown in Table 8.4, once the water level exceeds 63.00 m, storage capacity increases rapidly, while the water surface area continues to grow at a relatively steady pace. The difference between the estimated water level and the elevation of the dam location determines the corresponding dam height. The

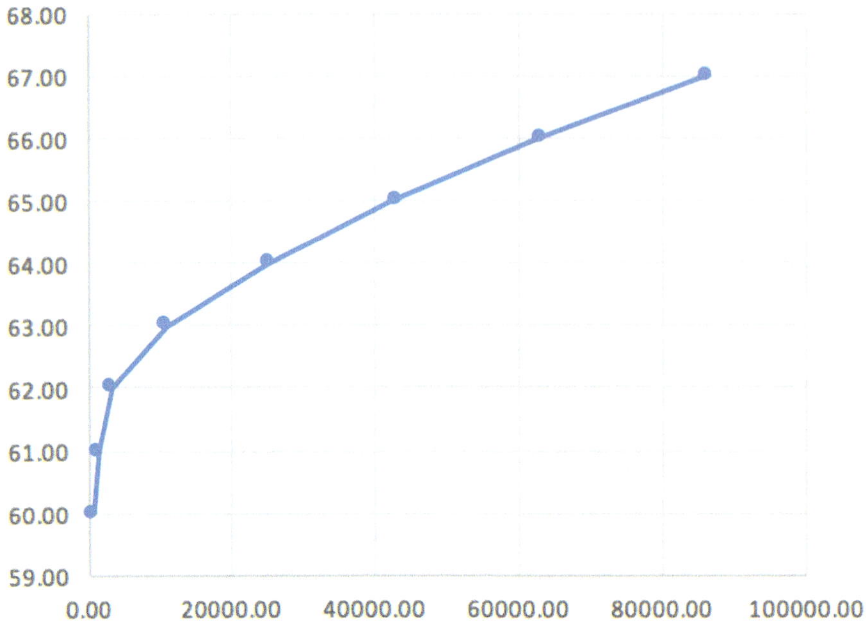

Fig. 8.22 Catchment D storage capacity curve

final dam height should be determined based on an assessment of the water body morphology at different water levels.

8.3.5 Water Body Morphology Regulation and Water System Generation

For Basin 4, once the primary constructed water body corresponding to Catchment Area D is determined, the parameterized simulating natural waterscape model can be used to design the next-level water body. It should be noted that the volume of water available to the next-level water body is the sum of runoff collected within the catchment and the overflow from the upper-level water body. Thus, the process of constructing the water system proceeds progressively from high to low elevations based on the topography of the basin, with ponds, marshes, pools, and lakes interconnected by water volume.

This study utilizes Grasshopper software running within the Rhino environment to compile operation sets and perform data computations. On one hand, algorithmic programming in Grasshopper enables accurate and efficient data computation; on the other hand, Grasshopper can process and output various types of data precisely, ensuring effective data integration and minimizing errors from manual calculations. Combined with Rhino, Grasshopper's computational results can be visualized and adjusted in real-time, facilitating iterative optimization. Based on the previously established design logic, and incorporating prior analyses and evaluations on water feature siting, runoff volume, site topography, and water body morphological indicators, parameters were selected (Table 8.5) and the initial data was input as constants into the model. The author developed four algorithmic modules for the simulating natural water feature design logic using Python within the Grasshopper platform. These included: the "Dam Height–Water Volume" algorithm, the "Water Body–Topography" algorithm, the "Water Volume–Earthwork" algorithm, and the "Water Morphology" algorithm (Fig. 8.23). These were integrated to form a parameterized design model for simulating natural water features. Due to space limitations, the "Dam Height–Water Volume" algorithm is used here to exemplify how the design model is implemented in Grasshopper. Based on site selection and runoff volume calculations, the available runoff volume is associated with the terrain data to determine dam height. However, since "Volume–Elevation" type algorithms are often complex and constrained, a direct algorithmic approach is impractical. Therefore, a solid subtraction method in Grasshopper is employed: the required water volume is compared against the available runoff, and the dam height input is iteratively adjusted until the designed water volume matches the available runoff within an acceptable margin of error. The specific algorithm process is as follows: a predefined water volume block is intersected with the terrain surface to obtain the resulting water body (solid), which is then used to compute volume and extract 2D contours. Based on the site analysis and overflow point, dam points (point sets) are defined. Using contour lines, a dam base line is drawn, which is then extended along the terrain direction to generate a reference surface at the water level elevation. This surface is scaled about the base line center to conform to the terrain, forming the

Table 8.5 Parametric design parameters for simulated natural water features

Name	Meaning	Data type	Control method	Symbol
Site topography	Represents the terrain trend of the site	Surface	Manual control	Sa
Dam elevation	Elevation of each dam body	Float	Manual control	Hd
Dam morphology	Shape of each dam body including location information	Curve	Manual control	Cd
Water body	Water enclosed by each dam	Entity	Automatically generated	Bw
Water body contour line	Planar contour line formed by each water body	Curve	Automatically generated	Cw
Water volume	Volume of the water body	Float	Automatically generated	Vw
Dam slope	Slope of the earthwork used to construct the dam	Float	Manual control	Sd
Dam body	Earthwork constructing each dam	Entity	Automatically generated	Bd
Earthwork volume	Volume of earthwork required to build each dam	Float	Automatically generated	Vd

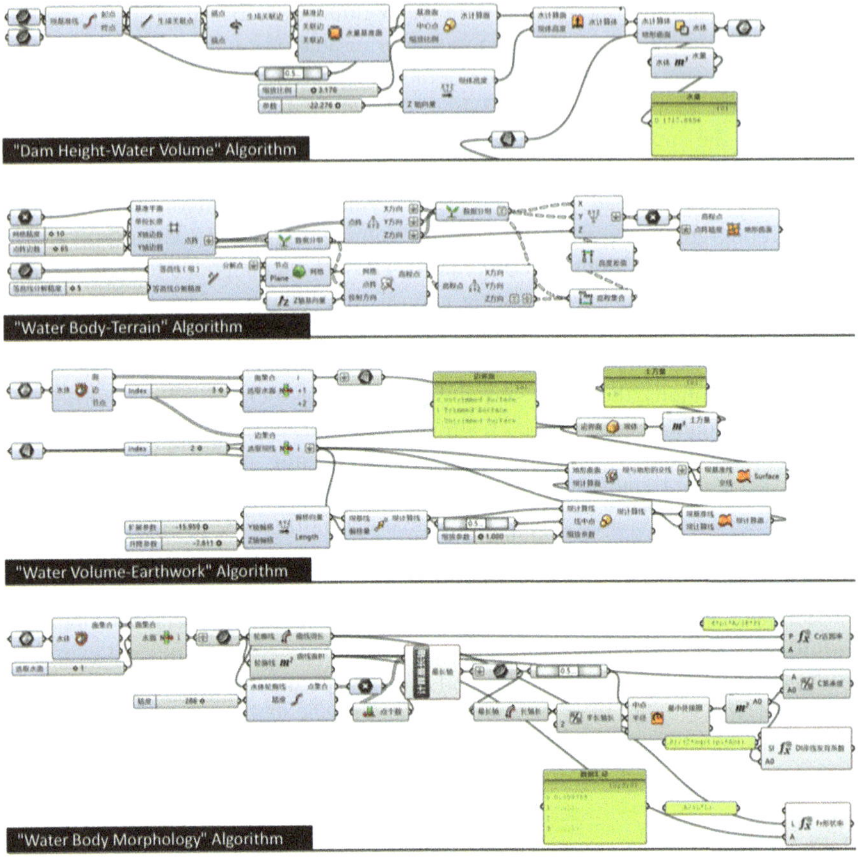

Fig. 8.23 Four algorithms

computational water surface. Based on previously obtained terrain data (terrain surface), and given the dam crest elevation (absolute height), the water surface is extruded upward to generate the final water body volume (Fig, 8.24).

Since a simulating natural waterscape is composed of a series of "individual" water bodies, constructing the overall water system requires integrating multiple individual designs. Based on the parameterized model for individual water bodies, the water system model is developed by associating water bodies within the same catchment area using shared water volume data. This requires the use of Grasshopper's tree data structure to simultaneously manage multiple water bodies. Individual data is converted into tree structure form, and through adjusting the water volume parameter, multiple water bodies are generated. With Rhino, the design outcomes from the parameter-controlled simulating natural water system can be visualized in real time. The interplay between water volume, dam height, and water body morphology constitutes a parametric modification scheme, which is fed back into

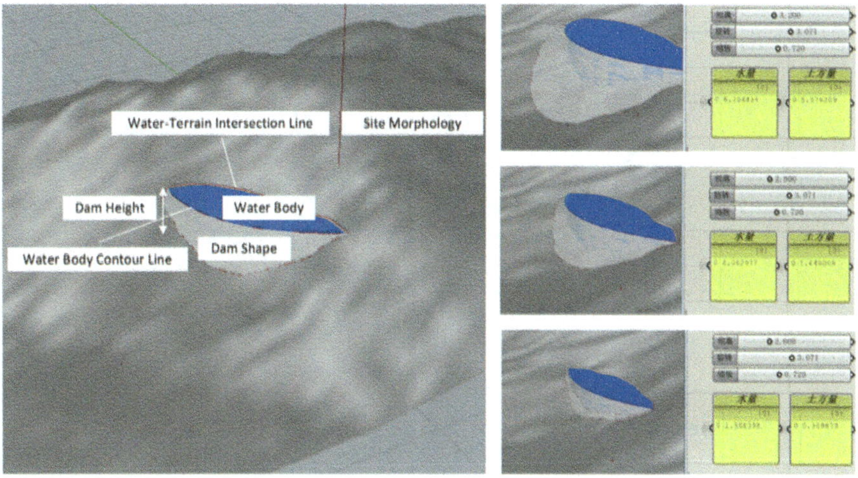

Fig. 8.24 "Dam Height-Water Volume" Algorithm Rule

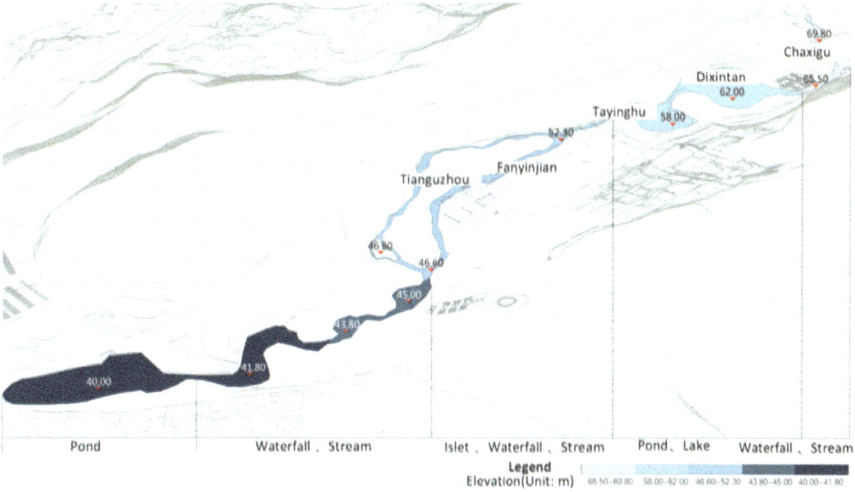

Fig. 8.25 Structural diagram of simulated natural water system in northern area of Niushou mountain scenic spot

the dam and terrain modules. After multiple rounds of adjustment and optimization, the final water system landscape design is produced.

Figure 8.25 shows the water system structure of the northern area of Niushou Mountain Scenic Area, created using the parameterized simulating natural waterscape model. It includes a variety of water features such as lakes, pools, creeks, streams, and waterfalls. Thanks to parametric analysis and computation, rainwater that would otherwise be directly drained into the municipal system is now collected

Fig. 8.26 Field photo of natural water system in northern area of Niushou Mountain scenic spot

and released gradually. The water system consists of three large water surfaces at varying elevations, connected by streams and cascades, and maintains perennial water presence through careful estimation and regulation. The project has been completed for over a year, with naturally beautiful water forms and babbling streams. An ecological water system has essentially formed. Based on systematic studies of Nanjing's rainfall and geomorphological conditions, this system utilizes natural rainfall and runoff with minimal human intervention, blending seamlessly into the surrounding mountainous environment. By fully leveraging the natural dynamics of surface water, the resulting water features align with natural processes, require no manual maintenance, and realize a sustainable simulating natural waterscape (Fig. 8.26).

References

1. Wang XR, Lin Q (2007) The meaning of nature. Chin Landsc Arch 1:7
2. Pan GX (2001) Landscape design art of Jiangnan. Southeast University Press, Nanjing
3. Chen Z (1988) Annotations on Yuan Ye (Garden Art). China Architecture & Building Press, Beijing
4. Strahler AN (1957) Quantitative analysis of watershed geomorphology. Trans Am Geophys Union 8(6):913–920
5. Shreve RL (1966) Statistical law of stream number. J Geol 74:17–37
6. Hu CH, Wang JX (2010) Runoff yield and concentration models and hydrological models in watersheds. Yellow River Water Conservancy Press, Zhengzhou

7. Wang DW (2008) Research on modern hydrology and water resources. China Water & Power Press, Beijing
8. Dong Y (2013) Analysis and calculation of rainwater resource potential in hilly small watersheds of central Sichuan. Yangtze River 9:8
9. Marsh WM (2006) An environmental approach to landscape planning (Zhu Q, Huang L, Yu KY et al, Trans.). China Architecture & Building Press, Beijing (Original work published 2006, p 172)
10. Niu WW (1987) Modern applied geography. Science Press, Beijing
11. 潘文斌，黎道丰，唐涛，等 (2003) 湖泊岸线分形特征及其生态学意义. 生态学报 2003(12): 29
12. George S, Robert MM (1990) Applications of fractals in ecology. Trends Ecol Evol 3:79–86
13. Pan WB, Li DF, Tang T et al (2003) Fractal characteristics of lake shorelines and their ecological significance. Acta Ecol Sin 12:29
14. Li XG, Jiang N, Wang HJ et al (2005) Dynamic changes in the shoreline morphology of lakes in the Taihu Lake Basin in recent 30 years. J Lake Sci 17(4):296
15. Jenson S K，Domingue J O. Extracting topographic structure from digital elevation data for geographic information system analysis. Photogramm Eng Remote Sens，1988，54(11) :1593-1600

Chapter 9
Vertical Design Model

Vertical planning and design in landscape environments must address a range of requirements, including spatial composition, surface drainage, vegetation planting, and the layout of facilities. It involves multiple site elements such as roads, water systems, and terrain, making the parametric vertical design model a representation of a complex system composed of numerous interrelated elements and integrated objectives. Vertical design at the landscape planning stage is of critical importance, directly impacting both the rationality and cost-efficiency of the overall scheme. Earthwork balance is a key indicator in vertical design and serves as the foundation for parametric vertical optimization. Through the regulation of cut-and-fill volumes, vertical design enables multi-objective optimization of the landscape environment. However, traditional vertical design methods, constrained by limited tools and approaches, often result in isolated treatment of subsystems such as terrain, roadways, water bodies, and open spaces. This segmentation prevents effective feedback among components, making it difficult to achieve integrated, optimized design outcomes. Additionally, the traditional process is often labor-intensive and time-consuming, with limited capacity for design iteration. The vagueness in design representation leads to imprecision in construction practices, which hampers both the realization of intended design effects and the control of project costs. With the advancement of computing technologies, vertical design expression has evolved from two-dimensional to three-dimensional modeling. The development of software such as Civil 3D has bridged the gap between design and construction, transforming the workflow of vertical design in landscape architecture. These tools have fostered the rise of interactive design practices and significantly improved methodological efficiency. The parametric vertical design model for landscape architecture, constructed on the Civil 3D platform, integrates all vertical elements within a unified digital terrain model, establishing dynamic relationships between these components. This chapter first explains the goals, content, and procedural framework of vertical design in landscape environments. It then discusses vertical design methods from two perspectives: parametric control of the design process and optimization of design methodology. Based on Civil 3D, the chapter develops a parametric vertical

Y. Yuan, *Parametric Landscape Architecture Planning and Design*,
https://doi.org/10.1007/978-981-95-8011-8_9

design model structured around four main elements—roadways, hydrological systems, site zones, and overall earthwork balance—clarifying the parametric components of vertical design and analyzing model applications through detailed case studies.

9.1 Vertical Design in Landscape Environments

The word landscape inherently contains land, highlighting the close relationship between landscape architecture and topography. In the practice of landscape planning and design, vertical design is a fundamental component. It aims to satisfy an array of integrated demands, including spatial organization, surface water drainage, vegetation planting, and the placement of buildings and structures. By appropriately modifying, optimizing, or transforming the original terrain, vertical design determines slopes, controls elevations, and achieves cut-and-fill balance. As a core component of vertical design, terrain serves as the foundational carrier for all other design elements in the landscape environment. Therefore, vertical design essentially defines the "structural framework" of the landscape topography.

9.1.1 The Value and Significance of Vertical Design

9.1.1.1 Shaping Spatial Form

As the foundational layer of landscape spaces, topography plays a crucial role in the formation of spatial morphology. "Topography" and "landform" refer to the three-dimensional undulations of the Earth's surface. In landscape environments, landforms include valleys, mountains, hills, grasslands, and plains, and can further be categorized into flatlands, convex areas, ridgelines, concave depressions, and basins. Beyond these classifications, landscape environments frequently involve a more commonly used category known as "micro-topography"—subtle undulations and depressions in the surface within a site-specific scope. Landscape spatial form is defined by enclosing elements, and topography serves as the most fundamental of these. All other design elements are attached to or integrated with the terrain, collectively forming spatial interfaces and enclosing diverse landscape spaces. Topography delineates space through three main components: the base surface, slope surfaces, and the skyline profile. By integrating vegetation, structures, and other components, it can be used to shape and define exterior spaces, thereby creating spatial forms with varying atmospheres (Fig. 9.1). Different types of terrain generate spaces with distinct characteristics, requiring designers to understand the spatial implications associated with each landform type.

Moreover, the shape and character of space are closely related to human visual perception. Topography affects how people perceive space, thus guiding the

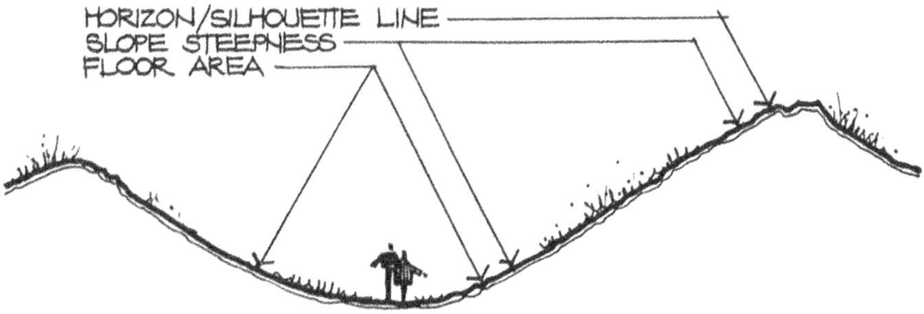

Fig. 9.1 Elements of spatial composition. Citation from Booth, N.K. (1989). Elements of landscape architecture design

experience and emotional response within landscape environments. Different landforms can evoke spatial atmospheres with distinct psychological qualities. The viewer's perspective plays a decisive role in the perceived sense of enclosure and spatial intimacy. In Basic Elements of Landscape Architecture Design, Norman K. Booth points out that the degree of spatial enclosure is determined by the size of the visual field, slope gradients, and the horizon line. Typically, the human visual field extends 40°–60° above the horizontal line and about 20° below. When the ratio of area, slope, and skyline reaches or exceeds 45° (i.e., a 1:1 ratio of height to length), the visual field is considered completely enclosed. Conversely, when the ratio falls below 18°, the sense of enclosure is lost (Fig. 9.2). Topography influences spatial perception and atmosphere, thereby affecting psychological responses. Gentle terrain does not obstruct sightlines, creating a visual sense of openness and comfort, evoking feelings of stability and vastness. In contrast, steep and rugged terrain can stimulate excitement and awe. This is exemplified by the Italian terraced gardens built into hillsides, such as Villa d'Este, Villa Lante, and Villa Farnese, and the classical French gardens, such as Château de Versailles and Château de Fontainebleau, which were laid out on flat terrain—each employing topography in distinctive ways to shape spatial ambiance. Additionally, topography contributes to spatial experience by directing sightlines. For example, by guiding views toward a specific focal point and constraining the visible range, designers can create a sequence of continuous landscape spaces. Therefore, in the design of landscape environments, it is essential to integrate design objectives with an in-depth understanding of how different topographic forms impact psychological perception in order to craft spatial experiences that are both intentional and distinctive.

9.1.1.2 Organizing Surface Water

Rainwater that falls onto the ground, aside from the portion that infiltrates the soil or evaporates, mostly flows downslope under the force of gravity, forming surface runoff. The volume, direction, and velocity of runoff are closely related to topography: the lower the terrain, the greater the accumulation of runoff; the steeper the

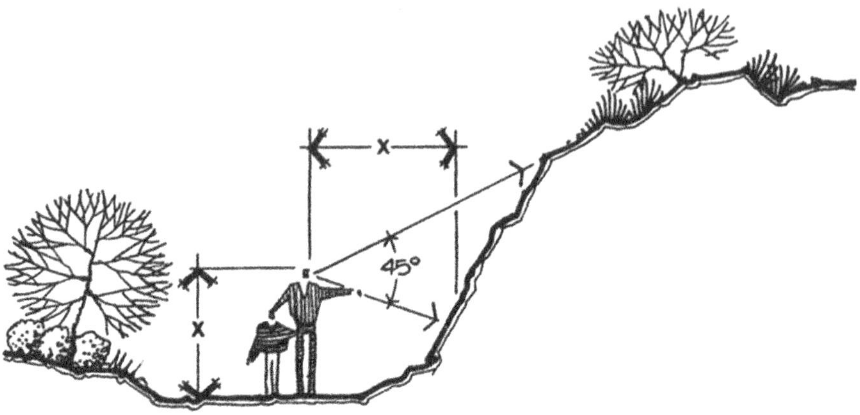

Fig. 9.2 When the field of view is 45°, the space achieves a sense of enclosure. Citation from Booth, N.K. (1989). Elements of landscape architecture design

slope, the faster the flow. Conversely, areas with little to no slope are prone to water accumulation. Therefore, vertical design plays a critical role in regulating surface drainage and directing runoff. One of the primary goals of drainage organization in landscape environments is to meet functional needs. First, as spaces intended for human use, sites must be designed to prevent waterlogging. Second, from a safety perspective, excessive surface runoff can result in severe erosion or prolonged saturation in geologically unstable areas, triggering hazards such as debris flows, landslides, or collapses—posing serious threats to public safety. Such risks must be minimized. Third, water storage and regulation should be addressed through vertical design: by directing and containing water within designated areas, designers can retain water during periods of excess and slowly release it during droughts. This regulation helps balance water resources, supports habitat improvement, meets the needs of plant growth, and reduces downstream impact. In addition to enriching the landscape visually, terrain also functions in the organization, management, and distribution of surface water. Therefore, beyond spatial form-making, the organization of surface water is a key consideration in vertical design.

There are two main types of drainage in landscape environments: natural surface drainage and engineered (organized) drainage. Both rely on gravity; thus, slope planning must fully account for the site's drainage conditions. In most areas, drainage is managed through natural surface runoff, where terrain itself serves as the organizing structure. Consequently, topographic design must prioritize drainage to avoid surface water accumulation. One of the defining features of drainage in landscape environments is its integration with topographic design, reducing the need for artificial channels or pipes and thereby lowering construction costs. This also eliminates the need for frequent maintenance, aligning with the principles of sustainable design. In certain cases, where site use demands more precise control, organized drainage systems—such as open ditches or subsurface pipelines—may be required. In general, rainwater runoff in landscape environments is relatively clean and can be discharged directly into nearby natural water bodies. However, runoff with higher levels of contamination or domestic wastewater must be treated to meet environmental standards before being released into natural systems or municipal infrastructure. Unlike drainage in built architectural settings, which often relies heavily on engineered systems, landscape drainage depends primarily on natural gravity flow. This makes vertical design especially critical in organizing and managing water movement across the site.

9.1.1.3 Satisfying Planting Requirements

In medium- to large-scale environments, the undulations of the terrain can alter both climatic and soil conditions, thus exerting a significant influence on the growth and distribution of plants. In addition to elevation—which imparts a vertical stratification to vegetation—slope aspect and gradient also create distinct distribution patterns. In the Northern Hemisphere, north-facing slopes generally receive shorter, less intense sunlight and exhibit higher humidity, favoring shade-tolerant, intermediate-layer plants; in contrast, south-facing slopes, exposed to prolonged and intense sunlight coupled with higher temperatures and increased evaporation, tend to support sun-loving species with certain drought-adapted characteristics. The east and west aspects represent transitional conditions between these extremes. Although differences in species distribution may be less pronounced on a smaller scale, variations in water and heat conditions can significantly affect plant growth. In many landscape environments, modifications to the terrain are focused on small-scale adjustments—that is, the creation of micro-topography.

In some regions of China, high levels of soil salinity and alkalinity can adversely affect plant growth. Besides methods such as increasing irrigation frequency to dilute soil salinity, the creation of micro-topography plays a critical role in salt isolation and neutralization. On one hand, as rainwater flows along the natural undulations of a micro-topographical surface, the resulting runoff can erode and thereby partially dilute the soil's salt content; on the other hand, constructing raised micro-topographic features with low-salinity soils in tandem with salt barrier measures can mitigate salt damage to plants, improve survival rates, and ensure high-quality growth. Furthermore, sculpting the terrain can effectively expand the available planting area. For example, a terrain with a slope of 5–10% may increase the surface area by approximately 2–3% and augment the soil volume, thereby providing a larger planting space. Since different plants have varying requirements—some thrive in sunny conditions while others prefer shade—a richly varied terrain can offer diverse, suitable environments for each species. Compared with a uniform terrain, a landscape with complex topographic variations can accommodate a broader range of plant species, not only achieving visually appealing plant landscapes but also fostering the development of well-structured and stable plant communities with notable ecological benefits.

9.1.1.4 Creating a Microclimate Environment

Topography influences airflow, causing variations in wind speed and direction, and interacts with water and heat to affect the microclimate of an area. According to ancient Chinese feng shui principles, the optimal arrangement for a residential site is "with a mountain at the back, facing water, and towards the sun." Mountains can shield a site from the harsh northern winter cold, while having a mountain at the rear helps to block cold winds. Facing water, especially in summer, allows the cool breezes from the water surface to lower temperatures and introduce moist air, thereby contributing to microclimatic regulation; in the temperate regions of the Northern Hemisphere, a southward orientation secures abundant sunlight. Generally, compared with flat terrain, north-facing slopes with inclinations of 2° to 5° receive approximately 25% less sunlight, while a 6° north-facing slope may receive around 50% less; conversely, south-facing slopes benefit from consistently better sunlight, maintaining relatively higher temperatures throughout the year. The interplay between topography and microclimate in landscape environments follows similar principles. Elevated forms such as ridges or mounds can block strong northwestern winter winds, thereby helping maintain site temperatures, while south-facing slopes, with their increased sun exposure, remain warmer all year. Additionally, topography can channel wind flow to adjust the internal climate of a site. For instance, saddle-shaped spaces such as valleys or depressions situated between highlands can direct wind passage, thereby enhancing site ventilation (Fig. 9.3).

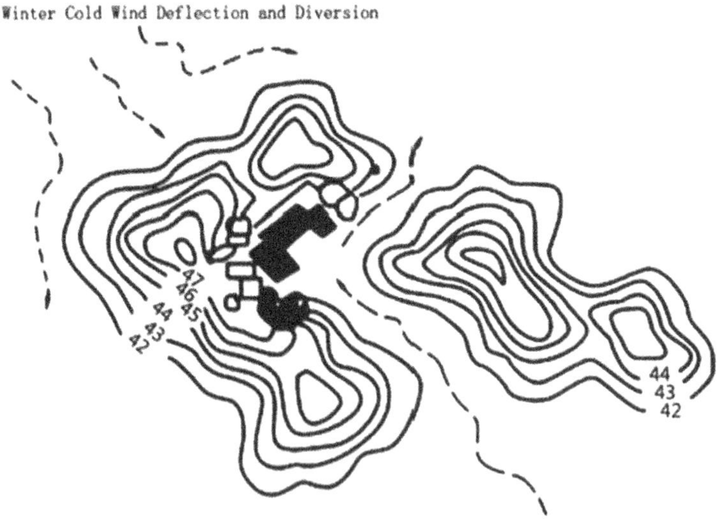

Fig. 9.3 The role of vertical design in shaping microclimates. Redrawn from: Han Yulin. Landscape architecture engineering [M]. Chongqing: Chongqing University Press, 2011: 18

9.1.2 Contents of Parametric Vertical Design

Vertical design is an essential component of landscape planning and design. According to Article 4.3.2 of the Park Design Code (GB 51192–2016), vertical control must include specific elevation indicators such as: peak or slope-top and slope-bottom elevations; main retaining wall elevations; maximum, normal, and minimum water levels; underwater topography and revetment top elevations; key turning, intersection, and slope-transition points on park roads; bridge deck elevations; elevation of roofs, interior, and exterior ground levels of primary buildings; interior and exterior ground elevations at each park entrance and exit; burial depths of underground utilities and structures; and ground elevations at key scenic viewpoints. These requirements are ultimately represented on drawings through contour lines and elevation annotations. Beyond the general content stipulated in design codes, modern vertical design integrates multiple objectives, including the generation of spatial form, surface water management, support for planting requirements, and the creation of microclimatic environments. The contour lines and elevation values shown on design drawings are derived from a series of analyses and calculations involving elevation, slope gradient and aspect, watershed and catchment delineation, and earthwork volume estimations.

9.1.2.1 Elevation Design

In the field of surveying, "elevation" refers to the vertical distance from a point to an absolute reference surface along a plumb line—commonly known as absolute elevation. Relative elevation, by contrast, measures the vertical distance from a point to an assumed benchmark surface, also referred to as a relative datum. Due to long-standing conventions, both absolute and relative elevations are often collectively referred to as "elevations" or "benchmark levels." To enable meaningful comparison and measurement, all surface points must reference a shared zero point, or standard baseline. Typically, the geodetic mean sea level is used as the reference datum (benchmark origin). In China, the official elevation reference surface is based on the average sea level of the Yellow Sea, with the national benchmark point established in Qingdao.In design practices, elevations may be expressed as either absolute or relative values and serve as the basis for vertical positioning.

During the schematic design phase of landscape planning, elevations must be established for key control points such as roads, topography, buildings, structures, and plazas. In the construction drawing phase, more detailed elevation data is provided for construction and on-site layout purposes. Vertical design in landscape architecture generally employs a combination of elevation control methods and contour line methods. The elevation control method involves marking elevation values at slope breakpoints and other critical locations, sometimes supplemented with directional arrows indicating surface water drainage flow. The contour line method uses contour lines to represent designed topographies, including roadways, plazas, parking lots, and green spaces. Drawing these contours requires reference to the elevation values of key terrain control points. In landscape planning and design, contour lines offer greater precision in representing terrain modulation. However, for elements like roads and hardscape areas, they may be less effective in expressing design intent. Therefore, these two methods are often used in conjunction. Figs. 9.4(a) and (b) illustrate examples of vertical design drawings from China's 1980s hand-drafted era and recent computer-generated drawings. Despite the shift in drafting techniques, both examples demonstrate the integrated use of elevation control and contour line methods in conveying vertical design strategies.

9.1.2.2 Slope Determination

Slope indicates the steepness of a surface and is typically defined as the ratio of vertical height to horizontal distance. It is one of the most critical factors in vertical design and, together with elevation, determines the landform of a landscape. Slope is closely related to drainage and plant cultivation; different slope gradients correspond to distinct land use strategies and spatial organization approaches. The four commonly used methods of expressing slope include the percentage method, the

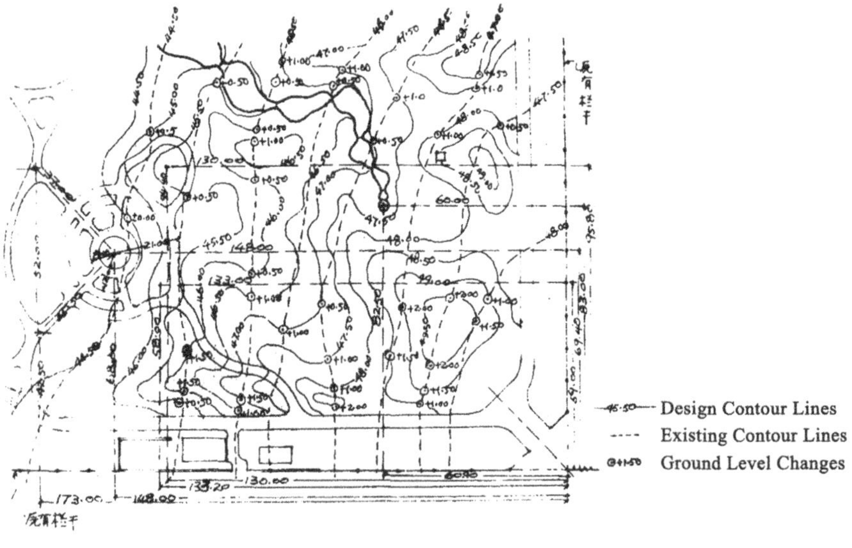

Fig. 9.4 (a) Partial topographic design construction drawing of a park, showing layout control points, both proposed and existing contour lines, as well as ground elevation changes at specific locations. Citation from: Li Jiale. Lecture six on garden planning and design knowledge: topography and plant arrangement techniques [J]. Beijing Gardening, 1989 (04). (b) Partial vertical design construction drawing of the northern area of Niushou Mountain scenic zone, Nanjing, illustrating both existing and proposed contour lines, revetment elevations, isobaths, and road design elevations. Citation from: Cheng Yuning studio profile

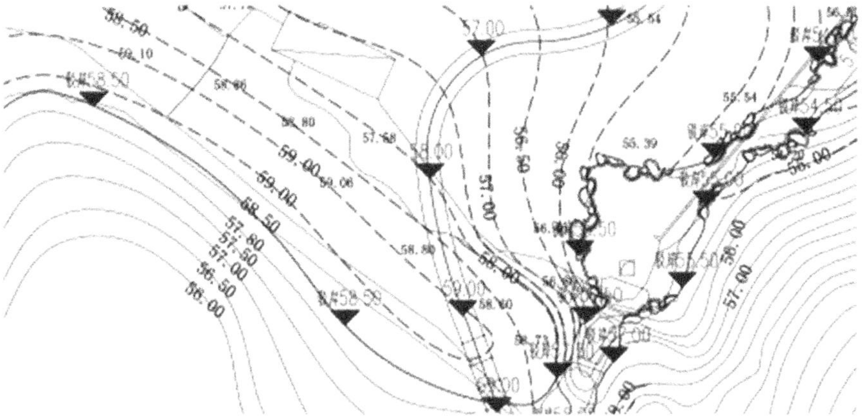

Fig. 9.4 (continued)

degree method, the mil method, and the fractional method, with the percentage and degree methods being the most prevalent. The percentage method expresses slope as the percentage of elevation difference over horizontal distance, calculated as: slope = (elevation difference / horizontal distance) × 100%. For example, a 3% slope means that for every 100 m in the horizontal direction, there is a 3-meter vertical drop. The degree method represents slope in degrees, calculated using the inverse tangent function: $\tan(\alpha)$ = elevation difference / horizontal distance, therefore α = arctan(elevation difference / horizontal distance).

The Commission on Geomorphological Survey and Mapping of the International Geographical Union has classified slope gradients for use in detailed geomorphological mapping. According to their standards: 0–0.5° is considered a plain, 0.5°–2° a very gentle slope, 2°–5° a gentle slope, 5°–15° a slope, 15°–35° a steep slope, 35°–55° a very steep slope, and 55°–90° a cliff. In landscape environments, slope must be fully considered in landform creation, road alignment, the design of water features, and the construction of buildings and structures. Construction should be avoided on excessively steep terrain to reduce environmental disturbance and prevent safety hazards. In vertical design, slope considerations mainly focus on terrain slope and the slopes of sites and roads; in some cases, the drainage slope must also be specified. Naturally accumulated soil, after a long period of settlement and stabilization, tends to form a relatively uniform surface with consistent slope—this is known as the natural inclined surface of the soil. The angle formed between this surface and the horizontal plane is called the soil's angle of repose. The angle of repose varies depending on soil type, moisture content, and other factors, typically ranging between 15° and 40°. In landscape planning and design, it is not advisable to excessively alter large areas of natural terrain; designed slopes should be as naturalistic as possible and take into account the relevant soil angle of repose. Overly steep slopes not only result in unstable landforms and increase the risk of disasters such as landslides but also make it difficult to maintain the intended design outcomes. When the slope of a terrain exceeds the natural angle of repose of the soil, appropriate slope protection, soil stabilization, or erosion control measures must be implemented. When designing sites and roads, the layout and routing should be flexibly adjusted according to surface slopes to ensure both safety and proper drainage. In particular, road design must conform to standard requirements for both transverse and longitudinal gradients to ensure safe vehicle operation. Slope design in landscape planning and design can refer to the Park Design Code (GB 51192–2016), which provides specifications for surface drainage slopes (Table 9.1). In low-lying areas where natural slope drainage cannot be achieved, subsurface drainage systems or culverts should be installed.

9.1.2.3 Earthwork Calculation

In landscape architecture construction, whether it involves building roads, plazas, buildings, or structures, or modifying and utilizing terrain, all operations are grounded on the earth's surface and depend on earthwork operations to be carried out. As such, earthwork is a preliminary and essential component in landscape construction. The investment and volume of earthwork often constitute a significant proportion of the overall cost of landscape projects. Therefore, the strict control and balancing of earthwork is an effective means of reducing construction costs. Although detailed earthwork quantity reports are not required during the conceptual design phase, the later stages of construction drawing design and implementation in landscape planning necessarily involve issues related to earth volume. Consequently, it is crucial to anticipate and manage earthwork quantities early in the design process to ensure continuity and controllability throughout.

Earthwork engineering includes estimating and balancing the quantities of excavation and fill for elements such as sites, roads, and water systems. The Vertical Planning Standard for Urban and Rural Construction Land (CJJ 83–2016) provides explanatory notes that list the quotas and balance standards for earthworks across various categories of urban construction land. Earthwork quantity is calculated by determining the difference in volume between the original terrain and the modified terrain. This includes two parts: the volume of excavation and the volume of fill. The sum of these two volumes constitutes the total earthwork quantity. Earthwork balance refers to achieving an equilibrium between the volume of excavation and the volume of fill within a defined area. Common methods for earthwork calculation include the contour method, grid method, cross-section method, average elevation method, Digital Elevation Model (DEM) method, and regional earthwork balancing method. Each of these is suited to different terrain conditions and varying

Table 9.1 Drainage slopes for various surface types (%)

Land cover type	Minimum slope
Grass	1.0
Sports turf	0.5
Planting surface	0.5
Paved area	0.3

requirements for calculation scale and accuracy. Among them, the grid method, cross-section method, and DEM method are the most commonly used in civil engineering practice.

9.1.3 Key Points of Vertical Design

9.1.3.1 Appropriateness and Minimization

Vertical optimization refers to the appropriate adjustment and transformation of topography in a measured and context-sensitive manner. In the context of vertical design for landscape environments, the fundamental principle is to act in accordance with local conditions. Landscape architecture projects often cover large-scale areas, and extensive earthworks or vertical modifications can lead not only to irreversible damage to natural ecosystems but also to significant waste of human and material resources. "Appropriateness" implies suitability and proportionality, while "minimization" is guided by the principle of reducing intervention to the lowest possible level. Emphasizing appropriateness and minimization in vertical design requires targeted evaluation, precise calculations, and thoughtful design to achieve intended goals with minimal environmental disturbance—ultimately protecting the ecological environment and optimizing project investment. At the same time, the principle of on-site earthwork balance should be upheld. On-site earthwork balance refers to achieving equilibrium between cut and fill within the designated design area, avoiding off-site soil sourcing and treating the introduction of foreign soil with caution. This principle plays a vital role in both controlling construction costs and facilitating ecological restoration. After conducting an ecological sensitivity assessment of the entire planning area, zones classified as ecologically sensitive—particularly in terms of vegetation and terrain—should be preserved without disturbance. Areas without recreational or functional needs do not require modification, as maintaining the status quo can offer greater ecological benefits. However, some zones with steep slopes that pose risks to visitor safety may require limited artificial intervention to ensure a secure environment and prevent accidents. Areas identified through site analysis as having low ecological sensitivity and high suitability for development serve as the primary spaces for visitor activities. In these areas, vertical adjustments should be selectively made based on design needs for spatial definition, drainage, and planting—such as in poorly drained lowlands, excessively flat and visually monotonous terrains, or areas lacking topographic variation. One of the critical aspects of vertical design in landscape planning is the scientific delineation of areas that require intervention and the development of appropriate and targeted design strategies.

"Appropriateness and minimization" also distinguish vertical design in landscape environments from that in urban land development. Although urban design also values the use of natural terrain, functionality is often prioritized. In contrast, landscape environments contain numerous ecologically sensitive areas where any terrain modification must begin with environmental protection. Furthermore, landscape architectural design seeks to express the unique character of place, and therefore does not impose rigid requirements on road slopes, elevations, or architectural positioning. Instead, it emphasizes following the natural terrain—adapting to elevation changes and responding to the site's conditions. This results in a more flexible approach to vertical design, which facilitates internal system balance and optimization.

9.1.3.2 Systematic and Complex Nature

The outcomes of vertical design are ultimately expressed through the form and configuration of the ground surface. Consequently, it is intricately linked with all landscape elements that depend on surface conditions, including roads, water systems, buildings and structures, open spaces, and plantings. Vertical design must meet the spatial and functional requirements of these elements, making it an essential component of landscape planning and design. Topography acts as a connective medium, dynamically integrating these various elements, and this dynamic interrelationship is reflected in the organization and distribution of earthwork volumes. The vertical design of the Jiangsu-Fukuoka Friendship Cherry Blossom Garden serves as a representative example to further explain the systemic nature of this process.

The Jiangsu–Fukuoka Friendship Cherry Blossom Garden is situated within the Ming Xiaoling Scenic Area of the Zhongshan Mountain Scenic Spot in Nanjing, Jiangsu Province. Established in 1996, it originally covered an area of 20 mu (approximately 13,333 m^2). Each year, the garden attracts numerous participants in Sino-Japanese friendship exchanges, making it a well-known cherry blossom viewing site in Nanjing as well as an important venue for friendship activities between Jiangsu Province and Fukuoka Prefecture. Due to years of neglect, the garden's landscape had deteriorated and its facilities fallen into disrepair. According to the client's requirements, the project involved enhancing the garden's landscape and expanding its area to over 60 mu (about 40,000 m^2). The author participated in this landscape improvement project alongside their advisor. One distinctive feature of the project was its emphasis on vertical design, which is presented here as a case study.

Located on the southern foothills of Zhongshan Mountain, the garden's base terrain is natural mountainous land sloping from north to south. The overall topography has some undulation, although certain local areas are flat and lack spatial variation. Additionally, the garden's original water system consisted of artificial hard embankments that were severely damaged, resulting not only in poor visual quality but also in significant leakage through the hard pond bottoms, making water retention difficult and causing ponds and channels to dry up. Furthermore, the eastern section of the site suffered from waterlogging, which adversely affected the growth of the cherry trees. To address these problems, one of the key design focuses was to reorganize the relationship between terrain and water through vertical design. The redesign began with the water system by removing the original hard embankments and pond bottoms, dredging, and reshaping the garden's hydrological features. HDPE geomembrane combined with soil covering was used to resolve water retention issues while allowing the planting of aquatic vegetation. By creating natural embankments, the design established beautiful water features such as cascades and floodplains. Next, localized microtopography was constructed. The earthwork generated during the natural embankment restoration was used to form microtopographic features around the water system, which not only alleviated the poor drainage problems hindering cherry tree growth but also enhanced spatial layering and harmonized with the surrounding mountainous environment. Finally, natural-style road drainage culverts were installed to prevent water accumulation on soft-ground roads caused by terrain modifications. Through a systematic vertical design approach, this project achieved a comprehensive integration of multiple objectives: water system restoration, terrain construction, drainage optimization, and planting enhancement. This can be aptly described as "achieving four goals with one effort" (see Fig. 9.5).

9.1.3.3 Earthwork Volume and Earthwork Balance

As stated in the classic gardening text Yuan Ye, "High mounds can be built up, low squares should be excavated." Here, "excavation" refers to the removal of earth, while "building up" refers to the addition of soil—essentially describing the "cut" and "fill" processes in terrain formation. Whether it is terrain construction, road building, or water system creation, the core focus of vertical design work lies in the balance between "cut" and "fill." In earthwork engineering, excavation and filling

Fig. 9.5 Jiangsu-Fukuoka Friendship Sakura Garden—Rrnovated Site Photo. Citation from: Cheng Yuning studio profile

generally coexist, and the sum of their volumes constitutes the total earthwork volume. Although seemingly straightforward, managing cut and fill often presents significant challenges for designers. On smaller-scale sites, it is relatively easier to analyze the original terrain and calculate vertical design and earthwork volume. However, when the terrain is complex and features varied landforms, accurately determining the extents of cut and fill becomes difficult. Designers may struggle to precisely delineate the areas of excavation and embankment. In traditional landscape architecture design processes, the quantities of cut and fill often remain vague concepts in designers' minds; only those with extensive field experience can truly visualize and manage these earthwork dynamics intuitively.

In landscape construction, terrain shaping, modification, and utilization form the most fundamental part of garden engineering. Whether building hills, managing water, laying ground surfaces, or planting, earthwork activities such as excavation, transportation, filling, piling, and leveling are involved. Therefore, earthwork is a major undertaking in landscape construction. Taking the Shanghai Botanical Garden as an example, the site's predecessor, the "Longhua Nursery," had a generally low elevation that required widespread raising of ground levels, resulting in nearly one million cubic meters of earthwork. Construction proceeded intermittently over 6 to 7 years. In principle, the volume of excavated soil and fill material generated within the site should be balanced and resolved onsite as much as possible, minimizing external impact—this is known as maintaining internal earthwork balance. When large amounts of excavated soil must be transported offsite as waste, and fill material must be sourced externally, such vertical shaping approaches not only risk causing significant artificial disturbance and damage to natural areas where soil is removed or dumped, but also impose high costs for long-distance soil transportation. Consequently, internal earthwork balance within the site is especially critical for environmental protection, cost reduction, and shortening construction periods. The term "earthwork balance" refers to an approximate equivalence between cut and fill volumes within a given area, representing a relative rather than absolute balance. Historically, landscape architects have focused on using terrain to shape space, guided by the principle of adapting design to local conditions while aiming to maintain earthwork balance in principle. However, constrained by the methods and technologies available at the time, it was difficult to effectively integrate qualitative and quantitative approaches in design, lacked technical controls during construction, and therefore struggled to regulate earthwork balance precisely.

9.2 Parametric Approach to Vertical Design

9.2.1 Parametric Regulation of the Design Process

Vertical design is a crucial component of landscape architecture planning and is carried out concurrently with the overall site planning. Based on collected data, vertical design first requires a clear understanding of the pre-design terrain and landform conditions. Then, by integrating design elements such as topography, roads, site layout, and water bodies along with the design concepts and requirements, control elevations are established and contour lines are drawn. Finally, earthwork volumes are calculated; if the earthwork is unbalanced, elevation adjustments must be made for each design element, followed by repeated earthwork volume calculations. This iterative process continues until earthwork balance is achieved. During this process, the various vertical design elements are treated in isolation and do not reflect the systemic relationships among them. As a result, achieving balanced and rational earthwork distribution within the system is difficult. Furthermore, this design process exhibits a linear characteristic with limited reversibility. Once the design is completed, real-time feedback on earthwork volume results is lacking. If earthwork remains unbalanced, elevations for the design components must be adjusted repeatedly until balance is reached.

With the advancement of information technology, traditional manual drafting and design have been replaced by computer-aided design (CAD) tools, greatly improving drawing and calculation efficiency. However, software such as AutoCAD remains primarily two-dimensional in nature, and the design process based on these platforms has not fundamentally changed; design components and stages remain largely independent. In contrast, Civil 3D, which shares the AutoCAD platform, is distinguished by its use of a three-dimensional digital model that links various design stages. Changes in any stage are dynamically reflected in related stages and presented as data. Three-dimensional model-based design enables not only dynamic interaction but also real-time feedback, significantly enhancing design efficiency and transforming the previously linear design process. As one of the BIM software platforms, Civil 3D also facilitates the integration of design and construction phases. Therefore, vertical design conducted using the Civil 3D platform can directly guide construction work, enabling full process control throughout landscape architecture planning and design.

9.2.2 *Digital Development of Design Methods*

9.2.2.1 Evolution of Representation Methods

Before the emergence of contour line representation, both Eastern and Western traditions, limited by technology and methods, relied on triangular symbols and text annotations to roughly and generally depict terrain undulations, rivers, and mountain locations (Fig. 9.6). Since modern times, contour lines have become the primary means of describing terrain. The development of surveying science and tools has gradually enabled more precise terrain depiction. However, with societal progress, two-dimensional terrain representation methods, such as contour lines, have increasingly struggled to meet the demands of design and construction. The advent of the digital era has likewise brought about a transformation in vertical design methods. The concept of digital terrain has facilitated more accurate terrain description and more efficient design workflows, making parametric landscape architectural planning and design possible.

Fig. 9.6 Kangxi map of the mouth of the Heilongjiang River, partial woodblock print. Collection of the first historical archives of China. Citation from: Li Xiaocong. "Surveying and versions of Kangxi's overview map of imperial territories." The Palace Museum Academic Quarterly, 2012, 30(1)

(1) Two-Dimensional Representation Methods

To this day, contour line representation and spot elevation representation remain the two main vertical representation methods used in landscape architectural planning and design, often combined to express vertical relationships within a site.

① Contour Line Representation

There are various definitions of contour lines. One can understand a contour line as a curve formed by connecting points of equal elevation, referenced from a fixed or temporary baseline elevation (i.e., zero elevation). Alternatively, contour lines can be defined as the horizontal projections of the intersection lines formed when a series of hypothetical, equally spaced horizontal planes cut through the terrain at a reference datum. Contour lines require scale and contour interval annotation to accurately interpret terrain features. Contour line representation combines elevation-marked contour lines with special symbols to reflect geomorphological characteristics. The earliest known contour map appeared in the sixteenth century in the manuscript map by Pieter Bruins, showing the 7-foot depth line of the Bay of Heitersheim. In 1791, D. Triel in France first used contour lines to depict terrestrial terrain. In the early nineteenth century, contour line terrain representation was primarily used in field surveying, but only by the twentieth century did its scientific and practical value gain widespread recognition, becoming the principal terrain representation method. Today, due to its scientific rigor and practicality, contour line representation remains the most fundamental, widely used, and accurate method for representing terrain on plan maps.

Contour lines are imaginary lines—they do not physically exist but serve as a human-created tool to describe the earth's undulating surface. On medium to large-scale topographic maps, contour lines are typically categorized into four types for readability: base contours (index contours), emphasized contours (primary contours), intermediate contours (secondary contours), and quarter contours (supplementary contours). However, general topographic maps usually show only two types: index and primary contours (Fig. 9.7). To avoid confusion, original terrain contours are often depicted with dashed lines, while design contours use solid lines. The contour line method offers a clear quantitative concept for terrain and geomorphology maps and can accurately reflect surface elevation, slope, surface area, and volume, thus possessing high practical value. Civil engineering designs and calculations such as route gradients, earthworks, and reservoir capacities can derive scientific basis through profile analysis generated from contour maps.

② Spot Elevation Representation

Spot elevations are points on a topographic map annotated with elevation data, indicating the height of specific points above or below a horizontal reference plane. Spot elevations on plan or profile views are used to express elevation. Typically, spot elevations complement contour lines in terrain description by marking special elevation changes on the plan, as well as elevations of peaks,

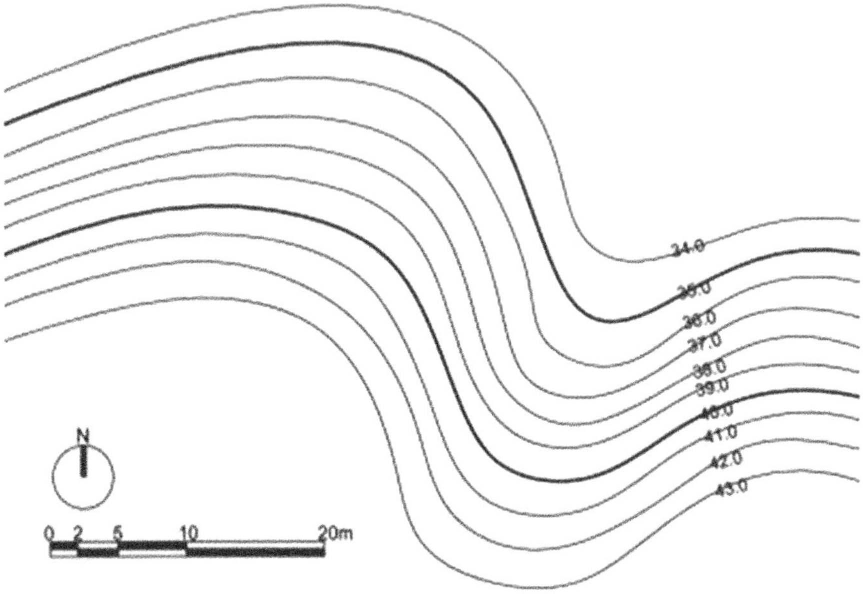

Fig. 9.7 Contour line representation

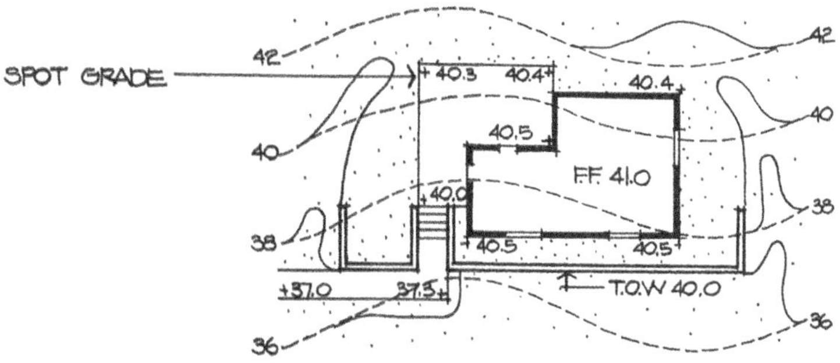

Fig. 9.8 The elevation point indicates a specific spot on the plane where there is a notable change in elevation. Citation from Booth, N.K. (1989). Elements of Landscape Architecture Design

depressions, and other features not clearly represented by contour lines. They may also serve as supplementary elevations located between two contour lines to enhance the reading accuracy of contours (Fig. 9.8).

Historically, due to limitations of surveying instruments, terrain elevation data was presented in two-dimensional drawings with numeric values reflecting elevation at discrete points. With advancements in surveying tools and technology, digital expression of drawings enables spot elevations to carry spatial coordinate attributes, thus becoming a form of three-dimensional representation. This serves as the foundation for generating Digital Terrain Models (DTM) and Digital Elevation Models (DEM), which will be discussed subsequently.

(2) Three-Dimensional Representation Methods
Whether using contour lines, spot elevations, or a combination of both to express the vertical conditions of a site, these methods essentially rely on two-dimensional graphics to describe three-dimensional spatial relationships. As a result, the process of translating from 2D to 3D is often fraught with uncertainty, ambiguity, and subjectivity. The emergence of Digital Terrain Models (DTMs) and Digital Elevation Models (DEMs) was revolutionary—they marked the beginning of truly three-dimensional representations of terrain, which significantly propelled the advancement of fields such as design, military applications, and engineering. In the context of landscape planning and design, DEMs form the foundational basis for parameterized vertical design.

① DTM
The concept of the Digital Terrain Model (DTM) was first proposed in the 1950s by American researchers Miller and R.A. Laflamme, who applied it in the alignment design of highways. A DTM is a simulated representation of continuous terrain constructed using coordinate points with known x, y, and z values in a given coordinate system. Here, x and y represent the horizontal (planar) coordinates, while the z value can convey a range of information—such as elevation, slope, or temperature—and even qualitative spatial attributes like soil types or geomorphological categories. Thus, a DTM can be understood as the digital expression of terrain surface morphology and attribute data—a form of digital description that integrates both spatial location and topographic properties. Conceptually, a DTM is a vector space of one or more dimensions overlaid on a two-dimensional geographic plane, incorporating geospatial positioning with descriptive terrain data. At its core, the DTM relies on 2D spatial referencing and digital encoding of terrain features. By inputting terrain undulation data into a computer, DTMs allow for more complete and accurate collection and storage of terrain information, providing a comprehensive, detailed, and precise reflection of both the spatial and attribute characteristics of the landform. Today, DTMs are widely applied, and with computers as operational tools, they can replace manual calculations in tasks such as earthwork estimation, slope analysis, and route selection. They also play a key role in the optimization of infrastructure designs—including highways, railways, and power lines—greatly enhancing both efficiency and accuracy.

② DEM
When the "z" value in a Digital Terrain Model (DTM) represents elevation, it constitutes a Digital Elevation Model (DEM), which is a digital method of describing and representing terrain. The most common way to express elevation in a DEM is by referring to height above sea level or relative elevation based on a reference plane. Therefore, DEMs are often also referred to as terrain models. In fact, a terrain model includes not only elevation attributes but also other sur-

face characteristics such as slope and aspect. Structurally, DEMs can take several forms, including gridded DEMs (GRID), Triangulated Irregular Networks (TIN), contour-based DEMs, surface DEMs, planar polygon DEMs, and spatial polygon DEMs. Among these, the most commonly used in landscape planning and design are the gridded DEM, the TIN DEM, and the contour DEM. These three types have relatively simple structures, which make it easier to establish topological relationships and to conduct visual analysis of the model (Fig. 9.9).

A gridded DEM represents terrain through a series of elevation values (z) arranged at regular intervals along the x and y directions. However, it has certain drawbacks: it can generate redundant data in areas with simple topography, it struggles to represent regions with highly variable elevation, and it cannot accurately depict key topographic features such as peaks, depressions, ridges, and valleys. Despite these limitations, the gridded DEM's data structure, computational methods, and data management are straightforward. The high degree of automation in data acquisition and its compatibility with remote sensing and raster-based GIS systems make it the most widely used DEM data structure at present. The Triangulated Irregular Network DEM connects elevation sampling points into a series of non-overlapping triangles that cover the entire surface to form an irregular mesh. It offers two major advantages: first, it overcomes the problem of data redundancy found in elevation matrices; second, it allows for the inclusion of significant surface data points, effectively utilizing key terrain features such as lines and points to represent complex topographies more accurately. This makes terrain analysis more convenient. However, the TIN DEM also has disadvantages, mainly in its more complex data storage and operations, which make standardized management more difficult. The contour-based DEM uses a series of ordered coordinate points to represent elevation. Each contour corresponds to a unique elevation value and can be considered a simple polygon or a polyline composed of multiple segments, each bearing elevation attributes. The advantage of expressing DEMs through contour lines lies in the

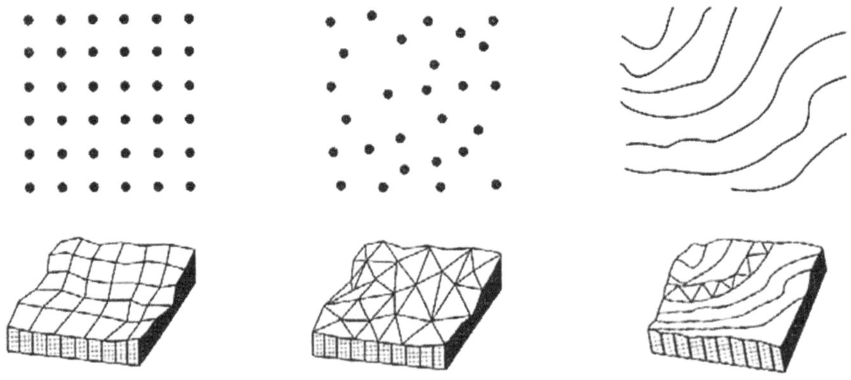

(a) Regular grid DEM **(b) Irregular Triangulated Network** **(c) Contour Line DEM**

Fig. 9.9 Three common DEM structures

intuitive nature of the representation and the relatively low demand for storage space. Nevertheless, this method has limitations: since contour-based DEMs only depict elevation at a limited number of levels, obtaining detailed elevation information within an area requires either increasing the density of contour lines or using interpolation techniques to calculate values for other locations. Fortunately, software tools allow for easy conversion among gridded, TIN, and contour-based DEMs, enabling designers to leverage the advantages of each structure as needed. For example, while contour-based DEMs are not suitable for slope or aspect analysis, they can be converted into gridded DEMs to perform such calculations.

Traditionally, terrain data structures have been two-dimensional, with elevation treated as a third-dimensional spatial attribute. The development of DEMs thus marks a significant and necessary shift in landscape planning and design—providing a foundational data model for parameterized vertical design. Compared with traditional topographic maps, DEMs—being a digital representation of surface terrain—offer the following distinctive advantages:

(a) Versatility in Representation: DEMs, once processed through computer software, can flexibly generate a variety of outputs including plan views, cross-sections, profiles, and 3D perspective diagrams, all tailored to different scales and formats. In contrast, traditional manually drawn vertical design maps are scale-dependent and must be redrawn for each change due to the independence of graphical elements.
(b) Precision Retention: Once stored digitally, DEMs retain their original resolution indefinitely. Their digital nature ensures consistent accuracy over time.
(c) Parameterization Enablement: As a digital format, DEMs support editable variables and serve as the base model for input parameters. They facilitate dynamic relationships among various design elements, enabling real-time updates and feedback—making parametric vertical design achievable.

9.2.2.2 Transformation of Design Approaches

Traditionally, once designers obtained the original topographic map, they would carefully study it, overlay tracing paper, and begin analyzing, sketching, and drawing upon understanding the site's terrain conditions. Interpretation of the terrain based on two-dimensional drawings relied heavily on the designer's experience, and thus the depth of experience directly influenced their understanding of the topography. Calculating earthwork volumes manually using the grid method was extremely inefficient. For instance, calculating the cut-and-fill volumes for a 1 km^2 site using a 20 m × 20 m grid would require computing the volume of 2500 grid cells—a process that is both time-consuming and complex, prone to errors. In large-scale grading projects, calculating earthworks could take several weeks. With the advent of computer-assisted calculations, the speed and accuracy of computations improved significantly. However, once the design scheme was modified, recalculating earthwork volumes typically meant starting over. Civil 3D software revolutionized this process by employing a triangulated mesh (TIN) method as the core of its earthwork calculation model. It generates a surface based on the road model and compares it with the existing ground DEM to compute cut-and-fill volumes. This two-surface comparison method greatly enhances accuracy. Moreover, Civil 3D integrates all parameters of vertical design into a unified model, such that any modification to one element is automatically reflected in the others. Consequently, even when vertical design schemes are adjusted, changes in earthwork volumes are immediately updated without the need for repeated manual recalculations.

Currently, vertical construction in landscape projects still relies primarily on the experience of on-site personnel, with layout lines and flagging being the common methods used. Traditional vertical construction drawings tend to be simplistic, typically presented as grid-based layout plans where elevation is marked at grid intersections. This method presents challenges in construction control, and the final built outcome often diverges significantly from the drawings due to its ambiguity and randomness. Traditional construction control methods are drawing-oriented, overlooking the operational convenience required during implementation. When dealing with large or complex sites, grid-based layout control becomes difficult to execute, often relying on the operator's subjective "feeling" rather than precise guidance—making accurate topographic construction largely theoretical. To achieve a satisfactory result, designers and contractors must repeatedly adjust the scheme on-site, yet it remains difficult to ensure comprehensive accuracy. In the face of today's rapidly advancing digital age, these conventional methods are increasingly outdated. The root of this lag lies in the fact that the translation from two-dimensional drawings to three-dimensional spatial understanding is performed cognitively by humans, without quantitative data to support the process. This introduces two major sources of error: the first is in the reading of drawings, which depends on the individual cognition and experience of the contractor, leading to varied interpretations; the second is during the implementation stage, where machine operation is guided by the operator's understanding of the drawings, making precise and accurate construction difficult. Not only does this make it hard to predict design outcomes with certainty, but it also hampers feedback and evaluation when comparing the results to the original drawings.

There are four primary methods for collecting DEM data: ground surveying, spatial sensors, digitization of existing maps, and digital photogrammetry. The advancement of data acquisition technologies has continuously improved DEM accuracy, making parametric vertical design increasingly precise and reliable. By directly importing completed DEMs into construction equipment, the precision of the construction process is ensured, as is the effectiveness of design implementation. From data input to output, DEM serves as the foundational object throughout the entire cycle of landscape planning and construction, becoming a critical component in achieving parametric design.

9.3 Construction of a Parametric Vertical Design Model in Landscape Architecture

As the final phase in the general layout design of landscape planning and design, vertical design is closely connected with the preceding stages such as site selection, road alignment, and water system development, and interfaces with the subsequent stages of detailed zoning design and construction documentation. It must not only meet the requirements of large-scale planning but also reflect the characteristics of detailed design. Software based on 3S technology, represented by ArcGIS, excels in large-scale planning and analysis but has certain limitations when applied to small-scale environments. In contrast, Civil 3D software, with its capabilities in precision, dynamism, and parametric modeling, complements ArcGIS and serves as the primary platform for constructing parametric vertical design models in landscape architecture.

According to the vertical design requirements of the general layout in landscape planning, the model mainly addresses three components: roads, water systems, and site terrain. The construction of a parametric vertical design model is guided by two fundamental principles. The first is the principle of system integration: roads, water systems, and site terrain within the same project area are treated as interrelated components within a unified system, where earthwork balance is achieved internally. The second is the principle of minimization, which aims to reduce disturbance to the site. This involves not only minimizing earthwork quantities for each individual component but also coordinating their integration within the overall system to achieve total earthwork balance. Considering earthwork volumes during the design stage carries important significance and value. Estimating and coordinating earthwork quantities helps designers identify optimal design solutions, effectively control construction costs, and significantly reduce redundant efforts during implementation.

The variables and parameters in a parametric vertical design model for landscape architecture can be categorized into two types. The first includes those that control terrain generation, such as elevation and slope. The second includes variables and parameters specific to the subcomponents of the design—for example, in road design, parameters such as longitudinal slope, cross slope, slope length, turning radius, vertical curve radius, and grade transition points. These variables and their respective ranges vary according to the specific project. Generally, elevation parameters change with the project context, while others are defined based on relevant design standards and regulations.

The parametric vertical design model in landscape architecture consists of four main components: vertical design of roads, vertical design of water systems, vertical design of site terrain, and, under the category of regulation and coordination, the overall earthwork balance (see Fig. 9.10). Table 9.2 lists the parameters involved in each subcomponent of vertical design, which will be further discussed in the following sections.

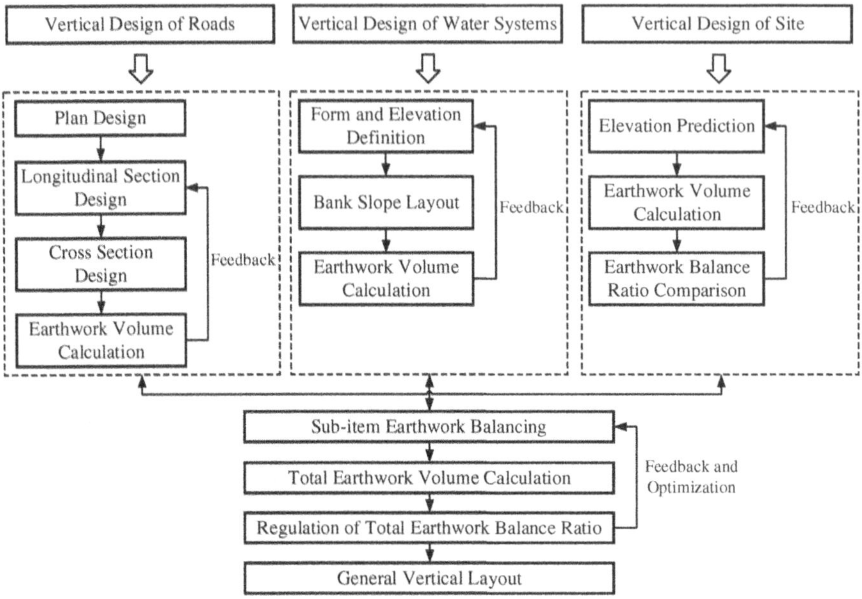

Fig. 9.10 Parametric vertical design model for landscape architecture

9.3.1 Road Vertical Design

Roads are typically classified based on their transportation function, traffic volume, and vehicle speed. Different types of roads have distinct functions, service targets, and technical requirements. In China, roads are generally divided into four main categories: highways, urban roads, accommodation roads, and country roads. Among these, accommodation roads are further categorized into industrial/mining roads and forest roads. The scale of landscape environments covers a broad spectrum, including both scenic areas near urban fringes and natural landscapes in remote suburbs. As such, roads in landscape environments often encompass all four types, making it difficult to classify them into a single category. Therefore, road design in landscape environments must integrate the technical standards of various road types while also considering site-specific characteristics (see Table 9.3). Typically, roads in landscape settings are classified by importance into four types: primary roads, secondary roads, tertiary roads, and pedestrian paths. Large scenic areas may contain all four types of roads, whereas smaller landscape environments may include only two or three.

The vertical design of roads in landscape architecture is closely tied to the road alignment phase, serving as its immediate follow-up. The alignment of a road is expressed through three interrelated components: the horizontal, vertical, and cross-sectional design. These three aspects are mutually constraining and complementary. Similar to highway and urban road design, the vertical design of roads in landscape environments generally includes four key components: the determination of the

Table 9.2 Parametric vertical design model for landscape architecture—parameter table

Item name	Sub-item name	Serial number	Parameter name	Unit
Vertical alignment of roads	Planar design	1	Design speed	Km/h
		2	Road length	m
		3	Horizontal curve	m
		4	Circular curve	m
		5	Transition curve	m
	Longitudinal profile design	6	Longitudinal slope	%
		7	Vertical curve	m
		8	Slope length	m
	Cross-section design	9	Roadway width	m
		10	Cross slope	%
		11	Slope face	%
	Earthwork calculation	12	Sampling line	–
Vertical alignment of water systems	Waterbody morphology design	13	Water surface form	–
		14	Underwater terrain form	–
	Bank protection design	15	Revetment slope	%
	Earthwork volume calculation	16	Original terrain	–
		17	Designed terrain	–
Site vertical design	Planar design	18	Construction area	m^2
	Vertical design	19	Elevation	m
		20	Site grading slope	%
	Earthwork volume calculation	21	Original terrain	–
		22	Designed terrain	–
Overall earthwork balance	Road vertical earthwork volume	23	Net earthwork volume	m^3
	Water system vertical earthwork volume	24	Net earthwork volume	m^3
	Site vertical earthwork volume	25	Net earthwork volume	m^3

road alignment, longitudinal profile design, cross-sectional design, and earthwork volume calculation. In landscape planning and design, the completion of these components enables the determination of the road's vertical elevation and the development of a compliant road design. The variables and parameters in a parametric road design model are derived from the parameters of these four components, collectively governing the generation of the road geometry.

9.3.1.1 Road Plan Optimization

A road is a linear structure, with its centerline forming a spatial curve. The road plan refers to the projection of this centerline onto a horizontal plane. Urban road planning is subject to numerous constraints, such as the layout of the road network, the

Table 9.3 Road design requirements listed in the 《Park Design Standard》 (GB 51192–2016)

<table>
<tr><th>Item</th><th colspan="5">Content</th></tr>
<tr><td>Article 6.1.1</td><td colspan="5">Park paths should be designed in accordance with the park's overall planned road network and classification, including the determination of path widths, horizontal and vertical alignments, as well as structural design</td></tr>
<tr><td rowspan="7">Article 6.1.3</td><td colspan="5">The width of park paths should be determined based on traffic requirements and must comply with the provisions specified in Table 6.1.3. Path width(m)</td></tr>
<tr><td rowspan="2">Park path classification</td><td colspan="4">Total Park area-A(hm^2)</td></tr>
<tr><td>$A < 2$</td><td>$2 \leq A < 10$</td><td>$10 \leq A < 50$</td><td>$A \geq 50$</td></tr>
<tr><td>Main path</td><td>2.0–4.0</td><td>2.5–4.5</td><td>4.0–5.0</td><td>4.0–7.0</td></tr>
<tr><td>Secondary path</td><td>——</td><td>——</td><td>3.0–4.0</td><td>3.0–4.0</td></tr>
<tr><td>Branch path</td><td>1.2–2.0</td><td>2.0–2.5</td><td>2.0–3.0</td><td>2.0–3.0</td></tr>
<tr><td>Minor path</td><td>0.9–1.2</td><td>0.9–2.0</td><td>1.2–2.0</td><td>1.2–2.0</td></tr>
<tr><td>Article 6.1.4</td><td colspan="5">Design requirements for the planar alignment of park paths:1.Park paths should integrate harmoniously with the terrain, water bodies, vegetation, buildings, paved areas, and other facilities, meeting the needs of both circulation and sightseeing, while forming a complete scenic composition;2.Park paths should create routes that orderly display garden landscape spaces or provide perspective lines for appreciating views ahead;3.The turning and connection of park paths should be smooth and continuous;4. For main paths accessible to motor vehicles, the minimum radius of horizontal curves shall be greater than 12 m</td></tr>
<tr><td>Article 6.1.5</td><td colspan="5">Design requirements for the longitudinal profile of park paths:1.Steps shall not be set on main paths;2.The longitudinal slope of main and secondary paths should preferably be less than 8%, and the length of a single slope section should not exceed 200 m. In mountainous areas, the longitudinal slope of main and secondary paths should be less than 12%; slopes exceeding 12% require anti-slip treatment. In regions with snow or ice, the road longitudinal slope should not exceed 6%;3.For branch paths and minor paths, the longitudinal slope is preferably less than 18%. For sections where the slope exceeds 15%, the pavement should have anti-slip treatment. If the slope exceeds 18%, the path should be designed as stepped stairs;4.For roads connected to plazas with relatively large longitudinal slopes, a transition segment with a slope less than or equal to 2.0% shall be provided at the connection;the slope of dedicated bicycle paths is preferably less than 2.5%. When equal to or greater than 2.5%, the maximum slope length must comply with the relevant provisions of the current industry standard "technical code for urban road engineering design" (CJJ 37)</td></tr>
<tr><td>Article 6.1.6</td><td colspan="5">The cross slope of park paths is preferably between 1.0% and 2.0%, and should not exceed 4.0% at maximum. In regions with heavy rainfall, a cross slope of 1.5% to 2.0% is recommended. For park paths and permeable pavements in areas with snow or ice, a cross slope of 1.0% to 1.5% is advisable. The longitudinal and cross slopes should not both be zero simultaneously</td></tr>
</table>

Citation from: Park Design Standards (GB 51192–2016)

width of planned red lines, and the locations of existing buildings along the street. As a result, the alignment of urban roads can only be adjusted within a limited range. In contrast, highways offer much greater flexibility in route selection. Road alignment in landscape environments resembles that of highways, requiring comprehensive consideration of terrain, geology, hydrology, and other natural factors. In a landscape planning and design process grounded in the principle of coupling, road alignment is completed during the general plan stage, which precedes vertical design. However, as part of a systemic design outcome, it is closely linked with vertical design: the basis for road vertical design lies in the result of the road alignment.

Straight lines and circular curves form the primary components of road plan geometry in landscape environments. On roads with relatively low driving speeds, the curved segments are composed solely of circular arcs. For roads designed to accommodate higher speeds, a transition curve is required between straight lines and circular curves to allow for a smoother vehicle transition. Hence, the horizontal alignment consists of a combination of circular curves and transition curves. Overly long straight segments lack geometric flexibility, tend to be monotonous, and are ill-suited to changes in terrain. They may also lead to driver fatigue, distraction, and speed beyond design limits, increasing the risk of traffic accidents. In areas with minimal topographic variation or in urban contexts, straight segments offer the benefits of shortest distance, ease of alignment selection, cost-effectiveness, and speed. However, in landscape environments, beyond safety considerations, the design aims to achieve a dynamic scenic experience and flexible spatial variation. Therefore, excessively short straight segments between curves are also undesirable. A winding and varied road plan effectively guides the viewer's gaze and enriches the scenic experience.

In landscape environments, road plans require further refinement based on the initial alignment, primarily through adjustments and constraints applied to the straight and curved segments in accordance with relevant standards and codes. Straight lines, circular curves, and transition curves collectively form the parameters for road plan design within the parametric vertical design model. The values of these parameters can be determined with reference to standards such as the Code for Design of Urban Road Engineering (2016 Edition) (CJJ 37–2012) and the Highway Route Design Code (JTG D20–2017) (see Tables 9.4 and 9.5).

Table 9.4 Calculation table for minimum lengths of tangent and circular curves

Design speed(km/h)		100	80	60	50	40	30	20
Minimum length of tangent curve	Calculated value(m)	166.7	133	100	83	67	50	33
	Adopted value(m)	170	140	100	85	70	50	40
Minimum length of circular curve	Calculated value(m)	83.3	67	50	41.7	33.3	25	16.7
	Adopted value(m)	86	70	50	40	35	25	20

Citation from: Urban Road Engineering Design Code (2016)(CJJ 37–2012)

Table 9.5 Minimum radius of circular curve

Design speed(km/h)		120	100	80	60	40	30	20
Minimum radius of circular curve (general value)(m)		1000	700	400	200	100	65	30
Minimum radius of circular curve (limit value)(m)	Imax=4%	810	500	300	150	65	40	20
	Imax=6%	710	440	270	135	60	35	15
	Imax=8%	650	400	250	125	60	30	15
	Imax=10%	570	360	220	115	–	–	–

Note: "The normal value" refers to the value adopted under regular conditions; "The limit value" refers to the value permissible under restricted conditions; "Imax" denotes the maximum superelevation adopted; "A dash (—)" indicates that the corresponding maximum superelevation is not considered

Citation from: Highway Route Design Specification (JTG D20–2017)

9.3.1.2 Generation of Longitudinal and Cross Sections

Upon completion of the longitudinal and cross-sectional designs, the road model is essentially finalized, yielding key outputs such as elevation data and earthwork volumes—making this a critical phase in parametric road design. These two sections largely determine the earthwork requirements for road construction and must be coordinated with the horizontal alignment during the design process.

(1) Longitudinal Section

The longitudinal section represents the vertical profile along the centerline of the road. Designing the longitudinal section involves shaping the vertical undulations of the road route. The longitudinal profile typically includes two primary lines: the existing ground line, which is an irregular polyline reflecting the elevation changes along the centerline based on surveyed points, and the design profile line, which is a geometric line determined by the designer after comprehensive comparison and analysis. The difference in elevation between corresponding points on the ground and design lines represents the construction height—indicating the depth of cut or the height of fill required. Where these two lines intersect, no earthwork is needed. This design process is pivotal, as the rationality of the longitudinal section directly impacts the volume of earthwork and the overall ecological disturbance. Therefore, in order to minimize earthwork and reduce site disturbance, the alignment should closely follow the natural terrain—striving for minimal deviation between the design and ground lines.

The longitudinal design comprises straight gradients and vertical curves. The straight sections, defined by elevation difference and horizontal length, are categorized as uphill or downhill slopes. To ensure vehicles can travel safely—climbing inclines with ease and descending without danger—a key constraint is the maximum longitudinal gradient. This gradient is a crucial design parameter, directly influencing driving speed and safety, route length, service quality, transport cost, and overall construction budget. It is necessary to determine the design speed of the road early in the process and select an appropriate range for the slope in accordance with design codes. In landscape environments, the maximum gradient is generally recommended to be below 8%. In hilly scenic areas, where driving speed is not a primary concern, local slopes may reach up to 12%. For bicycle paths, a slope of 2% is preferable and should not exceed 3%. For pedestrian pathways, slopes below 12% are considered comfortable; when exceeding 15%, steps should be introduced. In addition to the maximum gradient, landscape road design must also consider the minimum longitudinal slope and slope length. The minimum slope ensures proper drainage and prevents pipe blockages, typically no less than 0.3%. If slope lengths are too short, the road may become overly undulating, leading to discomfort for both drivers and pedestrians. Hence, slope length must be carefully balanced during design. Similar to how transition curves function in horizontal alignment, vertical curves are used to smooth the transition between slopes in the longitudinal profile. Based on the direction of slope change, vertical curves can be classified as sag curves (concave) and crest curves (convex), and are defined by their radius and horizontal length (see Table 9.6). It is also crucial to ensure that the longitudinal and horizontal alignments are coordinated for an integrated and functional road design.

(2) Cross Section

A road cross section refers to the profile perpendicular to the centerline of the roadway. It typically comprises elements such as carriageways, pedestrian walkways, and medians, with the total width of the cross section being the sum of the widths of these individual components. Urban roads feature complex traffic characteristics and functional compositions, accommodating motor vehicles, non-motorized vehicles, and pedestrians—each traveling at different speeds. Additionally, urban cross sections must account for utility corridors, drainage channels, traffic signage, and green space, resulting in relatively complex configurations. In contrast, roads within landscape environments possess distinct characteristics compared to urban roads or highways. First, their classification tends to be more diverse, often combining traits of both urban and rural road systems. Second, they serve multiple functions including tourism, logistical access, firefighting, and emergency evacuation, which typically require lower traffic speeds and lighter load capacities. Third, the primary users are visitors and tourists, and the modes of travel may include walking, bicycles,

Table 9.6 Minimum radius and minimum length of vertical curves

Design speed(km/h)		100	80	60	50	40	30	20
Convex vertical curve(m)	Normal value	10,000	4500	1800	1350	600	400	150
	Limit value	6500	3000	1200	900	400	250	100
Concave vertical curve (m)	Normal value	4500	2700	1500	1050	700	400	150
	Limit value	3000	1800	1000	700	450	250	100
Vertical curve length(m)	Normal value	210	170	120	100	90	60	50
	Limit value	85	70	50	40	35	25	20

Citation from: Urban Road Engineering Design Code (2016) (CJJ 37–2012)

electric carts, and motor vehicles. Given the relatively low traffic volume, cross-sectional design can remain simple: single- or dual-lane roads are often sufficient. Some roads may also include sidewalks on one or both sides (see Fig. 9.11). To accommodate drainage, roadways must be built with a slight transverse slope, known as crossfall or camber. In landscape environments, the cross slope should generally not be less than 3%. Another key parameter in cross-sectional design is the roadside slope or cut-and-fill slope. As landscape roads are often situated in mountainous or hilly terrain with significant elevation changes, construction frequently requires the formation of roadside slopes. These slopes can be categorized as either natural or engineered. To protect the environment and enhance visual aesthetics, natural slopes are preferable in landscape settings. However, the slope angle must not be excessively steep, as this could lead to safety risks such as landslides or collapses.

9.3.1.3 Earthwork Volume Control in Road Construction

Due to the functional requirements of safety, drainage, and accessibility, roads in landscape environments cannot fully conform to the natural terrain. Consequently, terrain modification—specifically the calculation and control of earthwork volume—is inevitable. In general, earthwork for road construction falls into three categories: full excavation, balanced cut-and-fill, and full embankment (see Fig. 9.12). Based on the road's horizontal alignment, longitudinal profile, and cross-sectional design, embankments are often constructed in valleys or depressions, while cuttings are applied through hills or steep terrains to reduce road gradients (see Fig. 9.13).

A commonly used method for calculating earthwork volume in design is the cross-sectional method. This involves dividing the road or site into multiple cross sections at regular intervals. Using the design standard cross section and the

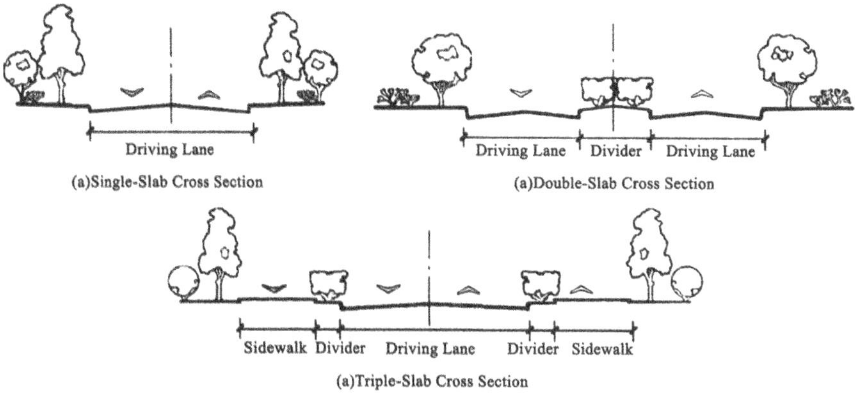

Fig. 9.11 Common cross-section types of roads in landscape and garden environments. Redrawn from: Han Yulin. Landscape architecture engineering [M]. Chongqing: Chongqing University Press. 2011:126

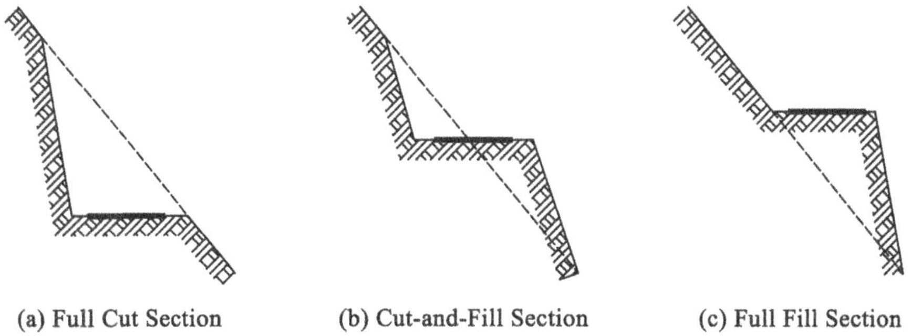

Fig. 9.12 Cut and fill methods in road construction

Fig. 9.13 Road cut and embankment

corresponding natural ground section, the enclosed area of each section is calculated. The average of the cut-and-fill areas between two adjacent sections is multiplied by the interval distance to obtain the volume between them. Summing the volumes of all sections gives the total earthwork volume. This can be performed using the average cross-sectional method or the prismoidal method. The formula for the average cross-sectional method is as follows:

$$V = 1/2\left(S_1 + S_2\right)L.$$

where S_1 and S_2 are the excavation or fill areas of two adjacent cross sections, and L is the distance between them.

Compared to the average cross-sectional method, the prismoidal method can achieve relatively higher accuracy, but the precision of both methods is generally limited, with errors usually around 10%. In areas with complex terrain, more sampling cross sections need to be established, which increases the computational workload. The parameterized vertical design model is constructed based on the Civil 3D software platform, where the core of earthwork calculation relies on the Triangular Irregular Network (TIN) method. This method generates a surface from the parameterized road model and compares it with the original terrain surface to calculate the earthwork volume with higher accuracy. Sampling lines are set to specify the locations of the calculation cross sections, and the density of these sampling lines determines the number of cross sections used in the calculation. Therefore, the higher the density of sampling lines, the more accurate the earthwork volume calculation becomes. For sites with significant terrain undulation, increasing the sampling line density can improve calculation precision.

The regulation of earthwork volume in road construction primarily involves several aspects. Firstly, the route selection should aim to pass through areas with relatively gentle slopes; most of this work is completed during the road alignment phase and further optimized in the horizontal design stage. Secondly, the longitudinal profile design must satisfy requirements for slope, as well as maximum and minimum slope lengths, while striving to make the design profile fit the existing terrain as closely as possible to minimize terrain disturbance and reduce earthwork. Thirdly, during cross-sectional design, slope treatment should maximize the protection of the original terrain, and reasonable slope gradients should be set to avoid excessive excavation or fill operations. Finally, after completing the previous steps, an earthwork volume estimation is conducted for individual components of the vertical design, and the results are fed back into earlier stages to make overall adjustments and optimizations, ultimately minimizing the earthwork volume in the detailed design.

9.3.2 Vertical Design of Water Systems

The vertical design of water systems in parameterized landscape architectural planning and design processes is similar to that of roads, both being based on the preceding design stages. This step involves the refinement and completion of water system design and can directly interface with the construction drawing phase. "Quasi-natural" waterscapes rely on terrain formation, meaning that the creation of water bodies must follow the terrain's configuration and elevation differences. Therefore, the construction of waterscapes is closely linked to vertical design. The "Code for Park Planning and Design" (GB 51192–2016) stipulates the key vertical control parameters related to water systems as: the highest water level, normal water level, lowest water level, water bottom, and the top of the revetment. The vertical design of water systems primarily focuses on regulating these key parameters to accomplish the overall design of the water system and to preliminarily estimate the earthwork volume required for water system construction.

9.3.2.1 Determination of Water Body Form and Water Surface Elevation

In parameterized vertical design, the form of the water body includes two components: the water surface form and the water bottom form. The basic form of the water surface is established in the earlier "Waterscape Construction" stage and is determined jointly by the terrain where the water body is located and the water level elevation. However, merely defining the water surface form is insufficient to complete the waterscape design. The vertical design of water systems in landscape architectural planning includes critical parameters such as water surface elevation (designed normal water level), water bottom elevation, and the water bottom morphology depicted by contour lines (see Fig. 9.14). Thus, in parameterized vertical design of water systems, it is necessary to combine the water surface form determined during the waterscape construction phase with the normal water level corresponding to the dam height, in order to design the water bottom morphology and achieve an integrated water system form that harmonizes with the surrounding terrain. The normal water level is defined as the elevation at which the water level at a given point equals or exceeds a certain height 50% of the time over one year or multiple years. According to Article 5.3.3 of the "Code for Park Planning and Design" (GB 51192–2016): (1) for artificial revetments without protective facilities, the water depth at normal water level within 2.0 m of the shore shall not exceed 0.7 m; (2) within 2.0 m around unprotected park bridges, stepping stones, and waterside platforms, the water depth at normal water level shall not exceed 0.5 m; (3) the vertical distance between the top of an unprotected revetment and the normal

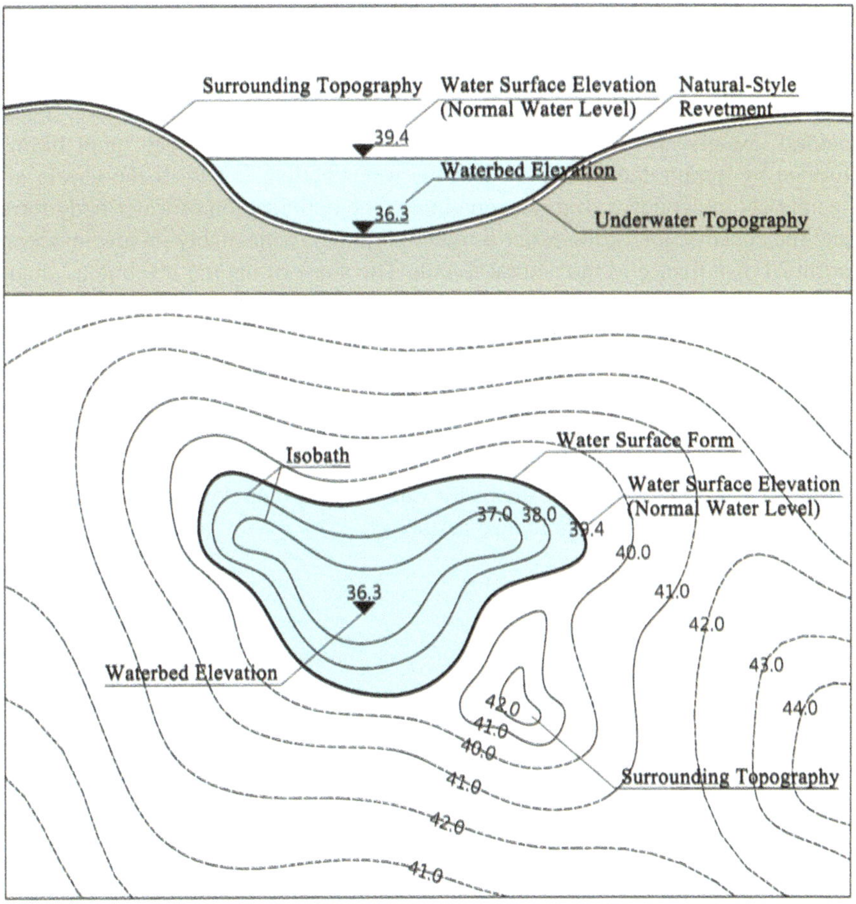

Fig. 9.14 Vertical design elements of water systems

water level shall not exceed 0.5 m. Generally, the normal water level of a "quasi-natural" waterscape is determined by the dam height, but the water depth generated in conjunction with the water bottom morphology must meet the regulatory requirements to ensure visitor safety. Contour lines depicting water depth, known as isobaths, function similarly to elevation contour lines by connecting points of equal depth with smooth curves, where every point on the line shares the same water depth. Therefore, isobaths can directly represent the depth of the water body or indirectly reflect the undulation of the water bottom terrain. Vertical design can achieve water depth regulation through the control of these isobaths.

9.3.2.2 Revetment Slope Control

A revetment, also known as a protective slope, is a structure used to safeguard water bodies in landscape environments. Revetments can be categorized by cross-sectional shape into formal and natural styles. "Quasi-natural" waterscapes commonly employ naturalistic rock revetments or gently sloping vegetated revetments, integrating the water body with the surrounding terrain by extending slopes into the water surface to achieve harmony with the natural environment. These "quasi-natural" waterscapes typically store water in natural or artificial "pool basins" formed by terrain depressions, where the water bottom is part of the terrain and seamlessly integrates with its surroundings. The optimization of water body forms and the construction of the water bottom morphology inevitably involve moderate artificial disturbance to the natural terrain. The slope of natural revetments should be selected based on the natural angle of repose of the different soil types to avoid safety hazards such as slope failure or collapse (see Table 9.7). Furthermore, the slope gradient of the revetment influences visitors' spatial perception: gentler slopes evoke a calming, soft atmosphere and tend to visually enlarge the water space; steeper slopes stimulate sensory excitement, create a sense of novelty and interest, and visually reduce the perceived scale of the water body (see Fig. 9.15). In summary, the design of revetment slopes for water bodies should comprehensively consider both safety and morphological aesthetics.

Table 9.7 Angle of repose of common soils

	Soil moisture condition			
Soil type	Dry soil	Moist soil	Wet soil	Soil particle size(mm)
Gravel	40°	40°	35°	2–20
Cobbles	35°	45°	25°	20–200
Coarse sand	30°	32°	27°	1–2
Medium sand	28°	35°	25°	0.5–1
Fine sand	25°	30°	20°	0.05–0.5
Clay	45°	35°	15°	≤ 0.001–0.005
Loam	50°	40°	30°	
Humus soil	40°	35°	25°	

Citation from: Internet

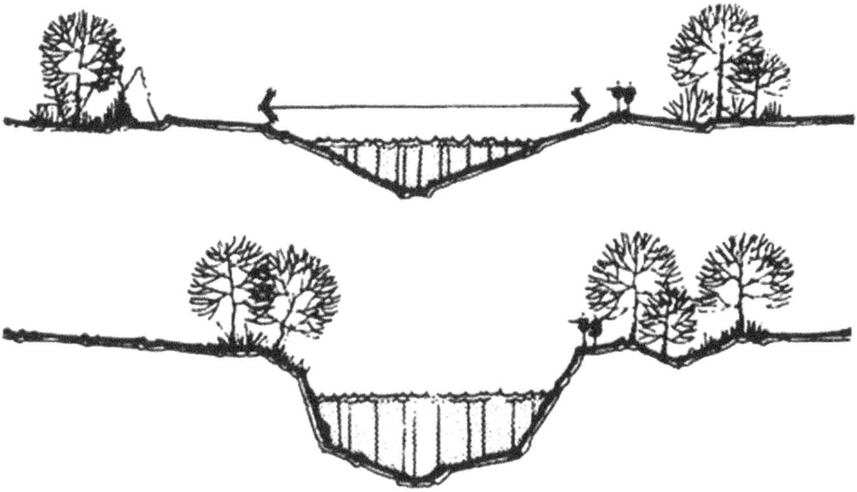

Fig. 9.15 The steepness of the surrounding revetments influences the perceived scale of the water body. Citation from Booth, N.K. (1989). Elements of Landscape architecture design

9.3.2.3 Earthwork Volume Control in Water System Construction

In the vast majority of cases, the form of designed water systems cannot be generated entirely by the natural terrain and requires local adjustments based on the original topography. The shape of the water system is formed through the creation of surface morphology. Therefore, water system construction inevitably involves earthwork operations, which arise from both morphological shaping and revetment slope grading. Similar to earthworks in terrain construction, the volume of earthwork depends on the degree of disturbance to the original terrain and is closely related to three key variables: the designed water depth, water body morphology, and revetment slope gradient. Compared to road construction, the calculation of earthwork volumes for water system construction in Civil 3D software is considerably simpler. It uses the "surface volume calculation" method, which determines the volumetric difference between the designed ground surface and the original terrain surface (see Fig. 9.16).

9.3.3 Site Vertical Design

The term "site" is commonly used in both landscape architecture planning and architectural design, but its meaning is broad. In a narrow sense, "site" specifically refers to the building's footprint area, including not only the building itself but also associated outdoor spaces such as plazas, parking lots, and activity areas. In this context, the site is dependent on the building and is generally referred to as the

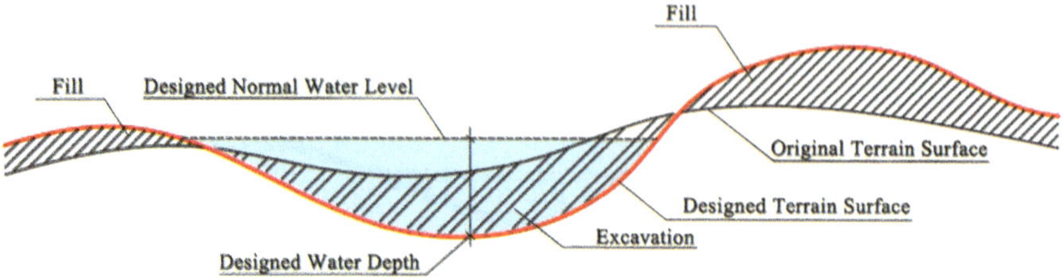

Fig. 9.16 Diagram of earthwork regulation in water system construction

"outdoor site, " a definition widely used in architectural design. In a broader sense, the "site" corresponds to land parcels that retain natural attributes or have been artificially developed. In landscape architecture planning, "site" usually refers to the latter — encompassing building plots and various artificially constructed land parcels. Site design consists of two key components: first, the plan layout, which aims to clarify functional arrangement and planar spatial needs; second, the vertical design, whose primary task is to shape the terrain to meet transportation, drainage, and vertical spatial requirements, while achieving earthwork balance with minimal cut-and-fill volumes. According to project requirements, combined with site topographical characteristics and construction technical conditions, it is critical to rationally manage the elevation of various site elements. This process holds significant ecological, economic, and intensification importance.

Construction in areas with excessively steep slopes not only causes considerable site disturbance but also presents safety risks. Therefore, it is necessary to select areas with appropriate slopes for development (see Table 9.8). In parameterized landscape architectural planning, the site location is determined during the project siting phase. Based on the requirements of vertical design, the vertical design stage primarily focuses on determining elevations and controlling earthwork volumes. This requires estimating and comparing site elevations based on existing topography and project needs, then regulating earthwork volumes accordingly.

9.3.3.1 Preliminary Estimation of Site Elevation

The elevation and planar layout of artificial facilities such as buildings and structures have a decisive impact on the earthwork volume of the site. Under the premise of meeting the functional requirements of these facilities, it is essential to reasonably determine the construction elevation in consideration of the existing site topography, so as to make full use of the terrain and minimize earthwork. Therefore, estimating the site elevation is a crucial task in vertical site design.

Once the project location is confirmed, the functions assigned to the site can be identified. At the same time, based on the projected scale requirements of the project location, a preliminary estimation of the project's land area can be obtained. Before estimating elevation, the plan layout must be roughly determined according to the anticipated site area. This stage represents a further design refinement involving

Table 9.8 Relationship between slope classification standards and construction feasibility

Category	Slope	Characteristics
Flat land	0%–1%	Poor drainage, potential for water accumulation
Gentle slope	1%–5%	The terrain is relatively ideal for construction use; however, attention should be paid to the potential visual fatigue caused by excessively large areas
Mild slope	5%–10%	Suitable for various forms of land use with good drainage, but attention should be paid to soil erosion
Moderate slope	10%–15%	Not suitable for large-scale construction
Steep slope	> 15%	Mostly unsuitable for construction

Citation from: Editorial committee of the architectural design data collection. Architectural design data collection [M]. second ed. Beijing: China Architecture & Building Press, 1994

detailed design of artificial facilities such as buildings and structures. Generally, design development in landscape architecture planning requires clarifying the style, plan form, and massing of construction units. Within the scope of the author's research, these aspects are not explored further. From the perspective of overall site arrangement, sites laid out along contour lines make better use of the existing terrain and require less cut-and-fill earthwork; conversely, sites arranged perpendicular to contour lines require significant leveling work, resulting in large volumes of excavation and fill and greater disturbance to the original environment. In practical application, site layouts can be flexibly arranged by segmenting the site based on contour distribution (see Fig. 9.17). On the basis of a basic planar form, three sets of preliminary site elevations are proposed by referencing the original terrain elevations. Because earthwork volumes differ under different site elevations, the final elevation is determined by comparing the earthwork quantities for these three sets. If necessary, more elevation options can be added to improve the accuracy of the estimation.

9.3.3.2 Control of Earthwork Volume in Site Construction

Earthwork balance refers to achieving a state where the volume of excavation approximately equals the volume of fill, or where the absolute difference between cut and fill is minimized. However, regarding total earthwork volume, both excavation and fill increase the overall earthwork amount. This is particularly evident in areas with significant terrain variation, where earthwork volume is greater than in flatter areas. Therefore, controlling earthwork volume in site construction must proceed from two aspects: first, selecting suitable site elevations that ensure excavation and fill volumes are approximately equal and minimized (see Fig. 9.18); second, optimizing the site's planar layout by preferentially choosing areas with relatively gentle topographic relief for construction (see Fig. 9.18). By calculating and comparing earthwork volumes corresponding to the three or more proposed site elevations mentioned above, the elevation with the least earthwork volume can be

(a) Parallel layout (b) Perpendicular layout (c) Freeform layout

Fig. 9.17 Site layout forms

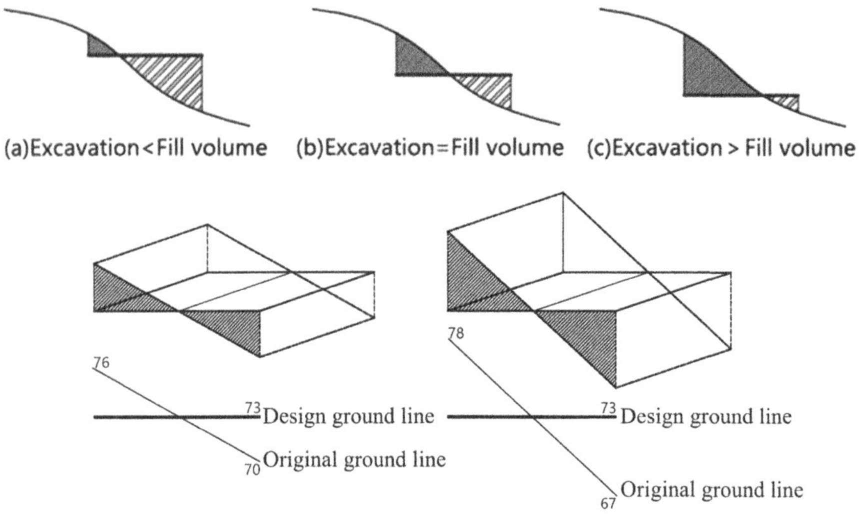

Fig. 9.18 The impact of slope on earthwork volume

determined as the final site elevation. Within Civil 3D software, the creation of 3D models enables precise earthwork calculations, and parameter adjustments can be used to achieve optimal vertical design.

9.3.4 Earthwork Balance

Earthwork balance corresponds to the final part of the parameterized vertical design model. It involves the overall regulation of earthwork balance based on the self-balancing of individual design components. This includes controlling the total earthwork volume, adjusting the degree of earthwork balance, and providing feedback and optimization for each sub-design. The objective of earthwork balance is to achieve a basic equilibrium between excavation and fill volumes within the design scope, thereby minimizing soil export and external soil import. This represents one of the key aspects of optimization in vertical design. Controlling earthwork balance requires compiling the excavation and fill volumes of each sub-component as well as the overall balance volume, and adjusting among the components to minimize the total earthwork volume while maintaining overall cut-and-fill equilibrium. If the analysis indicates that the target cannot be met, adjustments and optimizations must be made to the sub-designs until the optimal solution is generated.

9.4 Application of the Parameterized Vertical Design Model

In this section, the author presents a case study applying the parameterized vertical design model in the northern region of the Niushou Mountain Scenic Area, specifically the Xunzhen Dao (Seeking Zen Path) area. The Xunzhen Dao area is located in the southeastern part of the northern Niushou Mountain Scenic Area, known as the ancient Xunzhen Path and serving as the northern entrance to the Foding Palace (Fig. 9.19). The vertical design of this area follows the previously described procedures, addressing roads, water systems, and site design in sequence, and concludes with the regulation of overall earthwork balance.

9.4.1 *Road Vertical Design*

The road network planning is preliminarily completed during the road alignment stage. Road vertical design is a further refinement based on the finalized alignment. According to the road vertical design section of the parameterized vertical design

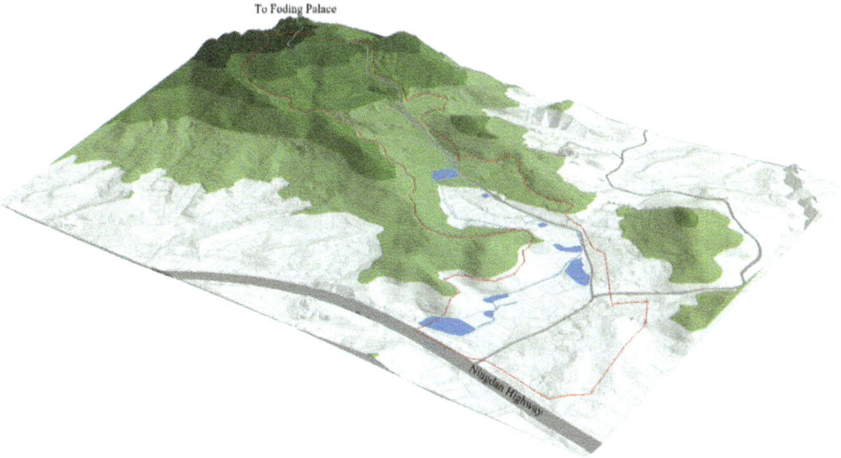

Fig. 9.19 Topography of the Zen-searching trail area in the northern region of Niushou Mountain scenic area, Nanjing. Citation: Cheng Yuning studio profile

model, the process consists of four parts: alignment optimization, longitudinal profile design and optimization, cross-section design and vertical model generation, and earthwork volume calculation. This is a progressive design process.

9.4.1.1 Alignment Optimization

(1) Alignment Refinement
Under given constraints, road alignment is initially calculated with the assistance of ArcGIS software, indicating the route direction, location, and relationships between roads. However, it does not fully meet the requirements of landscape architecture planning and design for road design. Further clarification of alignment parameters—such as horizontal curves, turning radii, and transition curves—is necessary to ensure the feasibility of the planned roads. Using Civil 3D software for road modeling, relevant standards and guidelines—such as the Highway Route Design Specifications (JTG D20–2017)—are referenced to reasonably determine parameters including design speed, road length, simple curves, circular curves, and transition curves. This allows optimization of the alignment path to meet road design requirements and standards. Fig. 9.20 illustrates the road alignment after optimization.
(2) Alignment Generation
Once the alignment is created, the software automatically generates detailed data such as primary and secondary stationing, curve points, and station links, enabling seamless integration with the construction design phase (Fig. 9.21).

9.4.1.2 Longitudinal Profile Design and Optimization

The dynamic longitudinal profile in Civil 3D refers to profiles created from surfaces, road models, and other longitudinal objects that update dynamically to reflect changes in the original objects, embodying the characteristics of parameterized design. The longitudinal profile generated along the road alignment surface is derived from the original terrain, reflecting terrain undulations along the road path. Fig. 9.22 shows the longitudinal profile of the main path of Xunzhen Dao corresponding to the original terrain. It can be observed that the path gradually rises with increasing terrain elevation; the ascent rate is relatively gentle in the first half but steeper in the latter half due to the rugged terrain. Overall, the path undulates gently with some localized increases in slope. In Fig. 9.23, the red line segment represents the original terrain profile along the path, serving as the baseline for the road design longitudinal profile. Compared with the green line segment in the same figure, the designed longitudinal profile follows the principle of minimal disturbance by largely adhering to the original terrain, optimizing areas with greater terrain variability, and controlling the road longitudinal slope within the standard range of 1% to 10%, with some sections reaching 11%.

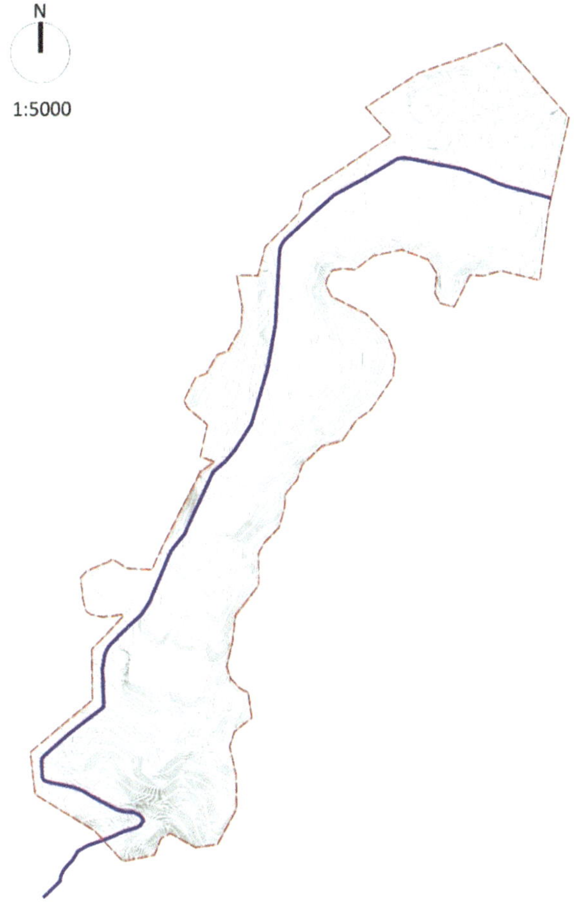

Fig. 9.20 Optimization of the route alignment in the Xunchan Path area

9.4.1.3 Cross-Section Design and Road Vertical Model Generation

Roadways are typically formed by extending a standard cross-section linearly along a designated path, or by combining multiple standard cross-sections along the same alignment. In Civil 3D, a parameterized road model is a three-dimensional spatial model that integrates the horizontal geometric characteristics of the alignment, the vertical profile, and the cross-sectional geometry. Therefore, a complete 3D road model requires the specification of path parameters, vertical profile parameters, and cross-section parameters. Once the first two are determined, the cross-sectional parameters must be designed to complete the model. In the context of landscape architecture planning and design, it is generally sufficient to define the road network structure within regulatory scope, road classification, and planned widths. Thus, excessive detail is not necessary in cross-section design. By specifying the basic form, roadway width, and side slope gradient, a basic vertical road model can be generated to meet landscape planning requirements. Fig. 9.24 shows the designed

Fig. 9.21 Generation of main and auxiliary station numbers, curve points, and station links in the road model

cross-section for the main road in the Xunzhen Dao area, consisting of two pavement panels, each 3.5 m wide, making a total road width of 7 m. The cross slope is 2%, and the designed side slope gradient is 30%. During actual construction, embankments or cut slopes may be required based on terrain conditions; the slope design herein refers to those embankment or cut slopes. Once the cross-section design is completed, the vertical road model is generated, as shown in Figs. 9.25 and 9.26. This model is built on a digital terrain model (DTM) and generated through the combined constraints of alignment, vertical profile, and cross-section parameters. These parameters and their associated data are interconnected, meaning any changes to one parameter are reflected in real time in the model through Civil 3D.

9.4.1.4 Earthwork Calculation for Road Construction

Accurate earthwork volume calculation is critical for selecting the optimal design scheme. Based on the Civil 3D platform, the parameterized vertical road model enables rapid computation of road construction earthworks, including quantities of

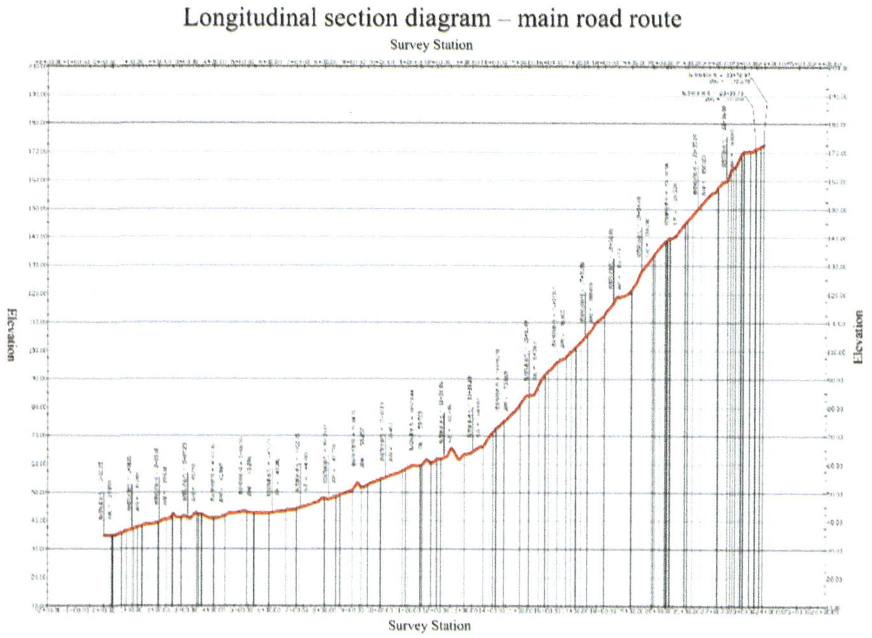

Fig. 9.22 Longitudinal profile of the path

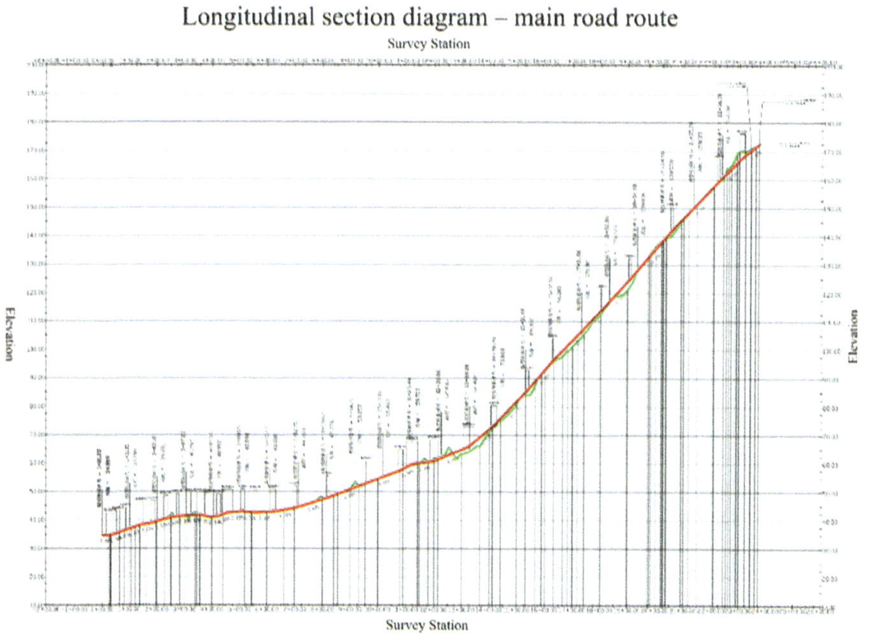

Fig. 9.23 Designed longitudinal profile

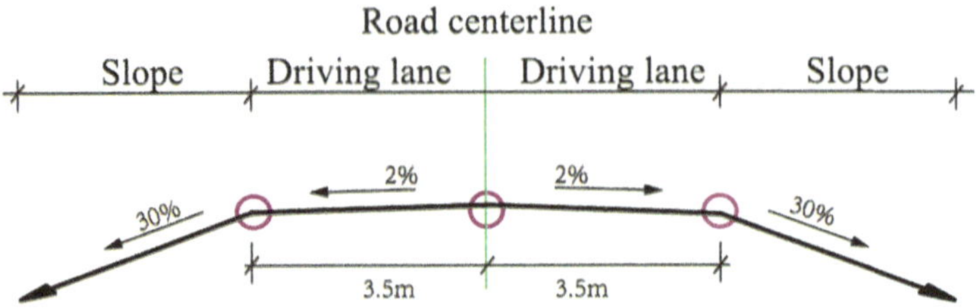

Fig. 9.24 Designed cross-section of the main road in the Xunchan Path area

Fig. 9.25 Generation of road vertical model

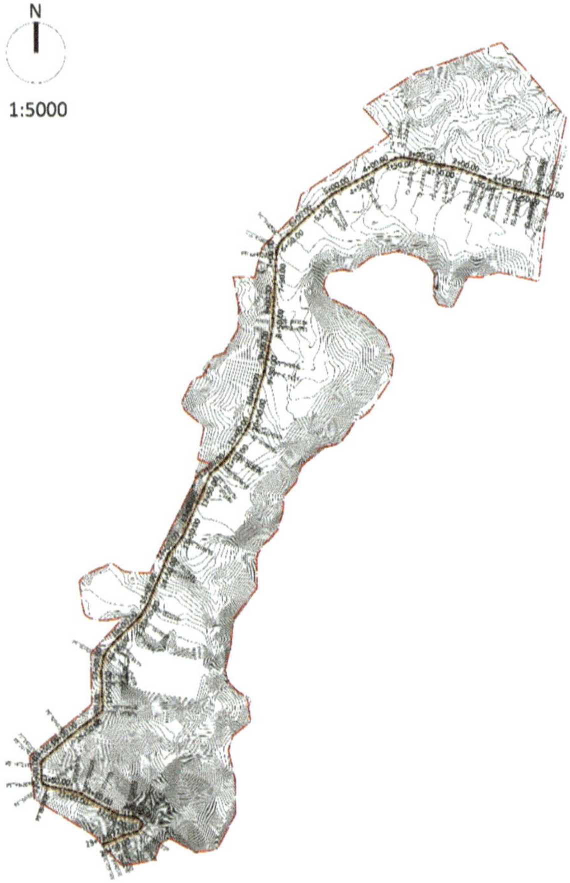

excavation, fill, and total earthwork volume. The core method employed is the triangulated irregular network (TIN) method, which compares the surface generated by the road model to the original terrain surface to calculate earthwork volumes with relative precision and efficiency. Civil 3D can automatically generate sample lines for road cross-sections according to specific station ranges or designated chainages.

Fig. 9.26 Road vertical design model—3D representation (Civil3D software)

Sample lines define the direction in which cross-sections are cut through a group of specified surfaces, and their density and width influence the accuracy of earthwork calculations. The denser the sample lines, the higher the accuracy—though this also increases data volume. Thus, appropriate sampling parameters should be chosen based on practical needs. Table 9.9 presents the earthwork calculation report for the main road in the Xunzhen Dao area, showing not only the areas and volumes of cut and fill, but also the net earthwork volume. This provides a detailed overview of the earthwork conditions for road construction and serves as the basis for earthwork regulation. From the data in this table, a preliminary judgment of earthwork balance can be made. If the balance cannot be achieved, the results must be fed back into the design process, where parameters of the road model are adjusted and optimized, and the earthwork recalculated until the design objectives are met.

9.4.2 Vertical Design of the Water System

In parameterized landscape planning and design, the basic horizontal layout of the water system is established during the water feature design stage. The vertical design of the water system focuses on shaping its three-dimensional form with the goal of achieving earthwork balance, making it an iterative and interactive process. The construction of the water system's vertical model begins with the optimization of its planar layout. Based on the existing terrain, a preliminary bottom profile is designed (Fig. 9.27) and a digital 3D model of the water system is generated in Civil 3D.Next, the water system model is integrated with the original terrain model, and suitable slope gradients are selected to design revetment slopes (Fig. 9.28). The

Table 9.9 Road earthwork volume report

Route: main road alignment; sample line group:SLG-main road sample lines; starting station:0 + 50.000; ending station:23 + 50.000

Station	Cut area(m^2)	Cut volume(m^3)	Reusable volume(m^3)	Fill area(m^2)	Fill volume(m^3)	Cumulative cut area(m^3)	Cumulative reusable volume(m^3)	Cumulative fill volume(m^3)	Cumulative net volume(m^3)
0 + 50.000	1.31	0	0	0.53	0	0	0	0	0
1 + 00.000	0.83	53.59	53.59	0.45	24.58	53.59	53.59	24.58	29.02
1 + 50.000	2.8	90.11	90.11	0.05	12.56	143.7	143.7	37.14	106.56
2 + 00.000	0.21	75.38	75.38	1.16	29.99	219.08	219.08	67.13	151.95
2 + 50.000	17.59	448.72	448.72	0.09	30.9	667.8	667.8	98.03	569.77
3 + 00.000	1.27	471.53	471.53	0.58	16.85	1139.33	1139.33	114.88	1024.45
3 + 50.000	7.45	219.63	219.63	0	14.4	1358.96	1358.96	129.27	1229.68
4 + 00.000	0	187.64	187.64	2.27	56.2	1546.59	1546.59	185.47	1361.12
4 + 50.000	0.2	5.08	5.08	0.02	57.33	1551.67	1551.67	242.8	1308.88
5 + 00.000	2.48	66.97	66.97	0.07	2.39	1618.64	1618.64	245.19	1373.45
5 + 50.000	0.4	72.55	72.55	0.39	11.57	1691.19	1691.19	256.76	1434.43
6 + 00.000	0.28	16.87	16.87	0.76	28.98	1708.06	1708.06	285.74	1422.32
6 + 50.000	0.06	8.35	8.35	1.62	59.72	1716.41	1716.41	345.46	1370.95
7 + 00.000	0.76	21.19	21.19	0.34	47.9	1737.6	1737.6	393.37	1344.24
7 + 50.000	2.85	90.24	90.24	0.26	15	1827.84	1827.84	408.36	1419.48
8 + 00.000	14.84	442.26	442.26	0	6.57	2270.11	2270.11	414.94	1855.17
8 + 50.000	0	370.97	370.97	0.87	21.8	2641.08	2641.08	436.73	2204.35
9 + 00.000	4.66	115.72	115.72	0.11	24.6	2756.8	2756.8	461.33	2295.46
9 + 50.000	0.14	120.01	120.01	1.03	28.57	2876.8	2876.8	489.9	2386.9
10 + 00.000	2.29	60.66	60.66	0	25.85	2937.47	2937.47	515.75	2421.71

10 + 50.000	2.47	119.04	119.04	0	0	3056.51	3056.51	515.75	2540.76
11 + 00.000	4.83	182.05	182.05	0	0	3238.56	3238.56	515.75	2722.8
11 + 50.000	1.55	159.18	159.18	1.38	34.75	3397.74	3397.74	550.51	2847.23
12 + 00.000	12	338.16	338.16	0	34.76	3735.9	3735.9	585.26	3150.64
12 + 50.000	56.64	1735.23	1735.23	0	0	5471.13	5471.13	585.26	4885.87
13 + 00.000	0	1416.01	1416.01	13.72	343.1	6887.14	6887.14	928.37	5958.77
13 + 50.000	0	0	0	21.35	876.84	6887.14	6887.14	1805.21	5081.93
14 + 00.000	0	0	0	9.23	768.48	6887.14	6887.14	2573.69	4313.45
14 + 50.000	0	0	0	9.82	474.33	6887.14	6887.14	3048.02	3839.12
15 + 00.000	0	0	0	6.69	412.73	6887.14	6887.14	3460.75	3426.39
15 + 50.000	0	0	0	28.81	890.01	6887.14	6887.14	4350.75	2536.39
16 + 00.000	0.21	5.12	5.12	1.06	749.9	6892.26	6892.26	5100.66	1791.6
16 + 50.000	0	5.21	5.21	12.16	330.57	6897.46	6897.46	5431.23	1466.24
17 + 00.000	0	0	0	19.81	793.36	6897.46	6897.46	6224.59	672.87
17 + 50.000	0	0	0	16.47	904.19	6897.46	6897.46	7128.78	−231.32
18 + 00.000	0	0	0	11.81	707.74	6897.47	6897.47	7836.52	−939.06
18 + 50.000	2.48	65.16	65.16	1.29	329.47	6962.63	6962.63	8165.99	−1203.36
19 + 00.000	0	61.8	61.8	30.45	793.48	7024.43	7024.43	8959.47	−1935.04
19 + 50.000	0.89	22.2	22.2	0.33	769.63	7046.63	7046.63	9729.1	−2682.47
20 + 00.000	11.47	309.43	309.43	0	8.72	7356.06	7356.06	9737.82	−2381.76
20 + 50.000	0	288.94	288.94	13.9	332.13	7645	7645	10069.95	−2424.95
21 + 00.000	0.9	22.33	22.33	5.34	483.29	7667.33	7667.33	10553.24	−2885.91
21 + 50.000	3.55	110.15	110.15	1.06	161.29	7777.48	7777.48	10714.53	−2937.05
22 + 00.000	3.76	182.65	182.65	7.66	217.94	7960.12	7960.12	10932.47	−2972.35
22 + 50.000	18.67	570.22	570.22	0.17	194.52	8530.35	8530.35	11126.99	−2596.65
23 + 00.000	80.28	2344.21	2344.21	0	5.05	10874.56	10874.56	11132.04	−257.48
23 + 50.000	10.79	2269.82	2269.82	0	0	13144.38	13144.38	11132.04	2012.33

Fig. 9.27 Design of water system morphology

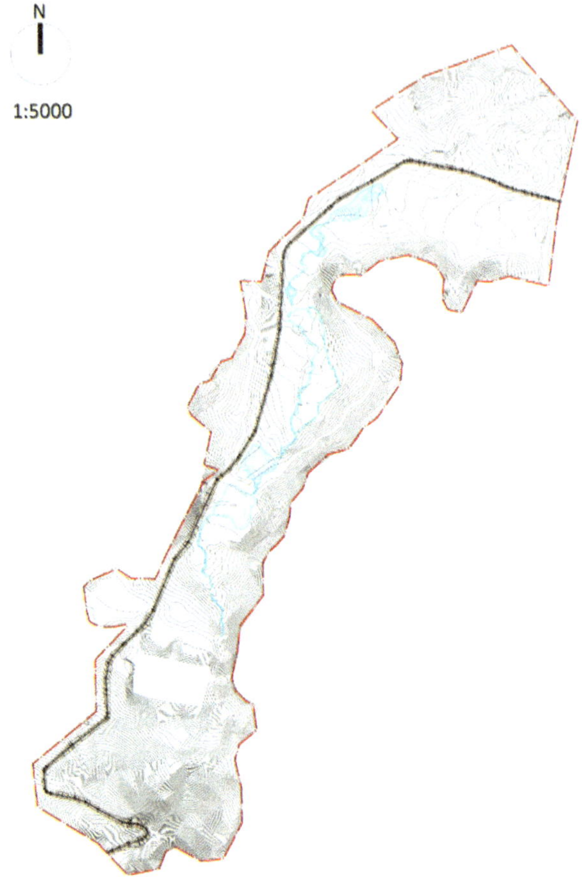

Fig. 9.28 Water system vertical model—3D representation (Civil3D)

Table 9.10 Earthwork volume calculation for water system construction

Volume summary						
Name	Cut factor	Fill factor	2D area (m^2)	Cut (m^3)	Fill (m^3)	Net (m^3)
Water system surface model	1.000	1.000	21392.61	31632.08	588.23	31043.85 < cut >

method for calculating earthwork volumes for water systems is essentially the same as that used for roads. The volume is derived from the difference between the water surface model and the original terrain surface (Table 9.10).

As with roads, the results of earthwork calculations can be used to provide feedback on the vertical design of the water system. Optimization of the water system's form based on this feedback can help achieve earthwork balance.

9.4.3 *Vertical Design of the Site*

During site selection, the specific location and potential construction area of the project are determined. Based on these conditions, the vertical design of the site is carried out using the digital terrain model within Civil 3D.

9.4.3.1 Site Vertical Model Generation

Site A is designated for the visitor center, with a construction area of 4070.8 m^2. A corresponding 3D site model must be constructed within the digital terrain model. First, the appropriate design elevation is determined based on the existing terrain elevations within the site. As shown in Fig. 9.29, the elevation in Site A ranges from approximately 41.0 m to 43.0 m, with the majority of the area lying between 41.0 and 42.0 m. Therefore, the design elevation for Site A is provisionally set at 41.50 m. Next, a site model with a design elevation of 41.50 m is created within the red line boundary. Where the site connects with the surrounding terrain, cut or fill is necessary. A slope gradient of 2:1 is selected based on design needs. Once the relevant parameters are set, the site model is generated (Fig. 9.30), which reflects the cut and fill conditions of the site relative to the original terrain at a design elevation of 41.50 m and with a 2:1 slope.

9.4.3.2 Achieving Earthwork Balance and Determining Site Elevation

Based on the three-dimensional digital terrain model (DTM) of the site, the Grading Volume Tool in Civil 3D can be used to quickly compute the cut and fill volumes for Site A. As shown in Fig. 9.31, when the site elevation is set at 41.5 m, the cut

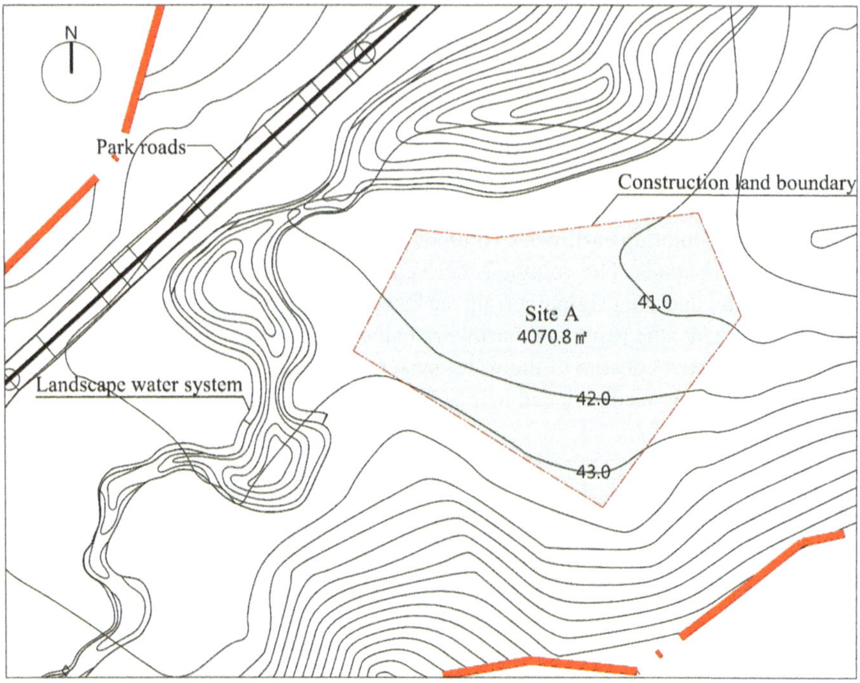

Fig. 9.29 Site A location assessment

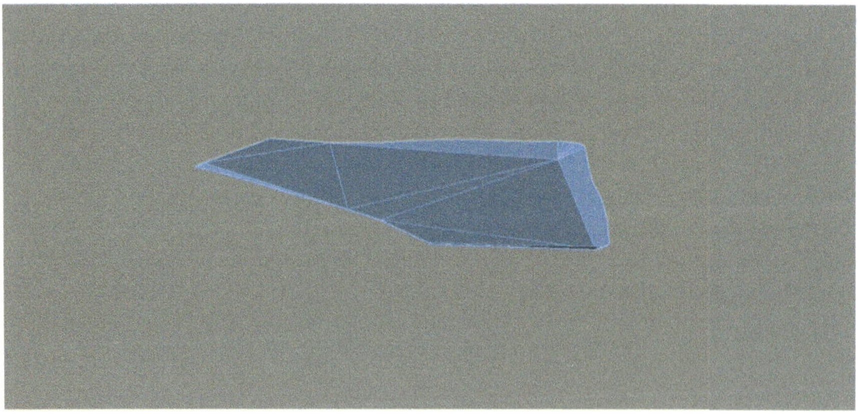

Fig. 9.30 3D digital model of site A

volume is 2094.84 cubic meters, while the fill volume is only 543.02 cubic meters. This indicates a significant surplus of 1551.82 cubic meters of excavated soil. The Grading Volume Tool offers an automated earthwork balancing function that adjusts the site elevation through calculation to achieve a balance between cut and fill

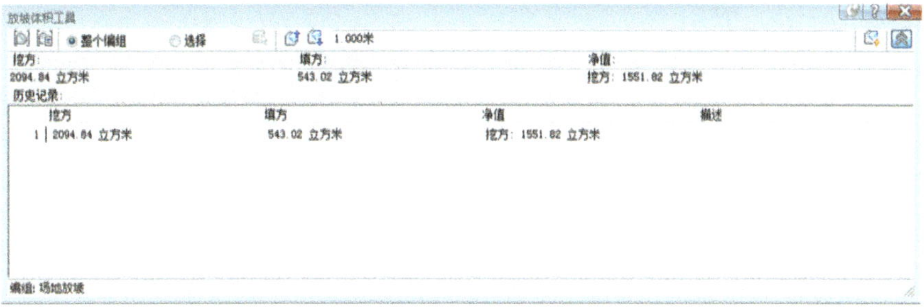

Fig. 9.31 Earthwork volume calculation for site A

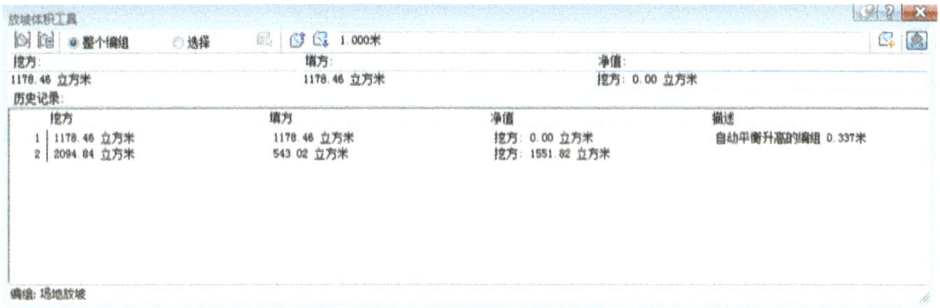

Fig. 9.32 Automated volume balance for site A

volumes. This tool provides a convenient and efficient means of achieving earthwork balance. It can calculate earthwork volumes and adjust elevations for a group of sites simultaneously, with real-time model updates that reflect elevation changes. As illustrated in Fig. 9.32, when the elevation of Site A is raised by 0.337 m, cut and fill volumes are balanced. Consequently, the final design elevation of Site A is set at 41.84 m. The tool can also be applied to multiple sites within the design area to regulate earthwork balance as required by the project. As shown in Fig. 9.33, Sites B and C are adjacent to Site A. These three sites can be grouped into the same grading set, and the Grading Volume Tool can be used to adjust the total earthwork balance across the entire group (Fig. 9.34), while simultaneously determining the final design elevations of all three sites.

9.4.4 *Earthwork Balance Optimization*

Based on the vertical digital model, Civil 3D automatically generates the earthwork quantities associated with each sub-component of the site design and calculates the net cut and fill volumes. As presented in Table 9.11, the initial design phase estimates approximately 5681 cubic meters of earthwork required for road construction. Meanwhile, both the water system and site development generate surplus

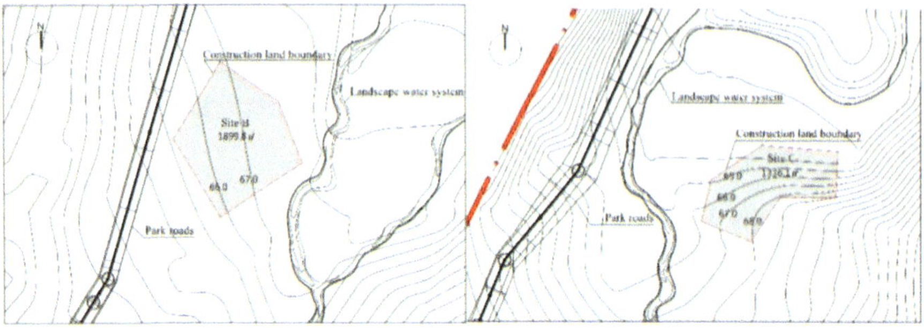

(a) Site B Location Assessment (b) Site C Location Assessment

Fig. 9.33 Location assessment of sites B and C

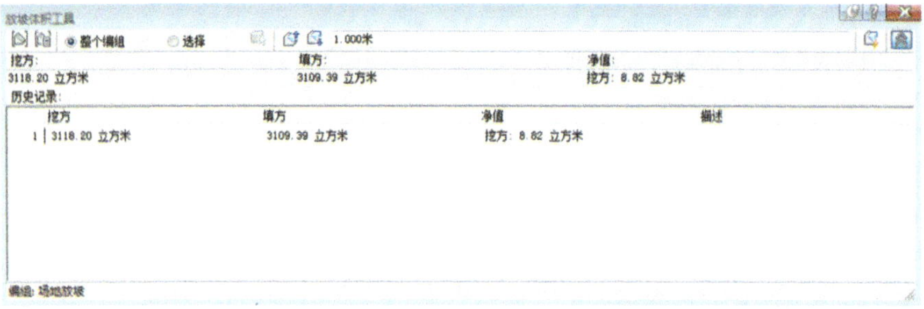

Fig. 9.34 Total earthwork balance for sites A, B, and C

Table 9.11 Earthwork volume summary table

Volume summary						
Name	Cut factor	Fill factor	2D area (m^2)	Cut (m^3)	Fill (m^3)	Net (m^3)
Road surface model	1.000	1.000	22605.21	5136.57	10817.58	5681.00 < fill >
Water system surface model	1.000	1.000	21392.61	31632.08	588.23	31043.85 < cut >
Site surface model	1.000	1.000	8978.69	3118.20	3109.39	8.82 < cut >
Total	–	–	52976.52	39886.86	14515.19	25371.66 < cut >

earthwork volumes, with the site essentially achieving self-balance, and the water system producing an excess of 31,043.85 cubic meters.

Therefore, the initial design does not meet the criteria for earthwork balance, and further adjustment and optimization of the design is necessary. Based on the second round of optimized design, the vertical digital model is updated accordingly, yielding revised earthwork calculations. Through iterative refinement and optimization, the earthwork balance is ultimately achieved, enabling the finalization of the vertical design results.

Chapter 10
Conclusion

With the advent of the digital age, emerging technologies and tools are being widely applied across the design industry. Research on "design science" has propelled the development of scientific design methodologies. Under the dual impetus of technological advancement and theoretical evolution, the form and substance of contemporary landscape planning and design have undergone profound transformation. Digital and parametric technologies, in particular, have exerted a significant and lasting influence on the field. Since the late 1990s, parametric design methods supported by computational technologies have flourished in the architectural domain, resulting in a wealth of practical applications. In contrast, research on parametric design within the discipline of landscape architecture has lagged behind. Much of the existing work borrows from architecture and related fields. However, although both disciplines belong to the broader category of human-environment sciences, they differ significantly in research focus, scale, scope, and design objectives and methodologies. How can landscape architecture—characterized by both scientific and artistic attributes—advance its planning and design methods through an integration of qualitative and quantitative approaches? What constitutes parametric planning and design in landscape architecture? Why are parametric methods applicable to this field? And what are the similarities and differences between parametric design in landscape architecture and in architecture? These were the core questions I grappled with throughout my doctoral research, which this book seeks to explore and address. This study begins by analyzing the underlying mechanisms of design, aiming to clarify the essence and operative logic of parametric planning and design from a methodological perspective. Reflecting on the overall research and writing process—from understanding the background, identifying research questions, defining concepts, elaborating theories, constructing systems, developing models, to conducting empirical studies—a relatively comprehensive logical framework and workflow have taken shape.

The book is dedicated to investigating the processes and mechanisms of landscape planning and design based on the discipline's unique characteristics, integrating both theoretical and practical dimensions. By incorporating parametric

Y. Yuan, *Parametric Landscape Architecture Planning and Design*,
https://doi.org/10.1007/978-981-95-8011-8_10

regulation techniques and applying quantitative methods, the research seeks to enhance the control and optimization of the design process and promote methodological progress in landscape planning and design.Thus, this study is neither a purely theoretical investigation grounded in analysis and induction, nor solely an applied technical study. Instead, it bridges theory and practice—starting from theoretical formulation, proceeding through practice-oriented feedback, and ultimately refining theoretical construction through empirical validation. From both theoretical elaboration and practical verification, this work conducts an in-depth exploration of parametric landscape planning and design, featuring several key innovations:

10.1 A Novel Perspective: Establishing a Parametric Landscape Planning and Design Process Based on the Principle of Coupling

The concept of "designing in accordance with local conditions" (因地制宜) has long been a foundational attitude toward nature and a well-accepted consensus in the design field. But what exactly is the internal mechanism of "designing according to local conditions"? What pathways enable its realization?

By introducing the concept of "coupling"—borrowed and adapted from other disciplines—into landscape architecture, this study offers a contemporary scientific interpretation of "designing according to local conditions." It aims to articulate the interrelation between design and site. Facing the current demands of landscape planning and design—ecological sensitivity, contextual specificity, minimal intervention, and sustainability—this research, guided by the principle of coupling, explores an actionable path toward achieving these objectives."Coupling" is not only a design strategy but also an embodiment of scientific design thinking. This research, grounded in the ontological foundation of landscape architecture, analyzes the value and significance of coupling in design, along with practical pathways for its realization. The coupling principle emphasizes the mutual adaptability between site and design, revealing their intrinsic connection and underlying the systemic formation of landscape planning and design. This connection serves as the foundational logic for generating parametric design solutions.Through the lens of the coupling principle, this study offers a renewed perspective on landscape planning and design processes. It clarifies the inner relationship between site and design, effectively breaking the "black box" of design and offering a transparent, structured outline of the process. In doing so, it lays the groundwork for a parametric approach to landscape planning and design.

10.2 Theoretical Innovation: Elucidating the Mechanism of Parametric Landscape Planning and Design Within a Systems Framework

Systems theory regards phenomena as organic wholes composed of multiple elements interconnected in a structured manner, using logic and mathematical models to explore both the configuration of components and their interrelationships. Applying a systems framework to landscape planning and design facilitates the integration of quantitative, logical, and mathematical approaches into the research process—approaches that form the very foundation of parametric design. A parameter is a variable with dynamic and mutable qualities, aligning with the inherent dynamism of landscape systems and thereby laying the groundwork for parametric implementation. From a systems perspective, the author examines the foundational logic and drivers behind parametric mechanisms in landscape planning and design and systematically investigates its core components.

Landscape architecture is a complex system involving multiple interacting elements. Its planning and design must meet diverse objectives while responding to sites that are equally varied and complex—incorporating factors such as water bodies, landforms, climate, flora and fauna, and cultural attributes. Each site possesses a unique character. Coupling site analysis with design through parametric methods is both a key focus and a challenge in this study. Furthermore, understanding how to establish and describe the parametric relationships among complex design elements within a dynamic system presents a significant research difficulty. The use of parameters and logical rules to describe the life processes of the landscape environment distinguishes parametric landscape design from its architectural counterpart. Within the system-based framework, this research constructs a parametric planning and design model grounded in the logic and structure of the landscape design process. By articulating the interrelationships among individual design elements and the system as a whole, the research transforms discrete, fragmented design factors into an integrated framework, rendering landscape planning and design more objectively grounded and scientifically structured.

Parametric approaches support the realization of multi-objective coordination and system-level optimization. The interdependence of parameters reflects the inherent relationships among design elements, corresponding to the internal logic of landscape planning and design. The author explores parametric implementation strategies for core components such as roads, water features, vertical design, and built projects, treating them as part of a unified system. By investigating the inner workings of landscape planning, this research achieves synchronized, multi-objective optimization of the environment.

10.3 Methodological Innovation: Exploring Parametric Landscape Planning and Design Through Integration of Qualitative and Quantitative Approaches

Scientific research is the process of investigating the essential nature and patterns of objective phenomena using appropriate methods, logic, and purpose. Common methods include surveys, experiments, and prototyping, with verifiability as the sole criterion for truth. In contrast, humanities research centers on the human spirit and culture, employing methods such as interpretive analysis, hermeneutics, and empirical investigation. Landscape planning and design is distinct from both scientific and humanistic research, demanding a unique form of design thinking and methodological approach. As both a science of art and an art of science, landscape architecture possesses dual attributes. Accordingly, it requires a methodological framework that accommodates both dimensions. This study begins with the internal nature of landscape architecture, supplementing traditional qualitative analysis with quantitative methods, thereby enabling the objective interpretation and evaluation of landscape environments. The integration of qualitative and quantitative methods is not only aligned with the discipline's dual nature but also forms the core of the parametric approach. The landscape environment includes both measurable physical components—such as hydrology, climate, vegetation—and intangible elements like culture and social values. Many of these are not easily quantifiable. A prerequisite for parametric design is the transformation of system components into numeric or abstract representations that can be processed computationally. Hence, a fundamental challenge is describing complex variables and converting qualitative information into quantitative metrics. Establishing scientific evaluation criteria for landscape components is essential to ensuring rational design outcomes. Given the complexity and diversity of evaluation indicators in landscape planning and design, integrating them into a coherent evaluation framework is another methodological challenge. As a multi-tiered system, effective evaluation requires defining hierarchical relationships and weighting schemes to ensure scientific validity and consistency.

Historically, landscape design has heavily relied on intuition with insufficient support from structured perception. The field has leaned toward qualitative approaches, often favoring emotion over rationality. Through integrated methodologies and scientific evaluation, this research establishes a structured combination of qualitative insight and quantitative rigor within a landscape systems framework. It explores pathways for implementing parametric design methods and enhances the scientific validity, precision, and practical feasibility of landscape planning and design.

10.4 Innovation in Outcomes: Development of a Parametric Evaluation and Planning Design Model Tailored to the Characteristics of Landscape Architecture

Building on the theoretical research into parametric design mechanisms, this study—guided by the principle of coupling—has preliminarily established a parametric evaluation and planning design model for landscape architecture. This model comprises six specialized modules: an ecological sensitivity evaluation model, a land-use suitability evaluation model, a project positioning model, a road alignment model, a quasi-natural waterscape construction model, and a vertical design model, collectively covering the primary components of landscape planning and design. Each model functions independently while also being interrelated, forming an integrated logical framework under the broader process of parametric landscape planning and design. As a part of the overall mechanistic investigation, the model incorporates parametric algorithm logic and has been tested and applied in practice, thereby laying a methodological and practical foundation for future programming and the development of domain-specific and integrated software platforms.

This parametric evaluation and planning design model is closely tied to real-world applications. Rooted in practice and refined through iterative use, it reflects a synthesis of theoretical research and practical validation. Developed progressively during the author's academic training and practice, the model has been applied and tested effectively, earning positive feedback from the professional community. By employing advanced technologies and quantitative methods for analysis, synthesis, and design, the parametric approach serves as a valuable auxiliary method in design processes. The parametric design methodology holds significant practical implications. In professional practice, a parametric approach grounded in the principle of coupling enhances the accuracy, rationality, and scientific rigor of landscape planning and design, reducing subjectivity and randomness, and substantially improving the efficiency of landscape environment planning, construction, and management. In education, it overcomes the limitations of ambiguity in traditional teaching methods, improves problem analysis and resolution skills, and diminishes the dependence on prolonged experience accumulation. For beginners, this approach offers a more accessible and effective path to mastering the internal logic of landscape planning and design. Applying the parametric landscape planning and design method based on coupling principles can therefore strengthen the core of landscape architecture as a discipline, improve and enrich design pedagogy, and provide students with more scientific and accessible ways to grasp design methodologies and internal planning logic.

As the discipline of landscape architecture continues to advance and research deepens, increasing attention is being given to the exploration of design methodologies and principles. This book focuses specifically on parametric landscape planning and design mechanisms, and the development of its associated evaluation and planning model represents an initial exploration of this area. It lays the groundwork for future specialized research and software development related to parametric design in landscape architecture. The models proposed in this book are built on principles of usability, accessibility, and operability, utilizing widely adopted and mature technologies and software platforms such as Analytic Hierarchy Process (AHP), ArcGIS, and Civil 3D. The research further explores the feasibility and potential for developing parametric design platforms based on these tools. For instance, ArcPy, a scripting module in ArcGIS, and RhinoScript in Rhino, can be used to realize "design by programming" using the Python language. Moreover, Grasshopper (a visual programming tool within Rhino) and the ModelBuilder in ArcGIS provide graphical algorithm editors (components) capable of combining operations to solve specific design problems and achieve defined design goals. Although programming tools and component-based editors serve the same fundamental function—algorithmic generation—they differ in their characteristics. Programming tools offer high flexibility and precision but are less accessible due to the technical demands of coding, making them less widely usable among landscape designers who are generally less familiar with programming languages. Also, rigorous logical syntax can suppress design spontaneity, leading to overly rigid or formulaic outcomes. Component-based editors (e.g., Grasshopper or ModelBuilder), on the other hand, are more approachable and user-friendly, yet they tend to be more generic and less targeted for the specific needs of landscape design, limiting their effectiveness and ease of use in this field.

Regardless of the platform, programming languages serve as the foundation of parametric design tools, playing an essential role. However, the generation of algorithmic logic remains the core of parametric design. Code produces objective outputs based solely on logical structures—unaffected by human will. Only through well-structured, purpose-driven algorithmic logic can design results truly align with the intended goals. That said, "design" remains the central, guiding force in landscape architecture, while "drawing" serves only as a tool—whatever form it may take—to support the realization of the design concept. The logical models proposed for each component of the planning process aim to establish a foundational framework for algorithm development. These logical models are designed for usability and built on mature software platforms. The next phase of work will focus on further coupling design stages, and either developing dedicated parametric design software based on existing platforms or exploring new platforms specifically tailored to the practical needs of landscape architecture. Whether developing domain-specific tools or custom software, such efforts will require cross-disciplinary collaboration—with landscape architecture researchers contributing domain knowledge and algorithmic logic, and experts in computer science, software engineering, and applied mathematics working alongside them to create targeted, effective tools.

Guided by the principle of coupling and centered on the parametric process of landscape planning and design, this study constructs a comprehensive Parametric Landscape Design System (PLDSM), which comprises six interrelated specialized models. The PLDSM articulates in detail the parametric workflow for landscape architecture design. The process begins with site cognition, which involves thorough site investigation and the collection and organization of relevant information. Quantifiable elements extracted from the site are then analyzed through algorithmic methods and processed via software computation. These computations transform raw data into variables and parameters, enabling the precise description of various environmental factors and accurately reflecting the actual site conditions. Based on the specific design type and project objectives, appropriate specialized design models are selected. These models establish a coupling relationship between design elements and site conditions. The previously generated parameters are input into algorithmic and auxiliary design software, ultimately producing a design outcome presented in the form of a fully data-driven model. In the design evaluation stage, digital analytical tools are used to assess and interpret the data model. The evaluation results can then be fed back into the design process, facilitating further refinement and optimization. This constitutes the preliminary framework of the Parametric Landscape Design System Model (PLDSM). Future work will focus on integrating this model with Landscape Information Modeling (LIM) systems to enhance the scientific rigor and informatization of landscape planning and design. Within the PLDSM framework, the data feedback loop is dynamic, enabling the generation and reconstruction of design results in the form of digital models within computational environments. This allows for faster and more precise landscape planning and design. The final data model output can be transmitted to conventional design media for visual representation and communication, or directly integrated into digital fabrication tools—such as intelligent robots, CNC machines, laser cutters, and 3D control systems—for automated construction, enabling the realization of complex forms and precise spatial positioning. Additionally, the model can be interfaced with architectural and urban planning systems that adopt similar digital technologies, thereby supporting data sharing and integrated design processes.

At present, designers can intuitively employ traditional tools such as drafting boards, rulers, and pencils, or work proficiently with digital software platforms like Photoshop, SketchUp, and AutoCAD. It is foreseeable that parametric and digital tools will become equally familiar and integral to landscape planning and design workflows. Like any method, design techniques will evolve and improve over time. As a supportive tool and methodology, the development of parametric design has already initiated a transformation in landscape architectural methods and design thinking. With continuous scientific advancement and deeper theoretical understanding, parametric tools will become a major driver in the evolution of design approaches and practices.

Nevertheless, design is fundamentally a creative activity, and cannot be separated from human thought. As current research suggests, computers cannot fully replicate the complexity of human cognitive processes—artificial intelligence cannot yet rival human intelligence. Consequently, landscape planning and design cannot be generated through logical computation alone. While the value of scientific and technological tools is fully acknowledged, it is equally important to avoid blind reliance on technology. We must neither follow trends uncritically, nor adopt a purely technocratic perspective, nor become subservient to tools. Scientific tools are not omnipotent; they merely offer designers more efficient and precise ways to understand and engage with the world. Ultimately, it is the synergistic collaboration between computational systems and human cognition, coupled with respect for objective laws and the full expression of human agency, that represents the rightful path for the future of landscape planning and design.

Appendix

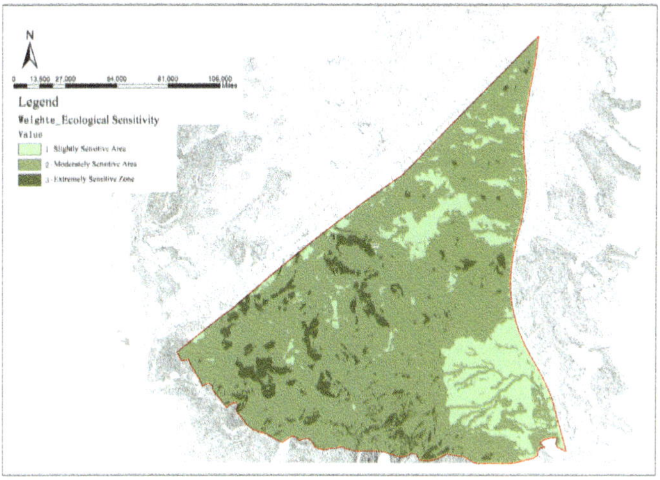

Ecological Sensitivity Zoning of the Northern Area of Niushou Mountain Scenic Area

Fig. A1 Ecological sensitivity evaluation model—drawing section

Y. Yuan, *Parametric Landscape Architecture Planning and Design*,
https://doi.org/10.1007/978-981-95-8011-8

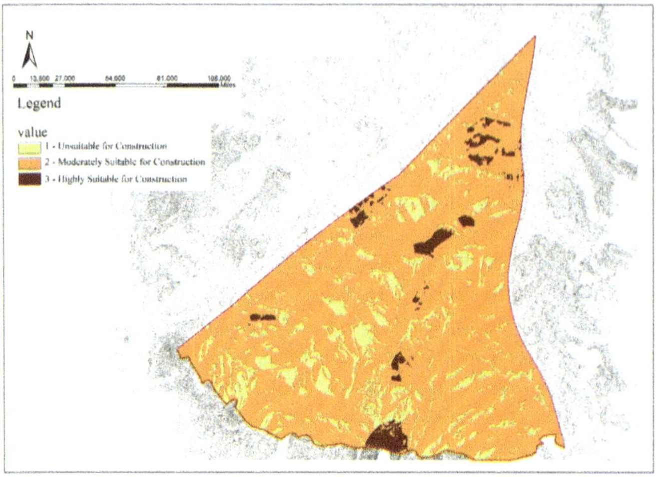

Land Use Suitability Zoning Based on Land Use Suitability Evaluation in the Northern Area of Niushou Mountain Scenic Area

Land Use Suitability Evaluation in the Northern Area of Niushou Mountain Scenic Area—Buildable Areas

Fig. A2 Land use suitability evaluation model—drawing section

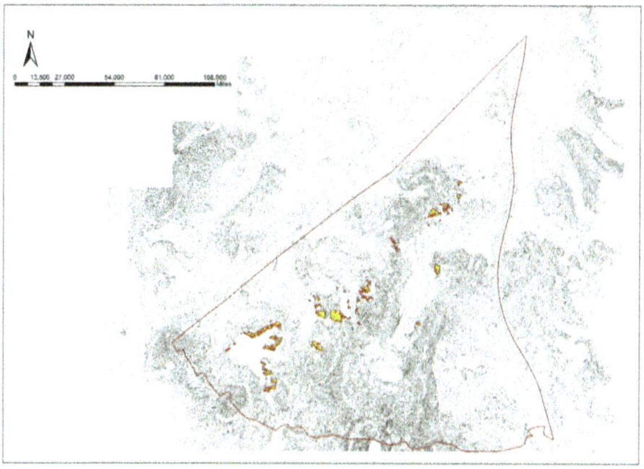

Figure 6-8 Secondary Site Selection Results of "Jingyi Mountain Lodge"

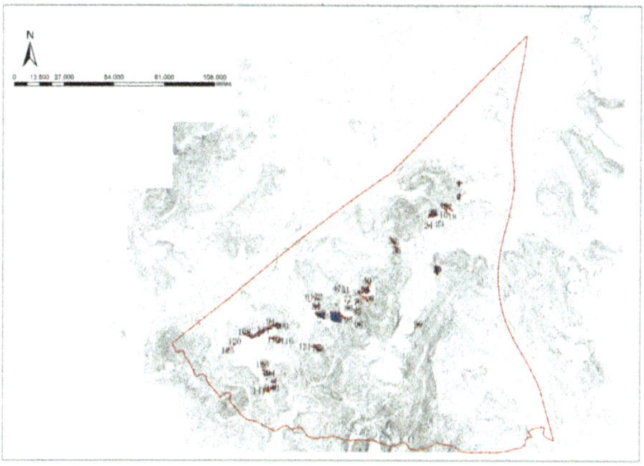

Generation of Alternative Land Options of "Jingyi Mountain Lodge" (For area statistics of alternative land, please refer to the Appendix)

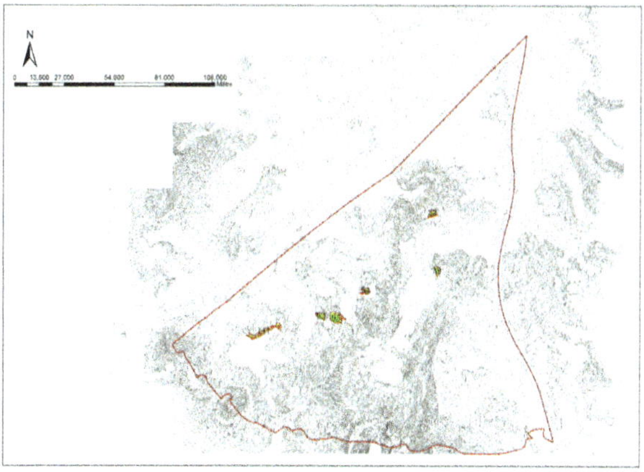

Land Screening Results (Area > 1500 m²)

Fig. A3 Project site selection model—drawing section

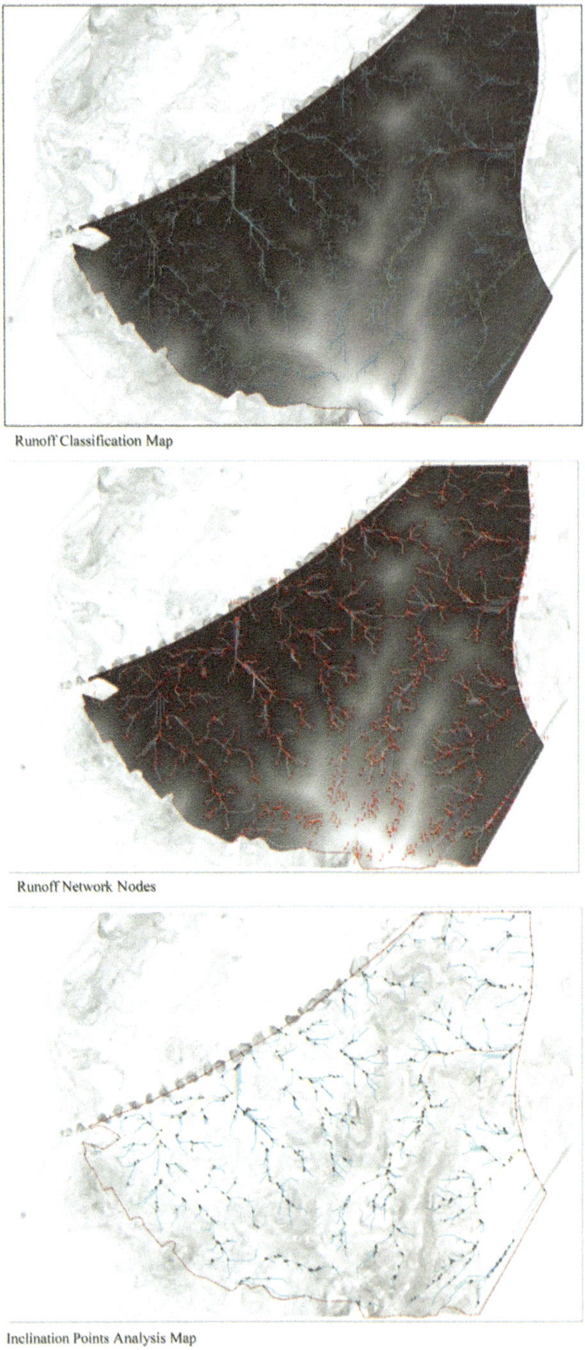

Fig. A4 Water feature creation model—drawing section

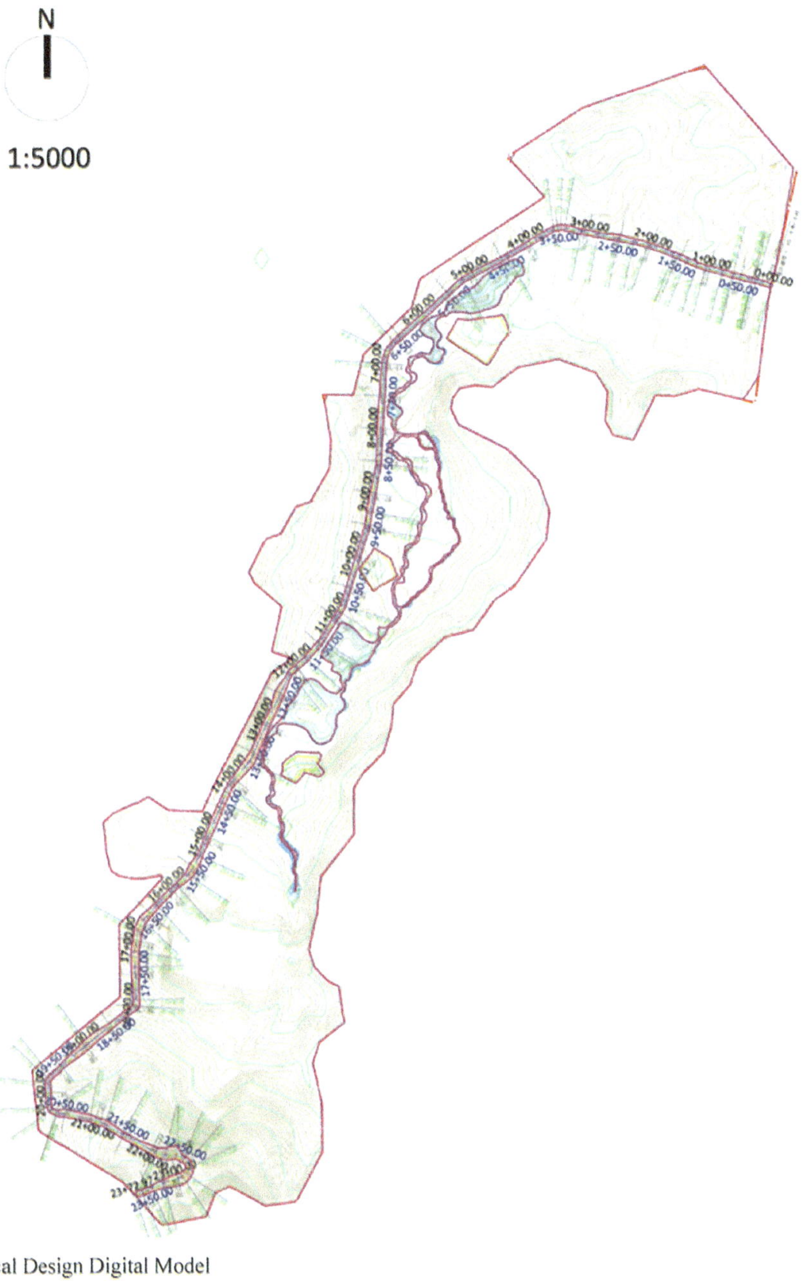

Fig. A5 Vertical optimization model—drawing section

Table A1 Statistics of optional land area for "Jingyi Mountain Villa"

ID	Gridcode	Shape_Leng	Shape_Area
1	10	126.84731000800	440.04905070900
2	10	19.95544772700	12.41075433090
3	10	107.58970954700	572.27325636900
4	10	28.16331209530	28.53875763060
5	10	8.72204340167	3.08760404587
6	10	122.80656127700	535.24890070100
7	10	82.66469180250	182.67120858300
8	10	21.77875915080	22.29521960580
9	10	23.74687842150	34.86455675210
10	10	83.49537655720	156.77491202300
11	10	3.77293209885	0.68468620628
12	10	3.77293209885	0.68468620628
13	10	10.40676307030	4.86050817743
14	10	6.00000000000	2.00000000000
15	10	4.00000000000	1.00000000000
16	10	4.00000000000	1.00000000000
17	10	3.77293209885	0.68468620628
18	10	18.44209931190	18.33593567830
19	10	3.77293209885	0.68468620628
20	10	33.62170937660	34.66422582980
21	10	3.77293209885	0.68468620628
22	10	348.93327290900	2027.20060802000
23	10	25.33427128460	21.54581979480
24	10	43.87946018670	86.83270218780
25	10	3.77293209885	0.68468620628
26	10	24.07713914850	16.49242276140
27	10	6.00000000000	2.00000000000
28	10	3.77293209885	0.68468620628
29	10	3.77293209885	0.68468620628
30	10	4.00000000000	1.00000000000
31	10	3.77293209885	0.68468620628
32	10	166.49975521900	353.54190214500
33	10	46.78156418690	105.20246430500
34	10	5.65885722219	1.36540561542
35	10	90.51480853110	220.66584483900
36	10	34.04794533110	49.23673728850
37	10	56.22654526340	114.69213130700
38	10	4.00000000000	1.00000000000
39	10	194.32297990200	1674.51898669000
40	10	41.00546467220	107.11404974000
41	10	36.29160316710	49.57325426490
42	10	8.17655938377	3.67598213255
43	10	64.62730680820	59.67532601580
44	10	7.62714463568	2.00429291651

(continued)

Table A1 (continued)

ID	Gridcode	Shape_Leng	Shape_Area
45	10	16.30133935170	10.40867321940
46	10	3.77291876978	0.68468098342
47	10	27.95060700030	49.37002536650
48	10	8.55330943599	3.29239717126
49	10	11.92865161660	7.85728809610
50	10	13.59866799320	9.40256702155
51	10	31.91320134000	48.57003963370
52	10	3.77291876978	0.68468098342
53	10	5.67373926129	1.34985382110
54	10	388.24733228600	1710.68169177000
55	10	5.67373926129	1.34985382110
56	10	7.62714463568	2.00429291651
57	10	7.07259600390	2.34680255875
58	10	28.34988150030	31.04097937050
59	10	3.77289210907	0.68467053585
60	10	18.71738567860	15.02298569310
61	10	40.38183946050	100.27895081400
62	10	4.00000000000	1.00000000000
63	10	8.20307525278	3.02052089572
64	10	4.00000000000	1.00000000000
65	10	7.62714463568	2.00429291651
66	10	17.04283854360	9.69274799153
67	10	102.69165409700	165.02020038300
68	10	32.68161536930	35.85739359260
69	10	14.26635734100	10.32481461020
70	10	97.58544377300	210.74837131400
71	10	27.91066372100	26.96961929090
72	10	71.27980877220	284.07837177600
73	10	5.65883949370	1.36538864300
74	10	33.86494108530	61.02276023480
75	10	44.37870147530	81.58557749730
76	10	64.93719307780	203.86003909300
77	10	74.02877997230	92.08643055710
78	10	3.77291876978	0.68468098342
79	10	3.77291876978	0.68468098342
80	10	3.77289210907	0.68467053585
81	10	12.65735510580	6.90594174713
82	10	3.77291876978	0.68468098342
83	10	61.24823118470	204.79249689400
84	10	165.24541177200	500.47185273300
85	10	28.41314055870	44.65755956250
86	10	13.48130966580	9.25007481873
87	10	7.06399849434	2.31673082709
88	10	10.48944550440	6.58970465511

(continued)

Table A1 (continued)

ID	Gridcode	Shape_Leng	Shape_Area
89	10	5.65897139651	1.36544527858
90	10	22.82343219360	28.22806672750
91	10	15.55313241590	10.73246503620
92	10	279.48312638400	2516.83282095000
93	10	47.74596247510	59.51155506820
94	10	11.49611555170	6.36024735495
95	10	548.35573322900	6342.33266278000
96	10	18.60318856580	18.21038393680
97	10	3.77291876978	0.68468098342
98	10	38.77700999650	60.97097333520
99	10	78.52977397260	238.11206945800
100	10	3.77291876978	0.68468098342
101	10	5.65897139651	1.36544527858
102	10	13.03132748720	5.37763203681
103	10	3.77294542898	0.68469142914
104	10	3.77294542898	0.68469142914
105	10	3.77291876978	0.68468098342
106	10	3.77293209885	0.68468620628
107	10	12.47082659820	5.41691998206
108	10	36.92372333730	68.38185234000
109	10	3.77293209885	0.68468620628
110	10	5.67365717991	1.34979403764
111	10	1142.43395376000	5439.58287053000
112	10	192.93716970300	680.56761865700
113	10	3.77291876978	0.68468098342
114	10	10.19627470030	5.40631569922
115	10	10.29773261210	6.51003295928
116	10	118.84671547300	605.34267500000
117	10	5.65896253214	1.36543679237
118	10	3.77291876978	0.68468098342
119	10	60.26349543510	93.05048081280
120	10	48.67756217250	76.93614140340
121	10	35.54748156820	48.15326631810
122	10	3.77291876978	0.68468098342
123	10	229.81537913500	1365.32717901000
124	10	7.58693980004	2.06416155398
125	10	80.59032570140	197.23139302100
126	10	6.98995563043	2.19129995257
127	10	5.65895366854	1.36542830616
128	10	94.74482273120	331.67655729500
129	10	11.92310508620	7.01674463972
130	10	16.39549788530	13.84911943040
131	10	8.17665039262	3.67609727383
132	10	31.46463904500	61.88251628730
133	10	23.19073874160	28.29501275720
134	10	222.37710920500	1189.31149707000

(continued)

Table A1 (continued)

ID	Gridcode	Shape_Leng	Shape_Area
135	10	23.48904722330	29.13173634190
136	10	128.64321316100	803.17386758000
137	10	14.38642052560	10.74560452250
138	10	15.07482142540	10.92661144580
139	10	14.48133212530	8.94047208689
140	10	262.15062686300	1206.19024239000
141	10	42.58542498330	97.30330718310
142	10	65.21452172150	104.55011982100

Table A2 Runoff classification statistics table

Category	Quantity	Category	Quantity	Category	Quantity
1	25 677	36	4	86	149
2	5 721	37	49	88	2
3	2 644	38	1	89	196
4	1 787	41	25	93	5
5	1 522	43	94	94	27
6	701	44	13	95	149
7	792	45	12	96	2
8	581	46	26	98	29
9	464	47	173	101	38
10	431	48	357	102	116
11	435	49	3	104	11
12	326	52	11	115	6
13	451	53	52	117	9
14	327	54	8	118	28
15	535	55	80	119	183
16	170	57	46	132	15
17	217	58	82	135	39
18	68	59	103	149	4
19	574	60	58	150	1
20	435	61	7	151	4
21	49	62	47	152	3
22	187	65	35	153	54
23	20	67	89	154	14
24	135	68	81		
25	97	69	16		
26	197	70	5		
27	122	71	15		
28	73	72	68		
29	91	73	29		
30	35	79	3		
31	11	80	8		
32	50	81	89		
33	228	82	29		
34	26	84	28		
35	17	85	54		

Table A3 Runoff node statistics table (No. 1–80)

OBJECT ID	Join_Count	Grid_Code	From_Node	To_Node	OBJECT ID	in_Count	Grid_Code	From_Node	To_Node
1	1	1	675	676	41	1	1	905	904
2	1	1	675	676	42	3	62	599	582
3	1	1	677	638	43	3	5	643	619
4	1	1	677	638	44	1	1	805	796
5	1	1	542	514	45	3	59	619	599
6	1	1	542	514	46	3	7	796	793
7	1	1	618	610	47	3	8	793	790
8	1	1	618	610	48	3	33	790	623
9	1	1	689	632	49	1	1	493	487
10	1	1	575	632	50	1	1	968	918
11	1	1	645	632	51	3	54	623	619
12	1	1	816	802	52	3	67	554	582
13	1	1	524	515	53	3	21	642	623
14	1	1	723	712	54	3	20	718	642
15	3	2	515	520	55	3	3	721	718
16	1	1	810	802	56	3	2	724	721
17	3	2	802	705	57	1	1	438	443
18	1	1	710	712	58	1	1	657	642
19	3	2	632	637	59	1	1	517	513
20	1	1	586	539	60	3	25	829	790
21	3	2	712	705	61	1	1	751	724
22	1	1	887	817	62	1	1	953	912
23	1	1	522	520	63	1	1	777	724
24	3	4	705	643	64	3	68	513	492
25	1	1	884	820	65	1	1	448	443
26	3	3	515	520	66	1	1	846	829

27	1	1	639	637	67	3	17	716	718
28	3	3	632	637	68	1	1	913	912
29	1	1	833	793	69	1	1	616	601
30	3	4	520	539	70	3	2	912	855
31	1	1	476	554	71	3	2	607	605
32	3	5	539	554	72	3	7	605	601
33	1	1	768	643	73	3	2	443	442
34	1	1	920	918	74	1	1	441	442
35	3	2	918	904	75	3	8	601	516
36	1	1	882	880	76	1	1	526	516
37	3	5	820	817	77	3	9	516	497
38	3	6	817	796	78	1	1	626	607
39	3	4	880	820	79	1	1	484	481
40	3	3	904	880	80	3	3	441	442

References

Academic Monographs

1. Wu LY (2001) *Introduction to Sciences of Human Settlements* [M]. China Architecture & Building Press, Beijing
2. Wang JG (2001) *Theories and Methods of Modern Urban Design* [M]. Southeast University Press, Nanjing
3. Cheng YN (2010) *Theories and Methods of Modern Landscape Design* [M]. Southeast University Press, Nanjing
4. Cheng YN (2015) *Place and Landscape: Selected Works of Cheng Yuning* [M]. China Architecture & Building Press, Beijing
5. Liu BY (1990) *Systematization of Landscape Engineering* [M]. China Architecture & Building Press, Beijing
6. Chen SH, Li SY, Lobel J et al (2009) *Digital Fabrication: Architectural Design, Computational Logic and Cognitive Theory* [M]. China Architecture & Building Press, Beijing
7. Li X, Zheng JQ (2006) *Design Methodology* [M]. Tsinghua University Press, Beijing
8. Wu JG (2007) *Landscape Ecology: Pattern, Process, Scale and Hierarchy* [M], 2nd edn. Higher Education Press, Beijing
9. Qiu JP, Wen TX et al (2010) *Evaluation Science: Theory, Methods and Practice* [M]. Science Press, Beijing
10. Tao QK, Wang YL (2001) *Operations Research and Systems Analysis* [M]. China Machine Press, Beijing
11. Wu JP, Li XW (2010) *An Introduction to the Development of Systems Science* [M]. Tsinghua University Press, Beijing
12. Du D, Pang QH, Wu Y (2008) *Modern Comprehensive Evaluation Methods and Selected Case Studies* [M]. Tsinghua University Press, Beijing
13. Zhu XL (2009) *Research on Subjective Evaluation Methods of the Built Environment* [M]. Southeast University Press, Nanjing
14. Qin SK et al (2003) *Principles and Applications of Comprehensive Evaluation* [M]. Publishing House of Electronics Industry, Beijing
15. Ma JF (1994) *Theory of Evaluation Activities* [M]. China Renmin University Press, Beijing
16. Peng N, Zhi WJ, Dai C (2009) *Dialogue between Phenomenology and Architecture* [M]. Tongji University Press, Shanghai
17. Wang YC (2007) *Principles of Landscape Ecological Planning* [M]. China Architecture & Building Press, Beijing
18. Tang GA, Yang X (2006) *Experimental Tutorial on GIS Spatial Analysis* [M]. Science Press, Beijing

Y. Yuan, *Parametric Landscape Architecture Planning and Design*,
https://doi.org/10.1007/978-981-95-8011-8

19. Cheng SG, Zhang CC (1999) *Environmental Impact Assessment and Planning* [M]. China Environmental Science Press, Beijing
20. Zhou QM, Liu XJ (2006) *Digital Terrain Analysis* [M]. Science Press, Beijing
21. Xiao DN, Li XZ, Gao J et al (2010) *Landscape Ecology* [M], 2nd edn. Science Press, Beijing
22. Huang GY, Chen Y (2003) *Theory of Eco-city and Planning Design Methods* [M]. Science Press, Beijing
23. Luo TQ, Wang M, Dai DX (2008) *Basic Theories and Methods of Modern Ecological Planning and Design* [M]. China Architecture & Building Press, Beijing
24. Liu K, Li TS (2004) *Ecological Planning: Theory, Methods and Applications* [M]. Chemical Industry Press, Beijing
25. Yang ZF, Xu LY (2008) *Urban Ecological Planning* [M]. Beijing Normal University Press, Beijing
26. Shen KN (2008) *Architectural Phenomenology* [M]. China Architecture & Building Press, Beijing
27. Zhou B (2007) *Principles of Architectural Design* [M]. Sichuan University Press, Chengdu
28. Dou HS, Jiang JH (2003) *China's Five Major Freshwater Lakes* [M]. University of Science and Technology of China Press, Hefei
29. Yan H (2012) *Site Planning in Architecture* [M], 3rd edn. China Architecture & Building Press, Beijing
30. Feng JZ (2010) *Artistic Conception and Space: Essays on Planning and Design* [M]. Orient Publishing Center, Shanghai
31. Wu JH (1999) *Landscape Morphology: A Comparative Study of Landscape Aesthetics* [M]. China Architecture & Building Press, Beijing
32. Institute of Linguistics, Chinese Academy of Social Sciences. *Modern Chinese Dictionary* [M]. Beijing: The Commercial Press, 1999.
33. Chen CC (2008) *Design Cognition: Cognitive Science in Design* [M]. China Architecture & Building Press, Beijing
34. Che WB (2001) *Dictionary of Contemporary Western Psychology* [M]. Jilin People's Publishing House, Changchun
35. Liang NJ (2003) *Contemporary Cognitive Psychology* [M]. Shanghai Educational Publishing House, Shanghai
36. Zhuang WM (2000) *Introduction to Architectural Programming* [M]. China Water & Power Press, Beijing
37. Pan GX (2001) *The Art of Jiangnan Garden Design* [M]. Southeast University Press, Nanjing
38. Chen Z (1988) *Annotations on Yuanye* [M]. China Architecture & Building Press, Beijing
39. Hu CH, Wang JX (2010) *Watershed Runoff and Hydrological Models* [M]. Yellow River Conservancy Press, Zhengzhou
40. Wang DW (2008) *Modern Research on Hydrology and Water Resources* [M]. China Water & Power Press, Beijing
41. Niu WW (1987) *Modern Applied Geography* [M]. Science Press, Beijing
42. Wang SM, Dou HS (1998) *Gazetteer of Chinese Lakes* [M]. Science Press, Beijing
43. Ren Y et al (2013) *Civil 3D 2013 Application Handbook* [M]. Tongji University Press, Shanghai
44. Wei SZ (2008) *Landscape Site Planning and Design* [M]. Southwest China Normal University Press, Chengdu
45. Zhang JH (2005) *Survey and Analysis Methods in Planning and Design Practice* [M]. China Architecture & Building Press, Beijing
46. Tan QM (2005) *Dimensional Analysis* [M]. University of Science and Technology of China Press, Hefei
47. Liu G (2002) *A Guide to The Structure of Scientific Revolutions* [M]. Sichuan Education Press, Chengdu
48. Lu MS (2012) *Qian Xuesen's Thought on Cognitive Science* [M]. Science Press, Beijing

49. Shen ZH, Mi HM (1995) *Design Process and Methods* [M]. Shandong Fine Arts Publishing House, Jinan
50. McHarg IL (1969) *Design with Nature* [M]. Natural History Press, New York
51. Norberg-Schulz C (1988) *Architecture: Meaning and Place* [M]. Rizzoli, New York
52. Watson GB, Bentley I (2007) *Identity by Design* [M]. Butterworth Architecture, Oxford
53. Alexander C (1964) *Notes on the Synthesis of Form* [M]. Harvard University Press, Cambridge, MA
54. Norberg-Schulz C (1980) *Genius Loci: Towards a Phenomenology of Architecture* [M]. Rizzoli, New York
55. Corner J (1999) *Recovering Landscape: Essays in Contemporary Landscape Architecture* [M]. Princeton Architectural Press, New York
56. Tzonis A, Lefaivre L (2003) *Critical Regionalism: Architecture and Identity in a Globalized World* [M]. Prestel, Munich
57. Frampton K (2007) *Modern Architecture: A Critical History* [M]. Thames & Hudson, London
58. Rapoport A (1982) *The Meaning of the Built Environment: A Nonverbal Communication Approach* [M]. University of Arizona Press, Tucson
59. Huber A (2004) *Region, Site, and Architecture* [M]. Birkhäuser, Basel
60. Turner T (1996) *Landscape Planning and Environmental Impact Design* [M], 2nd edn. E & FN Spon, London
61. Waldheim C (2006) *The Landscape Urbanism Reader* [M]. Princeton Architectural Press, New York
62. Peña WM, Parshall SA (2012) *Problem Seeking: An Architectural Programming Primer* [M], 5th edn. John Wiley & Sons, Hoboken, NJ
63. Tashakkori A, Teddlie C (1998) *Mixed Methodology: Combining Qualitative and Quantitative Approaches* [M]. Sage Publications, Thousand Oaks, CA
64. Marsh WM (2005) *Landscape Planning: Environmental Applications* [M], 4th edn. John Wiley & Sons, Hoboken, NJ
65. Naveh Z, Lieberman AS (1994) *Landscape Ecology: Theory and Application* [M], 2nd edn. Springer, New York
66. Birnbaum CA, Karson R (2000) *Pioneers of American Landscape Design* [M]. McGraw-Hill, New York
67. Ndubisi F (2002) *Ecological Planning: A Historical and Comparative Synthesis* [M]. Johns Hopkins University Press, Baltimore
68. White ID (2011) *The History of the Landscape Since the Sixteenth Century* [M]. Reaktion Books, London
69. Booth NK (1989) *Basic Elements of Landscape Architectural Design* [M]. Waveland Press, Prospect Heights, IL
70. Hall P (2002) *Urban and Regional Planning* [M], 4th edn. Routledge, London
71. Strahler AN, Strahler AH (1983) *Introducing Physical Geography* [M]. John Wiley & Sons, New York
72. Simon HA (1996) *The Sciences of the Artificial* [M], 3rd edn. MIT Press, Cambridge, MA
73. Hanna KC, Culpepper RB (1998) *GIS in Site Design* [M]. John Wiley & Sons, New York
74. Kuhn TS (1996) *The Structure of Scientific Revolutions* [M], 3rd edn. University of Chicago Press, Chicago
75. Carl Steinitz A. Framework for Geodesign: Changing Geography by Design [M]. Environmental Systems Research Institute Inc., 2013
76. Shannon McElvaney. Geodesign: Case Studies in Regional and Urban Planning[M]. Environmental Systems Research Institute Inc., 2012
77. Petscheck P (2008) Grading for Landscape Architects and Architects. Birkhauser, [M].Berlin
78. Mohsen Mostafavi, Ciro Najle. Landscape Urbanism: a manual for the machinic landscape [M].Achitecture Association, London, 2003
79. Stephen M. Ervin, Hope H. Hasbrouck. Landscape Modeling:Digital Techniques for Landscape Visualization [M]. McGraw Hill Professional, 2001

80. Christian Norberg-Schulz. Architecture:meaning and place:selected essays [M]. Rizzoli International Publications, 1988
81. Allen DW (2011) Getting to Know ArcGIS Model Builder [M]. ESRI Press
82. Bishop, Lange E. Visualization in landscape and environmental planning technology and applications [M]. London and New York, Taylor & Francis applications, 2005
83. Buhmann E., S. Ervin, M Pietsch(eds.) . In: Peer Reviewed Proceedings of Digital Landscape Architecture 2012 at Anhalt University of Applied Sciences. [M]. Wichmann, 2012
84. Buhmann E., S. Ervin, M Pietsch(eds.)In:Peer Reviewed Proceedings ofDigital Landscape Architecture 2013 at Anhalt University of Applied Sciences. [M]. Wichmann, 2013
85. Frederick Steiner. The living landscape:an ecological approach to landscape planning [M]. McGraw-Hill Inc, 1991
86. Charles A. Birnbaum, Robin Karson. Pioneers of American Landscape Design[M]. McGraw-Hill Inc, 2000
87. Sauer C (1925) The Morphology of Landscape [M]. in University of California Publications in Geography

Journal Articles

88. Cheng YN, Yuan YY, Cheng S (2013) Coupling-based strategies for landscape architecture design with reduced intervention [J]. *Chinese Landscape Architecture* 8
89. Cheng YN (2011) On the developmental trends of landscape architecture [J]. *Landscape Architecture* 2
90. Cheng YN, Yuan YY (2015) Landscape architecture under the background of contemporary science and technology [J]. *Landscape Architecture* 7
91. Cheng YN, Yuan YY (2015) Parametric design of quasi-natural waterscapes in mountainous environments [J]. *Chinese Landscape Architecture* 7
92. Yuan YY, Cheng YN (2018) Process, logic and model: An analysis of parametric landscape planning and design [J]. *Chinese Landscape Architecture* 10
93. Yuan YY, Zhu CH, Cheng YN (2018) Quantitative study on morphological characteristics of urban lake landscapes [J]. *Landscape Architecture* 8
94. Yuan YY, Chen YL, Cheng YN (2018) Parametric design of quasi-natural waterscapes based on logical construction and algorithmic implementation [J]. *Landscape Architecture* 6
95. Yuan YY, Cheng YN, Li Z (2017) Virtual and real integration: Teaching research on parametric landscape planning and design [J]. *Chinese Landscape Architecture* 10
96. Yuan YY, Cheng YN (2015) Parametric routing design of landscape environmental roads [J]. *Chinese Landscape Architecture* 7
97. Wu LY (2011) Reflections on architecture, urban planning and landscape architecture as first-level disciplines [J]. *Chinese Landscape Architecture* 5
98. Yang R (2013) On the development trajectory and characteristics of landscape architecture: What kind of landscape architecture does China need in the early 21st century? [J]. *Chinese Landscape Architecture* 6
99. Wu LY, Mao QZ (2002) Digital city and human settlements construction [J]. *City Planning Review* 1
100. Wu LY (1996) Lewis Mumford's academic thought and its implications for the discipline of human settlements science [J]. *City Planning Review* 1
101. Wu LY (2004) Human settlements science and landscape education [J]. *Chinese Landscape Architecture* 1
102. Chen JR, Chen Y, Song ML (2011) A review of Horst Rittel's design methodology [J]. *Art and Design (Theory)* 2

103. Li Y, Liu DM (2013) Analysis of digital design methods constrained by architectural design thinking characteristics [J]. *Architectural Journal* 10
104. Zhang WP (2009) Research and practice of parametric design [J]. *Urban Architecture* 11
105. Liu YC (2011) Parametric design: Methods, thinking and organizational models [J]. *Architectural Technique* 1
106. Huang WX, Xu WG (2013) Parametric and generative landscape architecture design: A case study of graduate design studios at Tsinghua University [J]. *Landscape Architecture* 1
107. Xu WG, Xu F (2010) Parametric design in Chinese architectural practice: An interview with Prof. Xu Weiguo and Mr. Xu Feng [J]. *Urban Architecture* 6
108. Chi ZW, Chen J, Zhang DS (2012) Progress in parametric design applications and implications for landscape architecture [J]. *Chinese Landscape Architecture* 10
109. Cao KZ, Zhu YF (2013) Inspirations of parametric diagrams for landscape planning and design [J]. *Landscape Architecture* 1
110. Kuang W (2013) Review and reflection on the development of parametric landscape planning and design [J]. *Landscape Architecture* 1
111. Yuan F (2012) From digital programming to digital fabrication [J]. *Time + Architecture* 5
112. Zhao M (2011) Exploration and practice of complexity theory: Digital generative practice and experimental teaching of Alisa Andrasek [J]. *World Architecture* 6
113. Wang SZ (2010) Overlay method and simplified research method [J]. *Chinese Landscape Architecture* 9
114. Wei LK, Zhang X, Xu Z et al (2012) Moving beyond the limitations of architectural digital technology [J]. *Architectural Journal* 9
115. Cai LH (2013) Conceptual genealogy and workflow of digital landscape planning and design [J]. *Landscape Architecture* 1
116. Cai LH (2012) Digital strategies in landscape planning and design [J]. *Chinese Landscape Architecture* 1
117. Li F (2010) Development and prospects of 3D landscape visualization technology based on GIS [J]. *Architecture & Culture* 9
118. Yang T (2012) Digital city and space syntax: A digital planning and design approach [J]. *Planner* 4
119. Ma JW (2010) An introduction to GeoDesign [J]. *Chinese Landscape Architecture* 4
120. Ma JW (2013) GeoDesign: Concepts, framework and cases [J]. *Landscape Architecture* 1
121. Tang YH (2010) GeoDesign: New thinking and new approaches [J]. *Chinese Landscape Architecture* 4
122. Yang YS, Li DH (2013) GeoDesign: Concepts, methods and practice [J]. *International Urban Planning* 1
123. Guo W, Hou XL (2010) "Philosopher of the land": American landscape architect Manning [J]. *Chinese Landscape Architecture* 3
124. Zhou X (2013) Suitability analysis methods for construction land within large urban green spaces [J]. *Modern Urban Research* 11
125. Chen LL, Wang Y (2009) Analysis and determination of building orientation for natural ventilation [J]. *Shanxi Architecture* 9
126. Dong Y (2013) Analysis and calculation of rainwater resource potential in small watersheds of central Sichuan hilly region [J]. *Yangtze River* 9
127. Zhao T (1995) Prospects of computer technology in landscape architecture: Conceptual framework of a comprehensive landscape information system (I) [J]. *Chinese Landscape Architecture* 11
128. Zhao T (1996) Prospects of computer technology in landscape architecture: Conceptual framework of a comprehensive landscape information system (II) [J]. *Chinese Landscape Architecture* 1
129. Bao RQ (2013) Strategies for computer-aided landscape planning and design [J]. *Journal of Beijing Forestry University (Social Sciences)* 1

130. Lu MS (2005) The formation and development of the qualitative–quantitative integrated method: In honor of Academician Qian Xuesen's 93rd birthday [J]. *Engineering Science* 1
131. Guo Y (2011) A methodological overview of contemporary design research [J]. *Landscape Architecture* 4
132. Li XG, Jiang N, Wang HJ et al (2005) Dynamic changes in shoreline morphology of lakes in the Taihu Basin over the past 30 years [J]. *Journal of Lake Sciences* 4
133. Liu S (2015) Research progress in digital landscape technology: An overview of the International Digital Landscape Conference [J]. *Chinese Landscape Architecture* 2
134. Yang PR (2005) Planning evolution of eco-cities abroad from 1900 to 1990 [J], vol S1. *Modern Urban Research*
135. Shi-yan LI, Yue-hou QU (2004) About Process Philosophy[J]. *Journal of Tsinghua University (Philosophy and Social Sciences)* 2
136. Xiangrong W, Jing L (2007) The Meaning of Nature[J]. *Chinese Landscape Architecture* 1
137. Zhao Ke Y, Li. (2014) Combining Qualitative and Quantitative: Meta-Synthesis Digital Urban Planning[J]. *Urban Development Studies* 2
138. Sun J, Cheng Y, Yan H (2004) Road Line-selecting Methods Based on GIS and Its Developing Tendency[J]. *Geomatics & Spatial Information Technology* 6
139. Pan W, Li D, Tang T et al (2003) The fractal character of lake shoreline and its ecological implication[J]. *Acta Ecologica Sinica* 12
140. Miller WR (2010) On the Definition of GeoDesign[J]. *Chinese Landscape Architecture* 4
141. Steinitz C (2001) Landscape Planning: A History of idea-Speech in Peking University(Part I) [J]. *Chinese Landscape Architecture* 5
142. Steinitz C (2001) Landscape Planning: A History of idea-Speech in Peking University (Part II)[J]. *Chinese Landscape Architecture* 6
143. Peter PETSCHEK (2013) LandscapeSMART[J]. *Landscape Architecture* 1
144. Girot C (2009) The Santa Gila Experiment[J]. *Chinese Landscape Architecture* 5
145. Flaxman M (2010) Fundamentals of GeoDesign[J]. *Chinese Landscape Architecture* 4
146. Dangermond J (2010) GIS: Designing Our Future[J]. *Chinese Landscape Architecture* 4
147. Joost G (2014) Modes of Design Research[J]. *Chinese Landscape Architecture* 7
148. Charles J (1997) Nonlinear architecture: New science = New architecture?[J]. *Architectural Design* 129
149. Appleton K, Lovett A. GIS-based visualization of rural landscapes: Defining "sufficient" realism for environmental decision-making[J]. *Landscape and Urban Planning*, 2003, 65.
150. Steven DE (1991) Model-based approaches to managing concurrent engineering[J]. *Journal of Engineering Design* 3
151. Strahle AN (1957) Quantitative analysis of watershed geomorphology[J], vol 6. *Transactions of the American Geophysical Union*
152. Shreve RL (1966) Statistical law of stream numbers[J]. *Journal of Geology* 74
153. George S, Robert MM (1990) Applications of fractals in ecology[J]. *Trends in Ecology & Evolution* 3

Conference Papers

154. Cheng Yuning; Yuan Yangyang (2012) Place cognition–based teaching approaches in landscape architecture design[C] //*Proceedings of the 2012 Conference on Landscape Architecture Education*. Southeast University Press, Nanjing
155. Yuan Yangyang; Cheng Yuning (2013) Site selection research for landscape planning and design projects based on coupling methods[C] //*Digital Landscape: Proceedings of the First International Digital Landscape Forum in China*. Southeast University Press, Nanjing

156. Shaoming L (2010) Research on the complexity and logical construction of human settlements from a multidisciplinary perspective[C] //*Architectural Digital Flow*. Tongji University Press, Shanghai
157. Margetts J, Barnett R, Popov N. Landscape system modeling: A disturbance ecology approach[C] //Shepherd, J.; Fielder, K. *Proceedings of the 13th Annual Australia and New Zealand Systems Conference 2007: Systematic Development—Local Solutions in a Global Environment*.
158. Steinitz C (2010) Landscape architecture into the 21st century-methods for digital techniques[C]. *Peer reviewed proceedings of digital landscape architecture*, Wichmann
159. Flaxman M (2010) Fundamentals of geodesign[C].*Peer reviewed proceedings of digital landscape architecture*. Wichmann
160. Jack D (2010) GeoDesign and GIS: Designing our futures[C].*Peer-Reviewed Proceedings of Digital Landscape Architecture*. Wichmann
161. Miller W R. Geo-based Design[C] //*Specialist Meeting on Spatial Concepts in GIS and design 2008*. CA: University of California Santa Barbara, 2008.
162. Ahmad M A, Abdullahi A A. The need for landscape information modelling (LIM) in landscape architecture[C]//*Digital Landscape Architecture Conference 2013*.Bernburg, Germany, 2012.
163. Andrew N. The place for information models in landscape architecture, or a place for landscape architects in information models[C] //*Digital landscape architecture conference 2014*. Bernburg, 2013.

Dissertations

164. Li, Meifang. *Application of Three-Dimensional Parametric CAD Technology in the Final Drive of Bulldozers*[D]. Master's Thesis. Shandong University of Science and Technology, Shandong, 2005.
165. Li, Tingting. *A Preliminary Study on Reflexive Regionalism: A Theoretical Extension of Critical Regionalism*[D].Doctoral Dissertation. Tsinghua University, Beijing, 2012.
166. Ai, Qiao. *Ecological Sensitivity Analysis and Evaluation of Scenic Areas Based on GIS*[D]. Master's Thesis. Southwest University, Chongqing, 2007.
167. Zeng, Huimei. *Zoning of Scenic Area Protection Based on Ecological Sensitivity Analysis and Landscape Resource Evaluation*[D]. Master's Thesis. Southwest University, Chongqing, 2013.
168. Zhang, Yan. *Ecological Suitability Evaluation of Urban Construction Land in the Junshan Area of Yueyang Based on Weighted Sensitivity Analysis*[D]. Master's Thesis. Hunan University, Changsha, 2012.
169. Wu, Wei. *A Preliminary Study on the Spirit of Place in Landscape Architecture*[D].Master's Thesis. Huazhong Agricultural University, Wuhan, 2005.
170. Zhu, Bin. *A Comparative Study of Place and Landscape*[D].Master's Thesis. Dalian University of Technology, Dalian, 2010.
171. Yao, Peng.*Representation and Construction Strategies of Materiality of Place in Modern Landscape Architecture*[D]. Doctoral Dissertation. Beijing Forestry University, Beijing, 2011.
172. Yuan, Yangyang.*Interpreting Place Information: Exploring Approaches to Landscape Character Formation*[D]. Master's Thesis. Southeast University, Nanjing, 2012.
173. Li, Zhigang.*Research on Parametric Vertical Design in Landscape Architecture*[D]. Master's Thesis. Southeast University, Nanjing, 2014.
174. Huang, Xiong.*Research on Road Route Selection Techniques Based on GIS Spatial Analysis*[D]. Master's Thesis. Changsha University of Science and Technology, Changsha, 2006.
175. Chen, Yuantao. *Research on Intelligent Road Route Selection Methods Based on GIS*[D]. Master's Thesis. Chongqing Jiaotong University, Chongqing, 2012.

Specifications and Standards

176. Regulations on Scenic and Historic Areas of the People's Republic of China.State Council of the People's Republic of China. Available at: http://www.gov.cn/gongbao/content/2016/content_5139422.html
177. Standard for Master Planning of Scenic and Historic Areas.Ministry of Housing and Urban-Rural Development of the People's Republic of China. Available at: http://www.mohurd.gov.cn/wjfb/201903/t20190320_239842.html
178. Standard for Detailed Planning of Scenic and Historic Areas.Ministry of Housing and Urban-Rural Development of the People's Republic of China. Available at: http://www.mohurd.gov.cn/wjfb/202007/t20200707_246203.html
179. Code for Park Design.Ministry of Housing and Urban-Rural Development of the People's Republic of China. Available at: http://www.mohurd.gov.cn/wjfb/201703/t20170301_230801.html
180. Code for Design of Urban Road Engineering.Ministry of Housing and Urban-Rural Development of the People's Republic of China. Available at: http://www.mohurd.gov.cn/wjfb/201607/t20160712_228082.html
181. Code for Vertical Planning of Urban and Rural Construction Land.Ministry of Housing and Urban-Rural Development of the People's Republic of China. Available at: http://www.mohurd.gov.cn/wjfb/201607/t20160722_228283.html

Online Resources

182. ESRI company website. http://resources.arcgis.com/zh-CN/home/
183. Website of the Ministry of Education of the People's Republic of China. http://www.moe.gov.cn/publicfiles/business/htmlfiles/moe/moe_823/201002/xxgk_82699.html
184. Public Meteorological Service Center of the China Meteorological Administration. http://www.weather.com.cn/static/html/ knowledge/20080507/721.shtml

Others

185. Remmen A, Jensen AA, Frydendal J. Life cycle management: a business guide to sustainability [R]. 2007
186. Report on the Application to Unify and Standardize Related Domestic Majors under "Landscape Architecture" and to Establish It as a First-Level Engineering Discipline [R]. Education Branch of the Chinese Society of Landscape Architecture, January 16, 2006.
187. Notice of the Ministry of Housing and Urban-Rural Development on Issuing the "2011–2015 Development Outline for Informatization in the Construction Industry." May 10, 2011. Ministry of Housing and Urban-Rural Development of the PRC

The manufacturer's authorised representative in the EU is Springer
Nature Customer Service Centre GmbH, Europaplatz 3, 69115 Heidelberg,
Germany. If you have any concerns regarding our products, please
contact ProductSafety@springernature.com

Printed and bound by CPI Group (UK) Ltd, Croydon, CR0 4YY
12/10/2026
02231738-0005